AF601868

Springer Series on Naval Architecture, Marine Engineering, Shipbuilding and Shipping

Volume 2

Series editor

Nikolas I. Xiros, University of New Orleans, New Orleans, LA, USA

The Naval Architecture, Marine Engineering, Shipbuilding and Shipping (NAMESS) series publishes state-of-art research and applications in the fields of design, construction, maintenance and operation of marine vessels and structures. The series publishes monographs, edited books, as well as selected Ph.D. theses and conference proceedings focusing on all theoretical and technical aspects of naval architecture (including naval hydrodynamics, ship design, shipbuilding, shipyards, traditional and non-motorized vessels), marine engineering (including ship propulsion, electric power shipboard, ancillary machinery, marine engines and gas turbines, control systems, unmanned surface and underwater marine vehicles) and shipping (including transport logistics, route-planning as well as legislative and economical aspects).

More information about this series at http://www.springer.com/series/10523

Nisith R. Mandal

Ship Construction and Welding

Nisith R. Mandal
Department of Ocean Engineering and Naval Architecture
Indian Institute of Technology Kharagpur
Kharagpur, West Bengal
India

ISSN 2194-8445 ISSN 2194-8453 (electronic)
Springer Series on Naval Architecture, Marine Engineering, Shipbuilding and Shipping
ISBN 978-981-10-2953-0 ISBN 978-981-10-2955-4 (eBook)
DOI 10.1007/978-981-10-2955-4

Library of Congress Control Number: 2016956183

Printed on acid-free paper

This Springer imprint is published by Springer Nature
The registered company is Springer Nature Singapore Pte Ltd.
The registered company address is: 152 Beach Road, #22-06/08 Gateway East, Singapore 189721, Singapore

Preface

This book aims to introduce various aspects of ship construction starting from ship types, material of construction, welding technology to accuracy control. This book is the outcome of my experience of teaching Ship Construction and Welding Technology, Design and Construction of Ocean Structures, Marine Construction and Repair Techniques as regular and elective courses in undergraduate and graduate curricula during about past 30 years in the Department of Ocean Engineering and Naval Architecture at IIT Kharagpur. While teaching and working in this field, I felt the lack of a suitable book covering the various basic aspects of ship types, its structural components, materials, and aspects of its welding and dimensional control. This inspired me to get on this job and provide the budding naval architects with a comprehensive book on ship construction and welding. The contents of the book have been logically organized and spread over 21 chapters.

It starts with introducing to the novice reader the various types of ships based on cargo type and functionality and also the basic characteristics of shipbuilding industry. It then goes on to describe the various loads experienced by the ship structure and thereby working out suitable structural arrangement to sustain these loads. This forms the background to the introduction of the types of framing system, basic structural components, structural subassemblies and assemblies. All of these are explained with necessary illustrations and details. The book then goes on to work out the midship sections of some of the most widely used ship types, explaining the design strategy based on functionality. The book also includes the aspects of structural compensation for unavoidable discontinuities in ship structure.

Next the book covers various aspects of material of construction. It includes material description, classification requirements and different methods of steel material preparation. Subsequently different methods of metal cutting, plate and section forming are introduced along with the concept of line heating for obtaining compound curvature plates.

The reader is then introduced to various welding techniques related to shipbuilding industry. Here different fusion welding methods, power sources, effect of welding process parameters, metal transfer mechanism are discussed in detail. The solid-state welding technique suitable for aluminum welding has also been incorporated. The formation of weld-induced residual stresses and distortion has also been explained in detail. It then goes on to present in-process distortion control and mitigation techniques such as heat sinking, thermo-mechanical tensioning, etc. suitable for ship structural units. Finally, the book introduces various possible welding defects that one is likely to encounter in welded structures and explains the nondestructive testing methods those are relevant to ship construction.

With all the construction done, it is necessary to have a suitable mechanism to know the ranges of variations in structural fabrication so that one can quantitatively target the end product accuracy. To address this aspect a chapter on accuracy control has been included in this book.

I believe the contents of this book should prove useful to the students of naval architecture and ocean engineering as well as the shipbuilding professionals.

Kharagpur, India
July 2016

Nisith R. Mandal

Acknowledgement

The author with all humility wishes to acknowledge the encouragement he got through teaching Marine Construction and Welding to several batches of B.Tech. students over these years in the Department of Ocean Engineering and Naval Architecture at IIT Kharagpur.

He would like to specially acknowledge the help provided by some of his research scholars and graduate students, namely Ms. Malabika Adak, Sri Pankaj Biswas, Sri Debabrata Podder, Sri Amith Gadagi, Sri Anudeep Joshi, Sri K.S. Kapaleeswaran, Sri Chandra Shekhar, Sri Sanyappa Pujari, Sri Chandra Shekhar, Sri P. Mohan Ravi Kumar, Sri Abhay Kumar, Sri Sharat Kumar, a Scientific staff in the Marine Construction and Welding Lab and Sri Atanu Pal, Technical Superintendent in working out various examples and conducting experiments involving various aspects of welding and welding distortion.

The author would like to put on record the encouragement and active support he got from Sw. Suddhidanandaji Maharaj of Advaita Ashrama, Kolkata, Prof. Hidekazu Murakawa and Prof. Ninshu Ma of Joining and Welding Research Institute, Osaka University, Prof. Purnendu Das of Strathclyde and Glasgow University and Prof. Sreekanta Das of Windsor University, Canada. The list is definitely not exhaustive, there are many who directly and indirectly provided support and guidance in compilation of this work.

The author is also very much grateful to the staff of Marine Construction and Welding Laboratory, Sri Brotin Dey, Sri Biplab Das, Sri Subhas Josef, Sri Bharat Karar and Ms. Salma for all the long hours they devoted spontaneously even after working hours in carrying out tedious welding experiments.

The author is indeed very much thankful to the Chairman and Managing Director of Cochin Shipyard Ltd. for kindly partnering with us to conduct full-scale distortion control and accuracy control experiments on live projects. Thanks are especially due to Sri S. Harikrishnan, Sri Thomas Mathew, Sri P.J. Varghese and all the workers involved in the work at Cochin Shipyard Ltd.

The help extended to the author in conducting various studies relating to shipyard production and welding by Mazagon Dock Ltd., Mumbai and Garden Reach

Shipbuilders and Engineers Ltd., Kolkata is duly acknowledged. These studies have greatly contributed to the text content of this book.

The author wishes to sincerely acknowledge the word processing and drafting support of Ms. Shreya Mallick. Finally, the author acknowledges with gratitude the support of Continuing Education Cell of IIT Kharagpur in preparation of this manuscript.

Without the sanction of Divine Will not even a blade of grass moves in this world. The entire work that has been done became possible only because of HIS wish and HIS mercy.

Nisith R. Mandal

Contents

About the Author

Prof. Nisith R. Mandal, B.Tech (Hons), Ph.D., FIE, FRINA is Professor at the Department of Ocean Engineering and Naval Architecture, IIT Kharagpur. Prior to joining the IIT in 1987, while working in Garden Reach Shipbuilders and Engineers Ltd. he was involved in the initiation of a modular method of ship construction. Subsequently he headed for higher studies and obtained his Ph.D. in Ship Production and Welding Technology from the Ship Research Institute, Gdansk, Poland in 1985. Since then he has been actively engaged in teaching and research in the field of welding techniques, distortion and line heating as applied to shipbuilding. He has undertaken several industrial consultancy as well as research projects in his area of expertise sponsored by various agencies like Garden Reach Shipbuilders & Engineers Ltd., Calcutta, Mazagon Dock Ltd., Mumbai, Cochin Shipyard Ltd., Department of Science and Technology, GoI, the Ministry of Surface Transport (Shipbuilding & Ship Repair Wing), GoI, Manipur Science and Technology Council, Maglev Inc., USA, Institute for Plasma Research, Ahmedabad, West Bengal Surface Transport Corporation, the Council of Scientific and Industrial Research, the Naval Research Board, etc. He has published over 75 research papers in peer-reviewed journals and for national and international conferences. He has authored three books, "Aluminum Welding", 2nd Edition, 2005, "Welding and Distortion Control", 2004, and "Welding Techniques, Distortion Control and Line Heating", 2009 published jointly by Narosa Publishing House, New Delhi and Alpha Science International Ltd., Pangbourne, UK. He has conducted several short courses for shipyard professionals on ship repair and maintenance, welding and distortion control in steel fabrication, aluminum welding in the marine industry, weld-induced distortion prediction and control, shipyard productivity—process evaluation and improvement, and distortion control in shipbuilding.

Chapter 1
Introduction to Ships

Abstract Over the years as international trade increased and also bulk transportation of goods became more and more necessary, various types of ships came into being depending on the type of cargo that needs to be carried. This trade will naturally involve very high volume of transportation of all kinds of items starting from bulk grain, ore, coal to crude, automobiles and various other kinds of farm and engineering products. Passenger ships and inter island ferries also play a very important role in transportation as well as tourism. These vessels, their outward features, i.e. the hull form may not be very different, but the internal structural arrangement will be very much dependent on the type of cargo the vessel needs to carry. The internal structural arrangement should be such that it will facilitate loading, stowage and unloading of the cargo. And needless to say, should ensure safe transportation of the cargo. It naturally implies that the structure will be able to sustain all the service loads and also at the same time the hull form should be hydrodynamically efficient.

It is well known that two third of this earth is covered with water and only one third is land. Also at the same time for the mankind to survive international trade of various kinds of goods will always remain inevitable. This trade will naturally involve very high volume of transportation of all kinds of items starting from bulk grain, ore, coal to crude, automobiles and various other kinds of farm and engineering products. Passenger ships and inter island ferries also play a very important role in transportation as well as tourism. Apart from this another very important sector is marine defence, which includes various kinds of naval craft, to name a few, frigate, corvette, seaward defence boat, landing craft, aircraft carrier, etc. In addition to all these craft, there are certain types of vessels which can be referred to as service vessels, like tugs, dredgers, pilot vessels, offshore support vessels, etc. Oceans are vast natural resource of food, minerals, energy, etc. Deep sea fishing is another area, where different kinds of fishing and fish processing vessels are used. Hence one can observe that the requirement of various kinds of ships will remain as long as mankind will be there in this earth.

N.R. Mandal, *Ship Construction and Welding*, Springer Series
on Naval Architecture, Marine Engineering, Shipbuilding and Shipping 2,
DOI 10.1007/978-981-10-2955-4_1

It is believed that one of the earliest modes of transportation was through water. People through natural instinct learnt the science of flotation and propulsion. For obvious reasons the ancient marine vehicles were all made of wood. As the metal making technology developed, slowly wood got replaced by steel as one of the most widely used material of construction of ships. Similarly the joining technology also graduated from riveted joints to welded joints. One of the good fall out of world war was significant contribution to the development of welding technology and understanding of phenomenon like brittle fracture.

Over the years as international trade increased and also bulk transportation of goods became more and more necessary, various types of ships came into being depending on the type of cargo that needs to be carried. These vessels, their outward features, i.e. the hull form may not be very different, but the internal structural arrangement will be very much dependent on the type of cargo the vessel needs to carry. The internal structural arrangement should be such that it will facilitate loading, stowage and unloading of the cargo. And needless to say, should ensure safe transportation of the cargo. It naturally implies that the structure will be able to sustain all the service loads and also at the same time the hull form should be hydrodynamically efficient.

1.1 Ship Types

Ships can be classified broadly under five general heads, Merchant ships, Naval Ships, Fishing vessels, Service vessels and Inland vessels. The major divisions are as follows:

Merchant Ships

- General cargo carrier
- Dry Bulk Carrier
  - Bulk carriers
  - Ore carriers
  - Combination carrier (OBO)
- Liquide Bulk Carrier
  - Crude oil carrier
  - Product tankers
  - Cryogenic liquid carrier (LNG/LPG/Ammonia)
- Container ships
- Roll On Roll Off (RO RO Ships)
- Passenger liners

Naval Ships

- Corvette
- Frigate
- Battle ship
- Landing craft
- Aircraft carrier

Fishing Vessels

- Fishing trawler
- Factory ship

Service vessels

- Sea going tugs
- Dredgers
- Offshore support vessels
- Pilot vessels
- Ice breaker
- Research vessels

Inland Vessels

- Propelled and non propelled barges
- Passenger boats and launches
- Harbour tugs
- House boats
- Floating restaurants

1.2 Basic Features

General Cargo Carrier

These ships are often referred to as go-any-where type of ships. These vessels can carry any type of packaged cargo of varying dimensions. The cargo can be in drums, bags, bundles, bales or individual pieces. These ships are always equipped with cargo handling gear of its own generally in the form of derricks or deck cranes. Thereby these vessels do not depend on the port facility as far as cargo loading and unloading is concerned. Hence, irrespective of the cargo handling facility available in a given port, these type of vessels can load/unload cargo in any port provided the water draft available is adequate for the ship. That is how it acquired the name mentioned in the beginning.

Structural features
These vessels are of single skin type with double bottom and generally have at least one tween deck. This is provided for facilitating cargo segregation and stowage. Decks and double bottom are longitudinally stiffened, whereas the side shells are transversely stiffened.

Hatches and Holds
The engine room is generally located in semi aft position, i.e. there is a cargo hold in the aft of engine room and the other holds are ford of engine room. Each hold is provided with one hatch opening, having hatch width somewhat less than the half deck width. For improving cargo loading and unloading, there can be multiple hatch openings in each hold. However this adds to the production cost of the ship.

Capacity/Speed
Capacity of general cargo ships are always on the lower side, generally not exceeding about 12,000 t. Similarly the speed of such vessels is also on the lower side about 10–12 knots. Since these type of ships carry various types of cargo of varying overall dimensions and sizes, the whole process of cargo loading and unloading becomes quite lengthy. Many a time it is observed that the port time of such vessels goes beyond 3 weeks. Hence if the capacity is increased, it will further increase the port time because of longer duration of cargo handling. The longer a ship waits in a port, it only spends money, only then it earns, when it sails with cargo. Hence it is necessary to see how to reduce port time, i.e. reduce loading/unloading time.

Ship's speed is one of the owners' important requirement as well as it is one of the important parameter with regard to ship's operation. Fuel consumption increases exponentially with increasing speed. This implies operation cost increases with increase in speed. On the other hand travel time reduces as speed increases, i.e. with higher speed, number of trips per annum increases. If the number of days saved in sailing due to increase in speed is comparable to the port time, then only speed increase becomes justifiable. In case of general cargo ships, where port time is of the order of 21–30 days or even more, merely saving few days of sailing time by increasing speed and thereby substantially increasing fuel consumption does not become economically viable. That is why the cruising speed of general cargo ships is kept on the lower side.

Dry Bulk Carrier
These ships are essentially meant for carrying dry cargo in bulk, like grains, pulses, sugar, ore coal, etc. Vessels other than those meant for carrying ore, coal or high density bulk cargo are generally referred to as Bulk Carrier. Other vessels are often termed as Ore Carrier, Coal Carrier or Combination Carrier.

For cargo loading these vessels need to depend on port facilities, however these ships are fitted with deck cranes for cargo unloading. Cargo loading generally takes place in a automated fashion through a combination of conveyor and hopper. Whereas unloading is done either by pumping in case low density cargo like

Fig. 1.1 Dry bulk carrier

grain/pulses or by using grab buckets for coal/iron ore. A typical dry bulk carrier is shown in Fig. 1.1.

Structural features

These are single decker vessels with top and bottom wing tanks with sloping bulkheads. The angle of inclination of these bulkheads is decided on the basis of angle of repose of the cargo to be carried. These wing tanks are either kept empty or are used for ballasting. The sloping bulkhead of top wing tank in some bulk carriers is hinged to the side shell structure. Therefore when it is in lowered configuration it behaves like a lower deck and the space can be used for carrying general cargo.

The wing tanks and the double bottom are longitudinally stiffened, however the side shell in the hold region is transversely stiffened. Though longitudinal framing system is a preferred mode of stiffening, still the side shell is transversely framed to avoid retention of cargo on the webs of the longitudinal stiffeners even after unloading of the cargo. Majority of structural failures are observed in the area where the side shell frames are connected to the sloping bulkhead of the bottom wing tank. The common mode of failure is detachment of the bracket from the sloping bulkhead primarily because of corrosion. The rate of corrosion in this zone is comparatively higher because of accumulation of dust and dirt from the bulk cargo along the welded joint. The dust and dirt being hygroscopic, it absorbs moisture and acts as electrolyte promoting galvanic corrosion. At the same time it is to be noted that the welded zone is also surrounded by the so-called heat affected zone (HAZ) having a different microstructure compared to the parent metal. Thus the HAZ has different electro-chemical potential from that of the parent metal causing the formation of galvanic cell leading to a higher rate of corrosion. Eventually under the normal service loads and due to gradual wastage of metal through corrosion, complete detachment of bracket takes place. This may lead to a cascading effect on the adjacent bracketed joints of the side shell frames.

The tank top plating in way of cargo hold needs additional strengthening in case of ore carriers. Because during loading this heavy cargo drops from a height over the tank top plating and also at the same time while unloading, the grab bucket will hit the tank top plating while emptying the hold. Both these actions will cause sever loading on the tank top structure causing erosion and possible deformation of the same.

The double bottom height of high density bulk cargo (ore/coal) carrier is substantially higher compared to the normal bulk (grain/pulses/sugar/etc.) carrier. This is done to raise the position of vertical centre of gravity (VCG) of the vessel in full load condition. The cargo being of high density (7–8 times compared to that of grain cargo), it will occupy less space in the cargo hold to attain the full load condition. Thus it will have two basic effects:

- Large part of the hold will remain empty. This may cause cargo shifting due to rolling of the ship leading to damage of the side shell. This may also lead to listing of the vessel because uneven shifting of cargo to one side of the hold. This permanent heel of the vessel will cause loss of course keeping stability. This will require continuous rudder angle to correct its course. This will cause additional fuel consumption because part of the power will get consumed in forcing the ship back in its path.
- The VCG position will be very low resulting in substantially high metacentric height (GM). This will make the ship stiff, i.e. it will have very high righting moment. In the event of some heeling of the vessel due to some external cause like wave action or gust of wind, the vessel will heel and will try to restore back to its upright condition very fast. Thereby it will roll further to the other side and thus it may experience quite severe rolling motion.

This brings in the concept of Combination Carrier (OBO carrier). Here the hold volume is made smaller by raising the tank top and providing inner side shells as shown in the Fig. 1.2. By doing so, both the above issues are solved. By raising the tank top the VCG position is raised and thus GM reduces, making the vessel tender. By putting inner side shells, the hold volume reduces and therefore hold remains full and shifting of cargo does not take place. The additional space that is created by the inner side shells is used to carry liquid cargo in return voyage. That means, outgoing voyage she carries, say, iron ore and in the return voyage she brings in edible oil.

Hatches and Holds

The engine room is located fully aft. That means all the cargo holds are forward of the engine room. Each hold has hatch opening with suitable hatch cover for water tight closing. The width of the opening is of the order of half the width of the main deck at midship. Deck cranes are located in between the holds on the main deck

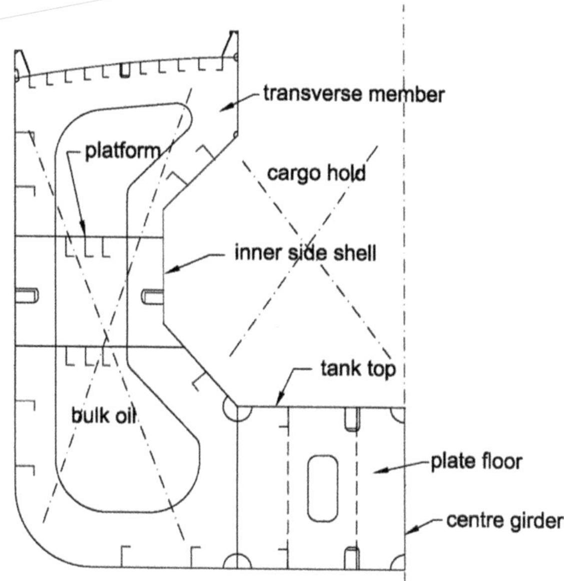

Fig. 1.2 Midship section of an OBO carrier

supported by transverse subdivision bulkheads. During loading or unloading extra care needs to be taken so as not to create a situation which makes alternate holds empty or full. This situation of alternate hold loading will cause development of severe shearing force in the structure in the vicinity of full and empty hold.

Capacity/Speed

Bulk carriers are available over a wide range of capacity, starting from about 20,000 t to about 80,000 t. Since both loading and unloading of cargo is well mechanised, the port time of such vessels are much less in comparison to that of general cargo ships. Hence depending upon the trade volume quite high capacity vessels are made. To increase the number of trips per annum, also one can go for somewhat higher speed of operation. The preferred speed for bulk carriers are in the range of 15–20 knots.

Liquide Bulk carrier

In these vessels liquid cargo is carried in bulk either at room temperature and ambient pressure or at cryogenic temperature and high pressure.

LNG/LPG

Cargo like liquid natural gas or liquid petroleum gas is carried under cryogenic conditions. These vessels are referred to LNG/LPG carrier. These are essentially extremely sophisticated vessels with appropriate containment system which can retain the cryogenic temperature of the cargo. This being extremely low density cargo, much of the hull remains out of water exposed to the atmosphere. This calls for additional design requirement against wind heeling moment. A typical LNG carrier is shown in Fig. 1.3.

Fig. 1.3 LNG carrier

In these liquefied gas carriers, the cargo is kept under positive pressure so as to prevent air entering the cargo tanks. This is done to ensure that only liquid cargo and cargo vapour are present in the cargo tank and flammable atmospheres cannot develop. The gas is carried always in liquefied condition and because of their physical and chemical properties, they are carried either,

- under pressures greater than atmospheric, or at
- temperatures below ambient, or a
- combination of both.

The gas carriers are therefore generally classified as,

- Fully Pressurised
- Semi-pressurised and refrigerated
- Fully refrigerated.

Natural gas mainly comprises of methane and ethane. LNG ships carry their cargo at −161 °C at ambient pressure. Whereas LPG is mainly propane and butane. LPG ships carry their cargo at −42 °C. It is often carried under pressure.

Crude Oil Tanker
One of the high volume liquid cargo is crude oil carried in crude oil tankers, shown in Fig. 1.4. Very high volume of international global trade is carried out in importing or exporting of crude oil. This gave rise to what came to be known as VLCC (Very Large Crude Carrier) and ULCC (Ultra large Crude Carrier).

Fig. 1.4 Crude oil tanker

Product Carrier and Chemical Tanker
The rest are various liquid products, carried in Product Carriers and chemicals, like various acids and alkalis carried in Chemical Tankers. A typical chemical tanker, Alcedo, 2800 t capacity is shown in Fig. 1.5.

Structural features
The liquid cargo carriers are all single decker vessels. The cargo in bulk is carried in the cargo hold directly. Irrespective of the type of cargo, all vessels have double bottom and completely closed main deck excepting in case of LNG/LPG carrier. Longitudinal framing system is adopted in the entire construction. In addition to the subdivision water tight bulkheads, these vessels depending on its size have one or more longitudinal bulkheads also. Longitudinal bulkheads are provided to reduce

Fig. 1.5 Chemical tanker

the free surface effect of the liquids in the cargo holds. These are generally fine form vessels.

Generally the crude oil carriers are of very high capacity with vessel length exceeding 300–400 m. The product carriers and chemical tankers are of smaller size compared to crude carriers. The length of product carriers is in the range of 220 m, and that of chemical tankers is 120 m. However the structural integrity and structural strength of all these vessels is very good as because there is no hatch opening, making it a perfect closed-section structure. In addition to this longitudinal framing system also contributes to the longitudinal strength of the hull girder. The crude carriers are full form ships having block coefficient 0.8 and above. However the product carriers and chemical tankers are of fine form.

Oil spillage caused by crude oil tanker disaster led to severe environmental pollution affecting ecosystem and coastal marine life. This eventually led to a regulation making it mandatory to have double wall construction for all sea going crude carriers. The idea is to have a second wall of defence in the event of any damage to the outer shell, thus preventing oil spillage. This double wall construction further contributes to the longitudinal strength. The same logic also applies to the product carriers as well as chemical tankers. All these vessels are of double wall construction.

Hatches and Holds

These vessels have cargo holds and each hold is serviced by inlet and outlet manifold for pumping in and out the liquid cargo. Naturally these vessels do not have any hatch opening. In LNG/LPG carriers, the cargo is carried in specialised containers with appropriate containment system and the whole assembly is placed in the holds. About one third of the container remains outside the deck level.

In product carrier, different holds are generally assigned to different products, e.g. different kinds of edible oils. Therefore each hold has its separate pumping system.

One of the typical feature of crude oil carrier is, the return trip is always empty, i.e. it does not carry any cargo. Hence it is in light ship condition and in this condition if it sails it may face various problems. It will be floating in light ship draft and hence good part of the propeller blade will stick out of water, causing a drastic drop in propulsion efficiency and increase in vibration. At the same time at lower draft there can be a case of drop in metacentric height, because of lower water plane area, thereby loss in stability of the vessel. Hence to overcome these problems, ballasting needs to be done to attain minimum required hull immersion. In the older days, there was no segregation and ballasting used to be done in cargo holds leading to contamination of ballast water with the left over crude oil in the holds. On discharging this contaminated ballast water, it led to huge pollution in the port waters. Because of this, MARPOL convention has made it mandatory to have well defined segregated ballast spaces, which will be used only for ballasting purpose.

Capacity/Speed

The capacity of crude carriers can be as high as 500,000 t or even more. Whereas the capacity of product carriers is generally in the range of 30,000–50,000 t and that of chemical tankers about 2,000–10,000 t. The capacity of gas carriers ranges from about 75,000–138,000 m^3 of gas. All these vessels have service speed in the range of 16–21 knots. These vessels can be operated in higher speed range because of the basic fact that cargo loading and unloading of these vessels are very fast. Therefore with increase in speed, one can achieve higher number of round trips per annum.

Container ship

The whole concept of container ship came into being to reduce the port time, i.e. cargo loading/unloading time. As one can observe in case of general cargo ship, the cargo loading/unloading time is very long, because each and every item of cargo needs to be individually handled. Here individual units of cargo can be packed in standardised containers and thus can be handled as a standardised unit of cargo. With this concept of containerisation of cargo, intermodal movement of cargo became possible, thereby increasing the efficiency of cargo transportation. Containers could be moved seamlessly between ships, trucks and trains. A typical container ship is shown in Fig. 1.6.

Every container is assigned an unique unit number. This can be used by any concerned person dealing with these containers to identify the container, the user and to track the container's whereabouts anywhere in the world.

Some facts about containers ships [1]

- Approximate lifespan of 26 years.
- The gantry cranes used for loading/unloading container ships can be 130 m tall and may weigh up to 2,000 tons.

Fig. 1.6 A typical loaded container ship

- The whole process of loading/unloading of containers and balancing the vessel accordingly by increasing/decreasing ballast are fully computerised.
- It is reported that CO_2 emission is less in container ships. On average it emits around 40 times less CO_2 that of a large freight aircraft and over three times less that of a heavy truck.
- Container shipping is estimated to be two and a half times more energy efficient than rail and 7 times more so than road.
- Weekly safety and fire drill sessions take place for all crew members when the vessel is at sea to make sure that they are practiced for any real emergencies.

Structural features

The fundamental distinct feature of container ship is that it has extremely wide hatch opening, making it a typical case of so-called open section structure. As the capacity went on increasing, it led to substantial increase in length overall, making the vessel more vulnerable to longitudinal and torsional loading. For smaller capacity container ships, the hull structure had to be strengthened against torsional loading whereas for larger ships longitudinal as well as transverse strength also became important. Adequate torsional strength was achieved by providing box girder arrangement for smaller vessels, however double wall construction became necessary for providing the required torsional, longitudinal and transverse strength for the larger, higher capacity vessels.

In case of hull construction with box girder arrangement for smaller container ships, longitudinal framing system is adopted in the box girder as well as in the double bottom. However, the side shell is transversely framed. The box girder provides for necessary torsional strength to the hull girder. Transversely framed side shell provides for necessary transverse strength as well as supports the box girder arrangement. As the vessel size increases, box girder becomes inadequate. To achieve the required strength against torsion and longitudinal bending, cellular construction is adopted. It can be considered as if the box girder is extended down to the tank top plating, thus obtaining the cellular nature of the structure.

The hatch covers are necessarily of very long span, because of the wide hatch opening. At the same time since containers are also stacked over the hatch covers, they need to be accordingly designed to withstand these loads.

Hatches and Holds

The prime most feature of container ship is its near open-deck construction. That means it has a very wide hatch opening to facilitate loading and unloading of the containers in the cargo hold. Putting the containers in the hold, one on top of the other is to be done vertically, same operation is required while unloading them. It needs to be kept in mind unlike in general cargo ship where, individual units of cargo are often moved transversely after lowering them in the cargo hold for proper stowage, in case of containers, this is not possible. If the containers need to be shifted inside the hold for adequate stowage, then the whole purpose of containerisation will be lost. It will be an extremely cumbersome job and will consume excessive time for handling and manoeuvring the containers. Hence it calls for hatch opening equal to the width of the hold.

The hatches are provided with hatch covers and stack of containers are also taken on top of the hatch covers. Proper lashing arrangement is done to secure the containers placed above the hatch covers as shown in Fig. 1.7. There are instances of losing containers due to sever storm causing violent motions of the vessel in deep sea.

Capacity/Speed

From the earlier days of container ships of about 1500 TEU in 1976, it has grown in excess of 12,000 TEU today, with some ships on order capable of carrying 18,000 TEU [1]. There are certain standard lengths of these containers. The most widely used sizes today, are the 20-foot and 40-foot lengths. The 20-foot container is referred to as a Twenty-foot Equivalent Unit (TEU). The capacity of a container ship is expressed in terms of TEU. It implies the number of such 20 foot containers that a ship can carry. Therefore one 40-foot length container will be equivalent to 2 TEU. With the market demand the container ships are growing in size and also becoming more fuel efficient. Some of the world's largest container ships are about 400 m long and about 55 m wide. In such vessels are powered by engines which weigh around 2,300 tons, the propellers 130 tons, and there are twenty-one storeys between their bridge and their engine room. Typically these vessels are 11,000 TEU and are manned by just about 13–14 people. Interestingly if that number of containers were loaded onto a train it would be about 71 km long.

Since all the container handling operations are fully computerised and automated, the port time is reduced to a minimum. Hence these vessels generally have higher speed of operation, in the range of about 20–25 knots.

Roll On Roll Off (RO-RO Ships)

With global trade of automobile import and export and also to facilitate tourism, the concept of RO-RO ships and RO-RO ferry (ROPAX) came into being. These

Fig. 1.7 Containers loaded on *top* of hatch cover

vessels are specially designed to carry wheeled cargo, which can be loaded and unloaded on their own wheels. These wheeled cargo can be passenger cars, trucks, trailers, railroad cars, heavy earth moving vehicles, etc. The ships are provided with ramps located at the stern or bow or at the port side of the ship. RO-RO ferry are widely used for passenger transportation between destinations separated by short stretch of sea, where building a bridge is not feasible. This provides a seamless movement of the passengers travelling from one point to the other. These car passenger ferries or RO-RO ferries typically have car decks for keeping the cars and restaurants and shops for the passenger to spend time while sailing. At the destination one boards the car and simply drives off. The passengers drive in and drive out of the ship. Thus it has very low port turn around time.

The ventilation of the cargo space in RO-RO ships is of additional importance because the cargo, i.e. wheeled vehicles which are loaded and also unloaded using their own petrol/diesel powered engines. Since hundreds of vehicles are stored in closed space, the exhaust from each, however small it is, may accumulate to a significant level. Hence the ventilation system of the cargo spaces is to be appropriately designed and it is kept completely segregated from other ventilation systems.

These ships can be further classified based on their dedicated usage, like Pure Car Carrier (PCC), Pure Car/Truck Carrier (PCTC) and Large Car and Truck Carrier (LCTC). A typical bow loading RO-RO vessels is shown in Fig. 1.8.

Fig. 1.8 A typical RO-RO ship with bow ramp

Structural features
One of the primary structural feature is, it is multi deck vessel and the decks need to have unobstructed passage from the ramp end to the other end of the ship till the bulkhead of the last hold. The earlier versions of RO-RO vessel did not have subdivision water tight bulkheads to satisfy the requirement of unobstructed passage of the vehicles. Thus any damage/leakage in the hold region anywhere along the ship's length will lead to complete flooding of the entire vessel.

The RO-RO ferry *Herald of Free Enterprise* capsized and sank shortly after leaving Zeebrugge port in Belgium in March 1987. It had a bow ramp. It was found that the bow door was not properly closed and the vessel left the port allowing water ingress, which eventually flooded the entire car deck [2]. The accident resulted in the deaths of 193 passengers and crew members. On 28 September 1994, the RO-RO ship *Estonia* was lost with more than 900 lives when the bow door got torn off by heavy seas. The car deck got flooded and the ship capsized within a few minutes.

Subsequently rules were enacted and now subdivision bulkheads are mandatory for RO-RO vessels. The transverse bulkheads are equipped with power-operated sliding doors. There should always be an inner door behind the bow door to act as a second line of defence [3]. Since these vessels do not require any hatch opening and at the same time there are multiple decks, therefore adequate global structural strength is easily achieved. The decks and double bottom are longitudinally framed whereas the side shell is transversely framed to provide for better load distribution from the car decks to the hull girder. The car decks need to be adequately strengthened to take the local load of the wheeled vehicles. In case of vessels meant for carrying extra heavy vehicles, accordingly appropriate strengthening of the decks is to be done.

The vehicles need to be properly lashed on to the deck to restrain the vehicles from any kind of movement inside the hold. As per IMO guideline [4], the suitable securing points are provided on the decks with:

- Longitudinal spacing <2.5 m
- Transverse spacing between 2.8 and 3.0 m

As per IMO guideline, the lashing shall consist of chain or any other device made of steel or other material with equivalent strength and elongation characteristics as given below,

- The minimum strength without permanent deformation of each securing point should be 120 kN. If the securing point is designed to accommodate more than one lashing (M lashings), the corresponding strength should be not less than M × 120 kN.
- Lashings should be attached to the securing points with hooks or other devices.
- Lashings should only be attached to the securing points.
- The angle between the lashings and the horizontal and vertical planes shall be preferably between 30° and 60°.

Hatches and Holds
RO-RO ships do not have any hatch opening for cargo loading. However it does have access opening at the bow or stern or in the side shell in the ford or aft region for vehicle loading/unloading through ramps. The transverse subdivision bulkheads in between the cargo holds have power operated sliding doors to provide for unobstructed access for vehicle movement and storage. There are multiple car decks depending on the size and capacity of the vessel. The car decks are connected to each other by internal ramps at either of the end of the decks.

To facilitate additional car storage in place of heavy and big vehicles, often movable car decks are installed in between the permanent decks.

Capacity and Speed
The PCTC generally have dual loading facility through stern ramp as well as side ramp for speeding up loading/unloading of thousands of vehicles. Often these ships are equipped with height adjustable decks to provide for required vertical clearance for extra high and heavy wheeled cargo. For example, a 6,500 unit RO-RO vessel with 12 decks can have three decks which can take cargo up to about 140 t having adjustable decks to increase deck height from 1.7 to 6.7 m. Since the cargo loading/unloading is very fast leading to low port time, these vessels generally have high cruising speed in excess of 20 knots.

Passenger Liners
With advent of aircraft, use of passenger ships for transcontinental passenger transportation has reduced, if not altogether eliminated. However in short distance inter island or main land to the neighbouring islands, passenger ships remained as one of the important mode of transportation of passengers. The advantage the passengers get in ships is, they can carry substantially larger amount of personal luggage as compared to that by air. Hence in passenger ships provision for storage of passengers' cargo is to be provided. Apart from this type of passenger transportation, passenger liners for tourism have remained quite popular. The prime purpose of these vessels is to give a sea sailing experience to its occupants with highest possible levels of comfort and safety. Therefore these ships are to be equipped with all kinds of facilities for passengers comfort and leisurely activities, like swimming pool to cinema halls, dance floors, shopping arcades to different playgrounds. A typical passenger liner is shown in Fig. 1.9. Unlike other ships, the aesthetic aspects of the ship are very important in case of passenger liners.

Some of the cargo ships also keep provisions to carry few passengers. Here off course, the passengers do not get the level of facilities as offered in passenger liners. Ships carrying more than 12 passengers need to satisfy all the statutory passenger ship requirements. On international voyages these vessels must comply with all relevant IMO regulations, including those in the SOLAS and Load Lines Conventions prescribed for passenger ships.

Structural features
Passenger ships are generally fine form ships, such that they can be efficiently operated at higher speeds. As can be seen in the Fig. 1.9, these vessels are often

Fig. 1.9 A typical passenger liner

provided with very high rake of the bow as well as high flare to make it aesthetically appealing and also to give more deck area. These are multi deck ships. The majority of the decks are above the main deck. There can be one or two decks below the main deck. The entire passenger accommodation is provided in these decks. In no case any of the passenger deck can be below the load water line level. This implies that the passenger accommodation deck below main deck should be clearly above the loaded draft.

One of the major structural issue in passenger ship is its extremely long superstructure. As per the classification rules a superstructure is defined as effective superstructure if its length is equal or more than 15 % the length of the ship. The structural significance is if the superstructure is effective, it takes part in longitudinal bending. Or in other words, it resists longitudinal bending, which implies, it contributes to longitudinal strength. It is well known that the stresses due to longitudinal bending will be higher on the members away from the neutral axis. Therefore the stresses in the structural members on the upper decks of the superstructure will be higher due to longitudinal bending of the hull girder. This would then require higher scantlings for the deck plating, deck longitudinals and other structural members. The direct effect of this would be the position of the vertical centre of gravity will go up. Whereas the position of meta centre, i.e. M does not change, as it depends only on the under water hull geometry. In a worst case scenario, this may lead to very low metacentric height (GM) or even a negative GM, both of which are not acceptable. This is a typical problem that a designer faces in passenger ships. At the same time it is also not advisable to provide for higher GM than what is required as per IMO guidelines. Because higher GM makes a vessel stiff, that means, it has high righting moment, causing the vessel to roll vigorously in the event of action of any external force. It makes the stay of the passengers and crew on board the vessel very uncomfortable. Other aspects would be higher fatigue loading.

Keeping in view the safety aspect as well as the comfort level, it is a challenging task for the designer to keep the GM at its minimum permissible value. This becomes specially difficult in case of passenger liners, as because the superstructure is very long and very high. Naturally the vertical centre of gravity position goes up significantly. One of the way, this can be counter acted is by putting permanent solid ballast at the double bottom region, thereby lowering the CG. However this may not be a very good solution as the solid ballast takes up a portion of the pay load.

The other solution to this problem is by making the superstructure discontinuous. The superstructure is made physically structurally discontinuous, thereby making it so-called ineffective superstructure. Structural discontinuity is incorporated over the entire length of the superstructure. Each length being less than 15 % LBP of the ship. These regions are connected to each other by expansion joints, which are flexible in nature. By doing this each part of the superstructure aligns itself with the bent shape of the hull girder undergoing longitudinal bending. The implication is, the superstructure in this condition does not resists longitudinal bending neither it contributes to longitudinal strength. Therefore the stresses developed do not depend on the distance from the neutral axis and also the stress levels remain on the much lower side. Hence the scantlings of the superstructure become low, reducing its weight substantially. Thereby the CG does not go up making the ship stiff.

On one hand this could be a good solution to the problem, at the same time it adds to different kind of problem in ship operation. All the piping connections running through these expansion joints need to be of flexible material. Over passage of time these flexible joints as well as the expansion joints may develop leakage and cause great inconvenience to the passengers on board.

It is therefore preferable to design with effective superstructure having scantlings able to withstand the hull girder bending load. These vessels like other ships also have double bottom. Longitudinal framing system is adopted in the decks shells and double bottom. The superstructure side shells are generally stiffened using transverse framing system whereas the partition bulkheads in the superstructure can be made of vertically corrugated plating.

Hatches and Holds

These vessels like any other vessel also have subdivision bulkheads, dividing the ship in number of holds. However these holds generally go empty, as no other cargo other than the passengers' personal effect are only carried. Some passenger accommodation are provided in the hold regions below the main deck, however they need to be always above the load water line. Below that deck level, the hold space up to the double bottom may remain empty. There are only access hatch and manholes in all the decks.

Capacity and Speed

The capacity of these ships is given in terms of number of passengers that it can carry. Large passenger liners have capacity to carry few thousands of passengers. One thing to be noted here is passenger ships for all practical purposes is also a cargo ship, only difference is its type of cargo. It is human cargo and to provide for even the bare minimum, it requires much voluminous space compared to that of any other cargo. Obviously to give a luxurious accommodation it will require several times more space. Hence in terms of dead weight, the capacity may not be very high.

Generally the speed of these vessels are on the higher side, though they are not always meant for point to point transportation. Depending on type of the vessel and the route, the speed may vary from about 15–25 knots.

References

1. Container Ship Design, © 2015 World Shipping Council.
2. Safety of RO-RO ferries, International Maritime Organisation. http://www.imo.org/OurWork/Safety/Regulations/Pages/RO-ROFerries.aspx.
3. IMO and ro-ro safety, International Maritime Organisation, Jan. 1997.
4. Resolution A.581(14), Guidelines for securing arrangements for the transport of road vehicles on RO-RO ships, International Maritime Organisation.

Chapter 2
Characteristics of Shipbuilding Industry

Abstract Shipbuilding industry has a very specific character as compared to that of any other manufacturing industry. Ship building involves usage of a wide range of equipment, materials and skills. The very size of ships makes it different from other industrial products. The huge size along with the required fittings and fixtures depending on the type of the vessel, it calls for a huge manhour requirement. It calls for a very wide variety of materials to be used in ship construction. As a very wide variety of materials are used it also calls for personnel with skills of various trades. It is not a mass production item, neither a show case item. It is a case of unit production. The ship builder gets only four inputs from the customer, i.e. the type of cargo, volume or weight of cargo to be carried, ship's route of operation and ship's cruising speed. Based on these, the builder needs to work out the entire design, build strategy, delivery schedule and cost of the ship. The ship builder also needs to guarantee that the vessel will deliver the required speed at the given loading condition.

Ships as a medium of large or bulk transportation will be ever present as long as human settlement is there on this earth. Global trade is going to exist and bulk import/export will remain economically and environmentally most viable through water/sea transport only. On the other hand as resources on land are getting depleted, more and more the focus is on offshore, be it energy, food, shelter, mineral, etc. Offshore units for oil and gas extraction, offshore energy firms, offshore fish firms, offshore mining, floating facilities, etc. are already there at different stages of usage/development. To support all these activities various types of specialised ships are also required. Lastly for a country's security and safety, a wide range of defence related ships are required. Thus it is very evident that ships will always play an important role as long as the human existence will be there in this earth. To cater to these needs shipbuilding industry will always exists. However for a given shipyard to flourish and exist, it needs to be competitive. This forces shipyards to improve productivity and thereby become more competitive. To make a product competitive one needs to understand the specific features of the product as well as that of the industry.

N.R. Mandal, *Ship Construction and Welding*, Springer Series
on Naval Architecture, Marine Engineering, Shipbuilding and Shipping 2,
DOI 10.1007/978-981-10-2955-4_2

Shipbuilding industry has a very specific character as compared to that of any other manufacturing industry. Ship building involves usage of a wide range of equipment, materials and skills. The very different character of this industry can be understood through studying the following aspects.

Product size
The very size of ships makes it different from other industrial products. The size of a ship may vary from few meters to few hundred meters. A small river boat may be 3 m in length whereas an ocean going crude oil tanker may have a length in excess of 300 m. Here it needs to be understood that such a huge product is not only to be manufactured but also to be launched and put into water unlike any civil structure. Other large industrial products like railway locomotives or aircrafts though not at all comparable as far as the size is concerned, are put into service using their own power. Whereas ships are built on land and need to be put into water for its operation. Hence it involves not only construction but also launching and its subsequent operation.

Manhour requirement
The huge size along with the required fittings and fixtures depending on the type of the vessel, it calls for a huge manhour requirement. The production time of a ship from start to end may be anywhere from about 1 year to about 3–4 years. Unlike other manufacturing industries shipbuilding is still very much labour intensive because of the very nature of the product. One of the major reason is ships are unit products, they are never a case of large series or mass production. The evaluation of exact manhour for a given ship is also very difficult, because of various overlapping and complex nature of activities that are involved in shipbuilding.

Wide variety of materials
The ocean going ships apart from having all the engineering requirements also need to have all the luxurious boarding and lodging facilities for ship's crew along with all the computerised navigational and communication equipment. Thus it calls for a very wide variety of materials to be used in ship construction.

Skills of various trades
As a very wide variety of materials are used it also calls for personnel with skills of various trades involving fitter, welder, piping, electrical equipment and fittings, main and auxiliary engine installation, electronic and navigational equipment installation, joinery work, air conditioning and ventilation, etc.

Unit production
It is not a mass production item, neither a show case item. It is a case of unit production. In shipbuilding, it is customer driven market, i.e. the customer tells what he wants and the builder is supposed to deliver satisfying the customer requirement. Thus in general each individual ship becomes more or less different from the other, hence the question of mass production never arises, making it a pure item of unit production. Whereas in case of say automobile industry, it is the manufacturer, who decides what he is going produce through his own market

survey. The customer cannot put his specific requirement. It is a pure situation of show case product. In such cases, once the design is made, tested and frozen, it goes for production. The production volume can be in millions. Hence one derives all the benefits of mass production, but in shipbuilding it being a unit product, the benefit of mass production is never achieved. Another distinct feature in ship building is there is no provision for prototyping.

Series production
One step above unit production is series production. There are situations where a customer, a shipping company may order for more than one vessel having identical specifications. These series of vessels are often referred to as sister ships. For medium sized to big vessels this series may be of the order of 5–10 vessels, whereas for smaller vessels, say fishing trawlers, it may be in the order of 50–60 vessels. At the same time if the number of ships required is on the higher side, it is very likely that the order for manufacturing them may also be spread over more than one shipbuilder. Hence again it becomes more or less a situation of unit production. Even if some vessels of identical specification is there, the second vessel in line is started after the 1st one has progressed, subsequently the third one and so on. By the time the 1st one is completed, delivered and put in service, the 2nd in line is in advanced stage of completion. Once the 1st one goes in service, often feed back from it may lead to some changes or modification in the internal fittings and other items. Also at the same time, the time lag between two identical operations of two sister ships are spaced so widely apart in time scale, that it looses all the benefits of series production, i.e. the fixtures and set up for identical component fabrication can not be kept waiting for such long periods.

Delivery schedule
The ship builder gets only four inputs from the customer, i.e. the type of cargo, volume or weight of cargo to be carried, ship's route of operation and ship's cruising speed. Based on these four information the builder needs to work out the entire design, build strategy, delivery schedule and cost of the ship. The builder signs a contract with the ship's owner mentioning among other things, the delivery date, the ship's cost and guarantees the speed requirement failing which the builder will be liable to pay heavy demurrage to the ship owner. The delivery schedule, ship's cost and speed guarantee all these are stated at the time of signing the contract when not even any design work has started. Hence to satisfy all these customer requirements becomes a challenging task for the shipbuilder.

Ship's speed
The cruising speed is one of the very important criterion for the ship owner. The sailing schedule and number of round trips that a ship can make over a period of time depends very much on its cruising speed. A ship with higher speed can make more number of trips, thus it can generate more revenue. However at the same time higher speed means higher power requirement resulting into higher fuel consumption causing escalation in operating cost. Thus the extra revenue generated by higher speed may get offset by the additional expenditure due to higher fuel

consumption. Hence the ships' owners are very careful about deciding on the ship's speed of operation. In the contract between the ship owner and the ship builder, the speed is one of the very important aspect which is mentioned with necessary demurrage clause. The ship builder needs to guarantee that the vessel will deliver the required speed at the given loading condition. Therefore determination of exact power requirement and selection of appropriate engine and propeller become very vital in ship construction.

All these various aspects give a very specific character to the shipbuilding industry as compared to any other manufacturing industry. Coordination of activities of huge workforce, materials all together pose a serious and difficult organisational problem to be handled in a shipyard.

Chapter 3
Structural Requirement

Abstract The basic challenge faced by the naval architect is to assess the uncertain loads that are coming on the structure. The solution to the structural requirement is not unique. It will have multiple solutions. One needs to figure out which one will be the most optimum one. A ship in a seaway experiences various kinds of motions causing dynamic loading on the ship structures. Along with this it is also subjected to static loading due to the self weight of the hull structure, cargo load, loading due to various machineries and equipment, buoyancy forces, etc. Thus it becomes an extremely challenging task to work out a structural solution satisfying this complex loading condition. All the loads coming on a ship structure on a seaway may be referred to as service loads. The response to these service loads is taken care of through four distinct strength considerations, longitudinal strength, transverse strength, torsional strength, local strength. Therefore the structural requirement calls for satisfying these four strength considerations and thereby takes care of the effect of service loads.

In determining the structural requirement of a ship, the basic challenge faced by the naval architect is to assess the uncertain loads that are coming on the structure. At the same time, the decisions like whether the deck plating thickness be increased by 1 mm to enhance buckling strength. What would be the effect of this on stability and on cost of construction. Likewise there are various other situations where the solution to the structural requirement will not be unique. It will have multiple solutions. One needs to figure out which one will be the most optimum one.

A ship in a seaway, depending on prevailing weather condition may experience various kinds of motions causing dynamic loading on the ship structures. Along with this the structure is also subjected to static loading due to the self weight of the hull structure, cargo load, loading due to various machineries and equipment, buoyancy forces, etc.

Loads coming on a ship structure can be classified based upon its variation over a period of time. Static loads are those which do not change over a short period of time. Stillwater loads, i.e. external hydrostatic pressure, buoyancy forces, lightship weight items, i.e. machinery, fittings and fixtures, piping, steering gear, fixed

N.R. Mandal, *Ship Construction and Welding*, Springer Series
on Naval Architecture, Marine Engineering, Shipbuilding and Shipping 2,
DOI 10.1007/978-981-10-2955-4_3

weapon systems, etc. and dead weight items, i.e. cargo, fuel, water, provisions, crew, etc. comprise the static load. This time period is considered to be far greater than the ship structures' natural frequencies of vibration or flexure. The static loads are shown schematically in Fig. 3.1.

Loads, having period slightly greater than the natural flexural periods of the ships primary structure can be referred to as slowly-varying loads. A ship hull gets subjected to variable hydrostatic and dynamic pressures because of wave action and also due to ship motion. These wave induced loads are slowly-varying loads and often they are referred to as quasi-static loads. The longitudinal distribution of the pressures on the hull may cause longitudinal bending of the hull, the transverse distribution may result into transverse distortion of the hull—called 'racking' and oblique distribution would cause combinations of bending, racking and torsion or twisting of the hull. The wave-induced loads causing longitudinal bending and hull racking are shown in Figs. 3.2 and 3.3.

In addition to wave induced loads, there are other 'slowly-varying' loads like, waves hitting on sides and foredecks, sloshing of liquids in tanks, shipping of green water on decks, etc. There are loads which can be classified as 'rapidly varying

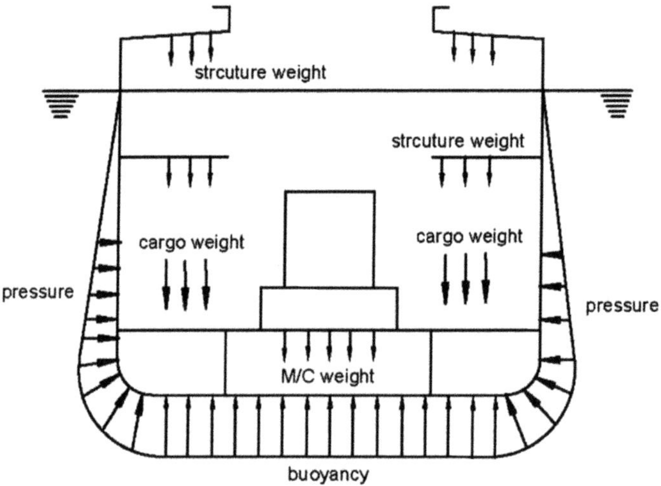

Fig. 3.1 Schematic representation of static loads on a ship structure

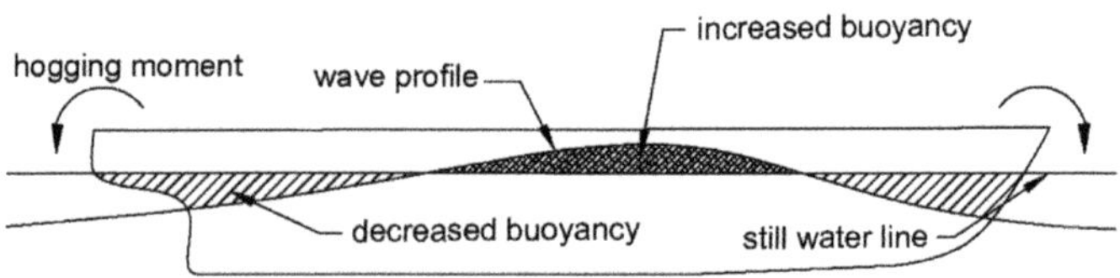

Fig. 3.2 Wave induced loading causing longitudinal bending due to hogging moment

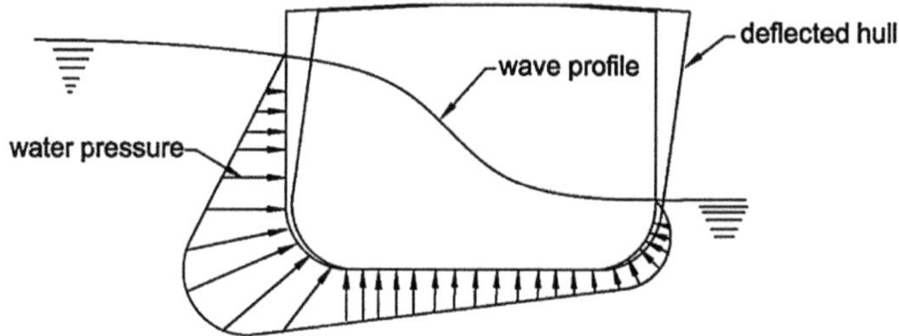

Fig. 3.3 Wave induced loading causing racking of ship hull

loads'. The timescale of variation of these loads is of the order of the natural periods of vibration or flexure of the ships' structure. This time period can be of the order of a second or less. These are the true dynamic loads on a ship structure. Examples are, *slamming* of forward bow causing localised bow damage, *springing*, a flexural resonance of the hull girder driving through waves causing rapid wave encounters. Other dynamic loading include mechanical vibrations within the ships structure caused by propeller and machinery rotations. Other sources of dynamic loads include combat loads in case of naval ships and underwater explosions. The dynamic loads generally are localised in nature.

Thus it becomes an extremely challenging task to work out a structural solution satisfying this complex loading condition. All the loads coming on a ship structure on a seaway may be referred to as service loads. The response to these service loads is taken care of through four distinct strength considerations as given below:

(i) Longitudinal strength.
(ii) Transverse strength.
(iii) Torsional strength.
(iv) Local strength.

Therefore the structural requirement calls for satisfying these four strength considerations and thereby takes care of the effect of service loads.

3.1 Longitudinal Strength

Ships, specially the ocean going ones are somewhat slender structure, i.e. it has a length to breadth ratio varying from about 3 to 6. Hence the longitudinal strength aspect becomes very important. Due to the difference in the weight and buoyancy distribution along the length, a ship experiences a longitudinal bending moment with its maximum at around midship region. With increase in ship's length, the bending moment also increases. Hence for larger vessels longitudinal strength requirement assumes more importance. This causes tensile and compressive stresses to develop in the hull girder. The deck and the keel plate being furthest away from the neutral axis, they experience maximum stress. To satisfy the longitudinal strength requirement, the structural arrangement is so designed such that the stresses developed at the furthest members from the neutral axis remain within the permissible limits. To avoid buckling or tensile failure it is always preferable to have longitudinal framing system in the deck as well as in the double bottom.

The structural members, which resist longitudinal bending, contribute towards longitudinal strength. For a member to effectively contribute towards longitudinal strength, it should be at least 15 % of the ship's LBP. Members which are shorter than this do not resist bending, instead align with the bent shape of the hull girder. It is explained through the following example:

Consider a beam of length L simply supported at the ends. Another beam is welded on top of it as shown in Fig. 3.4a. A bending moment is applied to it and the resulting deflected shape is shown in Fig. 3.4b. Now consider another identical setup of length L with the beam on top cut at intervals in small pieces as shown in Fig. 3.5a. Same bending moment is applied to it and the deflected shape is shown in Fig. 3.5b. Here one can observe that the deflection in the later case is more as the top beam is not resisting the deflection, instead the small pieces have aligned themselves along the bent shape of the lower beam.

The structural members which contribute to longitudinal strength in a ship structure are given below:

(i) Main deck plate
(ii) Deck longitudinal
(iii) Hatch side girder
(iv) Inner and outer side shell
(v) Inner and outer side shell longitudinal
(vi) Continuous lower deck plate
(vii) Lower deck longitudinal
(viii) Lower deck hatch side girder
(ix) Tank top
(x) Bottom shell
(xi) Tank top longitudinal
(xii) Bottom shell longitudinal
(xiii) Central girder
(xiv) Side girder
(xv) Sloping bulkheads of top and bottom wing tanks
(xvi) Wing tank longitudinal
(xvii) Longitudinal bulkhead.

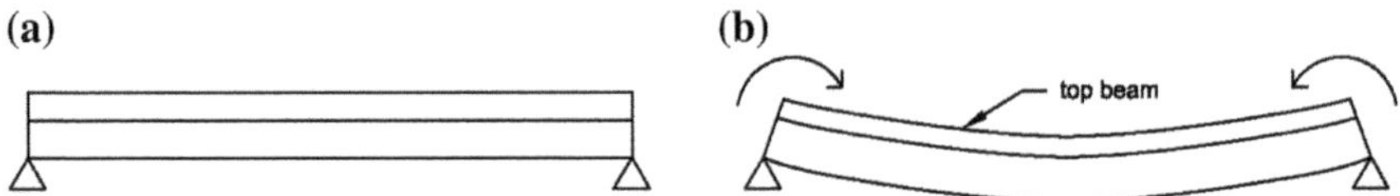

Fig. 3.4 *Top beam* welded to the *bottom beam*

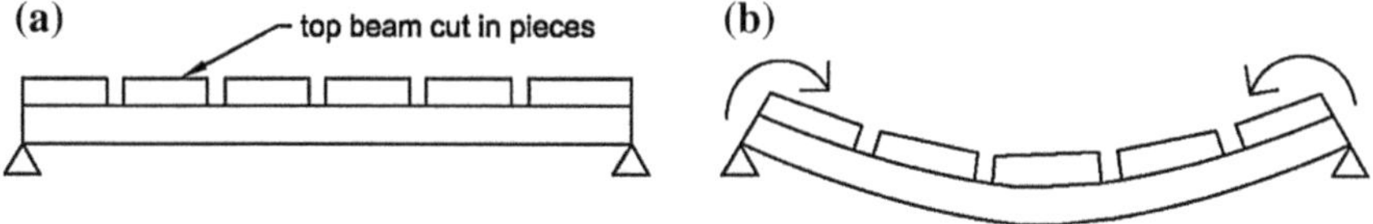

Fig. 3.5 *Top beam* welded to the *bottom beam* and cut in pieces

3.2 Transverse Strength

Due to external loads e.g. resulting from ships encountering oblique waves, the ship's hull is subjected to transverse loading. If the hull structure is inadequately strengthened for transverse strength, these transverse loads may cause deformation in transverse direction known as 'Raking' as shown in Fig. 3.3. To address this problem and to provide adequate transverse strength, adequate transverse stiffening members are provided. The structural members which contribute towards transverse strength of a ship are given below:

(i) Main deck plate
(ii) Deck transverses
(iii) Deck beams
(iv) Hatch end beams
(v) Continuous lower deck plate
(vi) Lower deck transverses
(vii) Subdivision transverse water tight bulkhead
(viii) Side shell frames and web frames
(ix) Wing tank transverses
(x) Plate floors and water tight floors.

3.3 Torsional Strength

The problem of torsional strength assumes significance particularly in case of container ships where deck openings are very large in comparison to the breadth of the vessels. It is almost a case of open deck ship. Unless adequate stiffening is done the ships may undergo torsional deformation leading to loss of directional stability. This is taken care of by providing box girder or by adopting double walled, i.e. cellular construction. This problem becomes more critical as the size of the vessel increases. However for other ships having smaller deck openings or no deck opening as in case of crude carriers, torsional strength aspect is not as significant because the normal transverse stiffening automatically provides for adequate torsional strength.

When the external forces do not pass through the shear centre axis of a ship hull cross section, it results in torsional moments. These moments may be caused by static or dynamic forces acting on the ship hull. The static torsion is due to the forces in still water condition, whereas the dynamic torsion is due to the wave induced forces. Static torsion in still water may be caused by non-symmetrical loading of cargo over port and starboard, however the overall cargo loading is such that the vessel remains upright. A ship in a seaway may experience non-symmetrical distribution of hydrostatic and hydrodynamic forces over port and starboard caused by wave action. This will give rise to dynamic torsion of the ship hull. Forces due to mass acceleration resulting from the ship's motion may also contribute to dynamic torsional moments. At the same time vibrations due to propeller shaft torque may also add to the torsional moments.

However, since cargo is generally loaded and unloaded keeping in view the symmetry of loading over port and starboard, the dominant forces causing torsional moments are due to the dynamic loading. The magnitude of these wave induced torsional moments depend on various factors; the hull form, wave height, heading angle and the position of the centre of twist which again depends on the structural arrangement of the hull cross section. Exact estimation of these forces and moments are not always possible, however classification societies provide guidelines and formulae for estimating these forces and moments for design purposes.

3.4 Local Strength

Local strength requirement assumes significance in the areas where the hull structure is subjected to localized loading like in the case of support bearings of propeller shaft, support structure for rudder stock, derrick and crane foundation on the deck, forward end structure subjected to slamming load, forward most deck plating subjected to load due to shipping in of green waters, etc. Local strengthening is done by providing additional stiffening members like stringers in the side shell and centre line wash bulkhead in the ford end construction, cant beams in the forward part of deck plating, plate floors at every frame space in engine room, increased scantlings and additional stiffeners in way of seatings of deck machineries, etc.

3.5 Structural Arrangement

The structural arrangement of a ship is laid out primarily on the basis of the type of cargo the ship will carry. For example the structural layout of the cargo holds of a bulk carrier will be very much different from that of a container ship or a crude carrier or a RoRo vessel. The structural layout should be such that it will facilitate efficient stowage, loading and unloading of the cargo. The structural arrangement also depends on the framing system adopted for structural design.

While deciding and laying out a structural arrangement the strength to weight ratio aspect must be kept into mind. A good design will provide for high strength to weight ratio, e.g. frame spacing should be so chosen that for the required strength the weight of the structure works out to be minimum. If the functional requirement permits then the framing system should be such which gives better buckling strength.

Wide Panel

The plate panels in between the stiffeners can be considered as either wide panel or long panel depending on framing system. The critical load that may lead to buckling for a wide panel with the loaded edges simply supported and the unloaded edges free as shown in Fig. 3.6 is given by Eq. (3.1),

$$P_{cr} = \frac{\pi^2 D b}{a^2} \tag{3.1}$$

Therefore the critical buckling stress for a wide panel can be written as given by Eq. (3.2),

$$\sigma_{cr} = \frac{P_{cr}}{b \times t} = \frac{\pi^2 D}{t^3}\left(\frac{t}{a}\right)^2 \tag{3.2}$$

i.e.

$$\sigma_{cr} = \frac{\pi^2 E}{12(1-\gamma^2)}\left(\frac{t}{a}\right)^2 \tag{3.3}$$

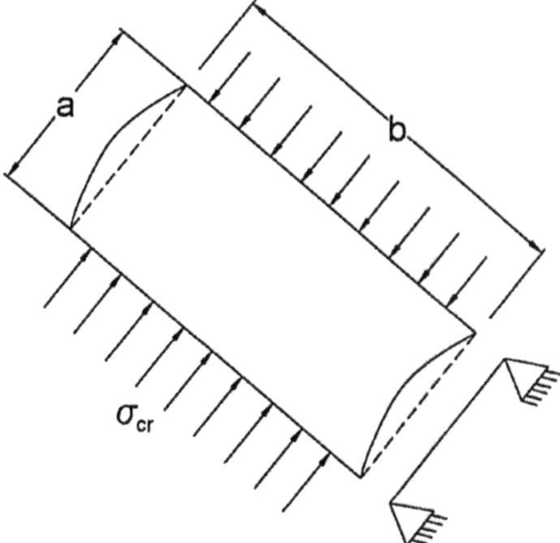

Fig. 3.6 A wide panel subjected to buckling load

where,

E modulus of elasticity
D flexural rigidity = $\frac{Et^3}{12(1-\gamma^2)}$
t plate thickness
a plate length
b plate width
γ Poison's ratio

Long Panel
A long panel under compressive loading exceeding critical buckling stress is shown in Fig. 3.7. The buckling mode shape is shown considering all the four edges to be simply supported.

The critical buckling stress in case of this long panel is given by Eq. (3.4),

$$\sigma_{cr} = \frac{\pi^2 a^2 D\left(\frac{m^2}{a^2} + \frac{n^2}{b^2}\right)^2}{tm^2} \tag{3.4}$$

where, m and n are half waves in each direction in the buckled shape. Here m = 3 and n = 1.

Minimum value of σ_{cr} is obtained with one half wave in transverse direction, i.e. n = 1. Therefore the critical buckling stress as given by Eq. (3.4) becomes,

$$\sigma_{cr} = \frac{\pi^2 D}{a^2 t}\left[m + \frac{1}{m}\left(\frac{a}{b}\right)^2\right]^2 \tag{3.5}$$

$$= \frac{\pi^2 D}{b^2 t}\left[\frac{m^2 b^2 + a^2}{abm}\right]^2 \tag{3.6}$$

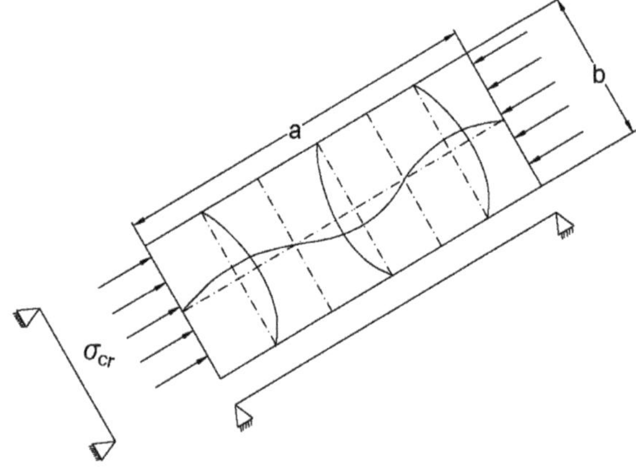

Fig. 3.7 A long *panel* subjected to buckling load

Let

$$\left[\frac{m^2b^2 + a^2}{abm}\right]^2 = k, \quad \text{i.e.} \quad k = \left[\frac{mb}{a} + \frac{a}{mb}\right]^2 = \text{buckling coefficient} \tag{3.7}$$

By setting $\frac{dk}{dm} = 0$ in Eq. (3.7), one obtains, $m = a/b$.

It signifies that the critical buckling stress attains the lowest value when the number of half waves in the longitudinal direction equals the aspect ratio of the panel. Therefore substituting m = a/b in Eq. (3.7) the buckling coefficient for the case of lowest critical buckling stress becomes 4.

Therefore the critical buckling stress for a plate of width b in terms of this buckling coefficient k can be written as given by Eq. (3.8),

$$\sigma_{cr} = 4\frac{\pi^2 D}{b^2 t} \tag{3.8}$$

3.6 Framing System

The basic structural component for ship construction is stiffened plate panels. They can be either flat or curved. These stiffened panels are prefabricated. The entire hull structure along with the superstructure is basically an assembly of these stiffened pate panels. The stiffening is always done in orthogonal direction, i.e. there will be stiffeners in both longitudinal and transverse direction. Longitudinal stiffeners and girders are those which run along the length of the ship and the transverse stiffeners, frames and beams are perpendicular to the length, i.e. along the transverse direction.

The system of stiffening is referred to as framing system. It is either longitudinal framing or transverse framing. However in a ship structure there can be a combination of both, e.g. decks and double bottom are stiffened with longitudinal framing system whereas the side shells are transversely stiffened, i.e. transverse framing system.

3.6.1 Longitudinal Framing System

In this system of stiffening, the longitudinal members often referred to as primary stiffeners are closely spaced and the secondary stiffeners, i.e. the transverse members are widely spaced as shown in Fig. 3.8. These transverse stiffening members are generally of higher scantling. In longitudinal framing system, the transverses are provided at about 3–4 frame spaces to reduce the unsupported span of the longitudinals and thereby provide support to the longitudinals as well as to provide for transverse strength to the hull girder.

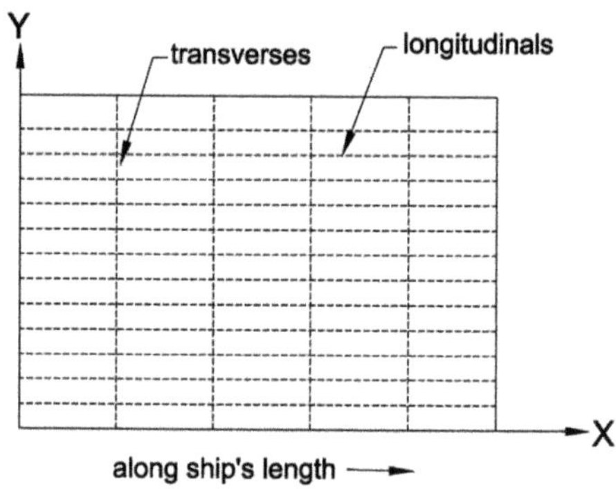

Fig. 3.8 Longitudinal framing system

In large ocean going ships, the dominant loading is due to longitudinal bending. The loading situation becomes worse when the ship's length becomes equal or more than the length of the encountering waves in the sea. This results in the hull girder experiencing very high longitudinal bending stresses. Longitudinal framing system is therefore preferable for ships where dominant loading is due to longitudinal bending (flexure). Hence to keep the bending stresses within the permissible limits, necessary section modulus of the midship section can be easily achieved with longitudinal framing system. The primary role of longitudinal members is to resist the longitudinal bending stress due to hogging and sagging moments as shown in Figs. 3.2 and 3.9 respectively.

Due to the sagging moment as shown in Fig. 3.9, the deck plating near midship region experiences compressive loading. It may lead to local buckling of the deck plating. However the longitudinal framing system having primary members as deck longitudinals contributes towards buckling strength of the deck structure.

A part of the longitudinally framed stiffened panel is shown in Fig. 3.10.

Here the plate panels in between the stiffeners behave like long panels. The critical buckling stress for such panels as given by Eq. (3.8), is obtained by substituting b = s,

$$(\sigma_{cr})_l = 4\frac{\pi^2 D}{s^2 t} \tag{3.9}$$

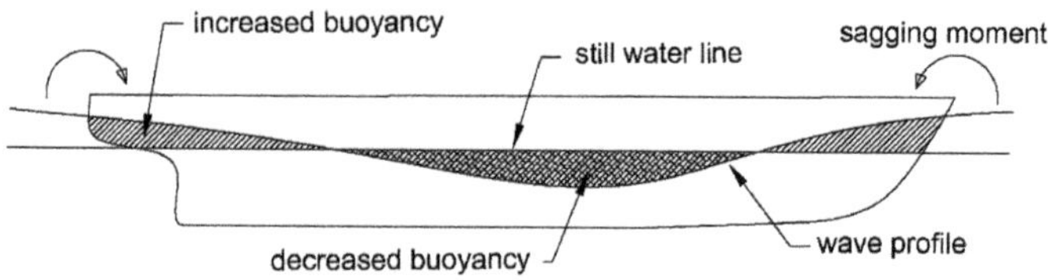

Fig. 3.9 Longitudinal bending due to sagging moment

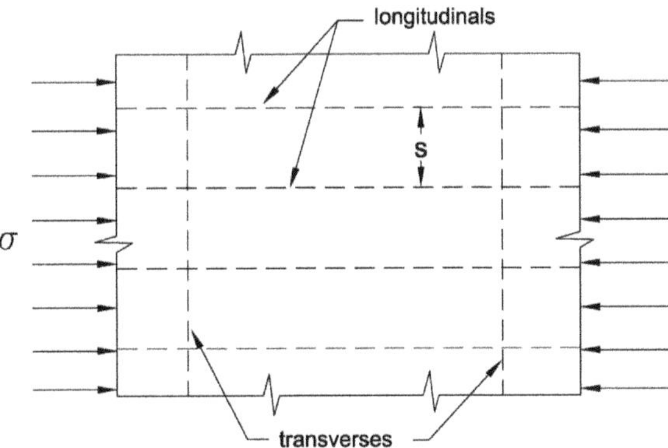

Fig. 3.10 Part of longitudinally framed stiffened *panel* under compressive loading

Fig. 3.11 Typical longitudinally framed stiffened *panels*

Typical longitudinally framed stiffened panels of a ship are shown in Fig. 3.11. One can also observe the buckling that has taken place in the plate panel. This is caused by weld induced stresses.

3.6.2 Transverse Framing System

In transverse framing system, the transverse frames or beams are the primary stiffeners. They are closely spaced with widely spaced longitudinal girders or stringers supporting them as shown in Fig. 3.12.

Transverse framing is used mainly in shorter vessels and submarines where loading is primarily hydrostatic and longitudinal bending load is not significant. At the same time it is used from functional point of view where longitudinal framing system may interfere with cargo loading, unloading and stowage. Many of the large ocean going ships have what may be termed as mixed framing system, e.g. the decks, bottom shell and inner bottom are longitudinally framed whereas the side shells are transversely framed as can be seen in general cargo carrier. Also in dry bulk carrier, the side shell in the hold region is transversely framed, whereas the rest of the structure has longitudinal framing system.

In transverse framing system, the plate panels in between the transverse stiffeners behave like wide panels, having aspect ratio $s/B<1$ as can be seen in Fig. 3.13.

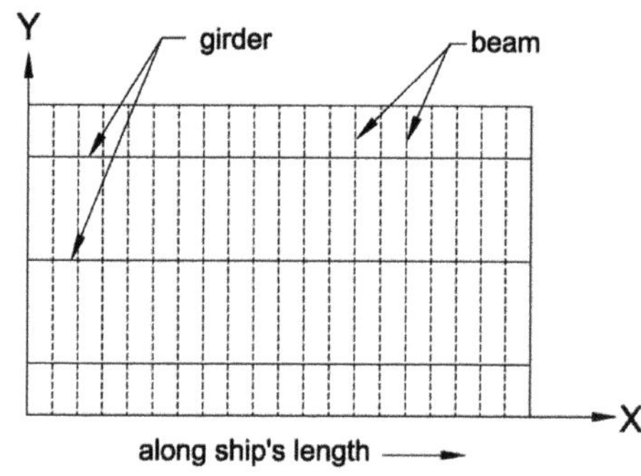

Fig. 3.12 Transverse framing system

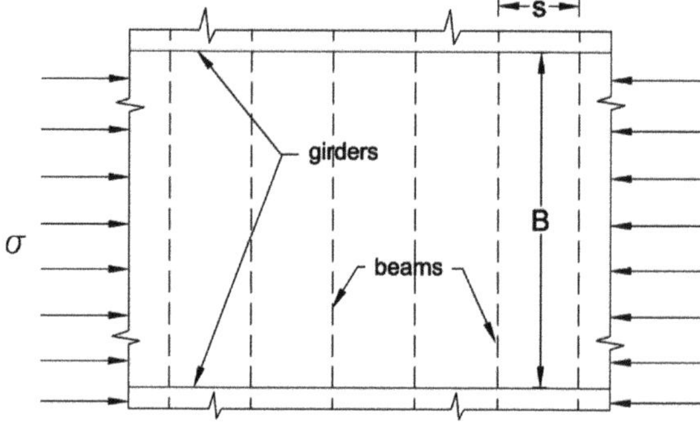

Fig. 3.13 Part of transversely framed stiffened *panel* under compressive loading

For such cases, m = 1. The critical buckling stress for wide panels as given in Eq. (3.5) can be re-written as,

$$(\sigma_{cr})_t = \frac{\pi^2 D}{s^2 t}\left[1 + \left(\frac{s}{B}\right)^2\right]^2 \tag{3.10}$$

In transverse framing system the longitudinal girders will be there providing for required longitudinal strength as well as provide support to the transverse primary stiffeners. The spacing of these girders is generally 4 times or greater than the spacing of the primary stiffeners, i.e. $\frac{s}{B} \geq \frac{1}{4}$.

Therefore the critical buckling stress in transverse framing system will be given by,

$$(\sigma_{cr})_t = \frac{\pi^2 D}{s^2 t}\left[1 + \left(\frac{1}{4}\right)^2\right]^2$$

i.e.

$$(\sigma_{cr})_t = 1.13\frac{\pi^2 D}{s^2 t} \tag{3.11}$$

Comparing the critical buckling stress given by Eq. (3.9) for longitudinal framing system with that of transverse farming system as given by Eq. (3.11), one can see that the critical buckling stress in case of longitudinal framing system is about 3.5 times higher than that of transverse framing system.

$$\text{i.e.}\quad (\sigma_{cr})_l = 3.5(\sigma_{cr})_t$$

Therefore to provide for the required tensile and buckling strength, the scantlings of the stiffening members as well the plate thickness in case of transversely framed stiffened panels is more compared to that in longitudinal framing system. Hence the weight of transversely framed panels works out to be more than longitudinally framed panels having same load bearing capacity. Therefore the strength to weight ratio in case of longitudinal framing system is higher than that of transverse framing system.

Preferred framing system

The above discussion shows that the longitudinal framing system is preferable to transverse framing system for the following reasons:

(i) $\sigma_{cr(L)} > \sigma_{cr(T)}$

The critical buckling stress in longitudinal framing system is about 3.5 times higher than that of transverse framing system.

(ii) $W_L < W_T$

Weight of stiffened panel with longitudinal framing is less than that of in transverse framing system. Therefore the strength to weight ratio in case of longitudinal framing system is higher than that of transverse framing system.

Chapter 4
Basic Structural Components

Abstract A ship is an assembly of various types of structural components. All these structural components provide the basic strength and support to the ship's shell structure. These components broadly fall under either longitudinal or transverse components. These can be in the form of basic structural members, e.g. deck longitudinal, side shell frame, etc. or prefabricated components, e.g. plate floor, wing tank transverse, etc. In ship construction an orthogonal form of stiffening arrangement is followed. The stiffening members are never arranged in an arbitrary fashion nor in any oblique direction. The reason being, the various structural components are prefabricated and they are subsequently aligned and welded in position. Brackets play a major role in making the connections between various components and thus providing for the load path. The primary failure cause of these brackets is buckling. To prevent such failures, flanged brackets or using curling, providing higher corrosion margin or elimination of lighting holes and various cutouts in highly loaded areas are used.

A ship is an assembly of various types of structural components. These can be in the form of basic structural members, e.g. deck longitudinal, side shell frame, etc. or prefabricated components, e.g. plate floor, wing tank transverse, etc. All these structural components provide the basic strength and support to the ship's shell structure. These components broadly fall under either longitudinal or transverse components. However brackets are not so classified. They play a major role in making the connections between various components and thus providing for the load path.

4.1 Longitudinal and Transverse Members

In ship construction an orthogonal form of stiffening arrangement is followed. The stiffening members are never arranged in an arbitrary fashion nor in any oblique direction. The reason being, the various structural components are prefabricated and

N.R. Mandal, *Ship Construction and Welding*, Springer Series
on Naval Architecture, Marine Engineering, Shipbuilding and Shipping 2,
DOI 10.1007/978-981-10-2955-4_4

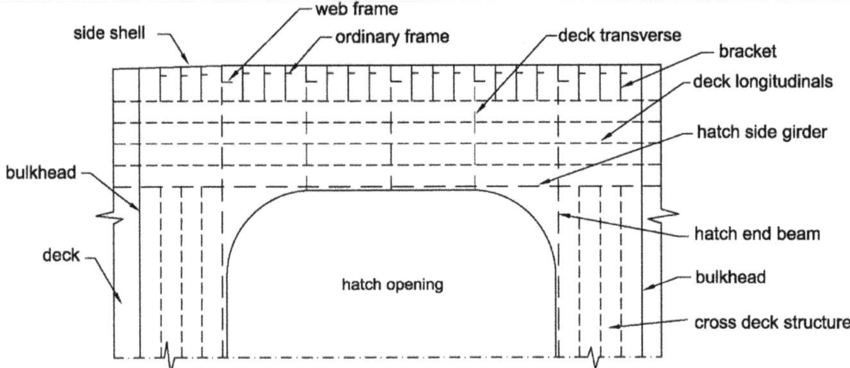

Fig. 4.1 Stiffening arrangement of deck and side shell of a general cargo ship

they are subsequently aligned and welded in position. This joining of two components, say two pieces of deck panels with longitudinal framing, will involve butt welding of the panel edges as well as welding of each longitudinals.

In transverse framing system, it is often observed that at a wider interval frames of higher scantlings, referred to as web frames are used as shown in Fig. 4.1. The spacing of these web frames are of the order of 3–4 times the frame spacing. The upper limit of the spacing is given by classification rules. The rest of the frames are so-called ordinary frames of lesser scantling. In general cargo ships and also in dry bulk carriers, transverse framing system is adopted in the side shell of the hold region. The web frames are placed in the same frame positions of deck transverse and plate floors in the double bottom. Thereby in bulk carriers the top wing tank transverse, web frame in the side shell, bottom wing tank transverse and plate floor in the double bottom form a ring like structure. Similarly in general cargo carrier, the deck transverse, web frame in side shell and plate floor in double bottom forms the so-called ring structure as shown in Fig. 4.2. This provides support to the longitudinals as well as transverse strength to the hull girder.

4.2 Girders and Transverses

Stiffening members of higher scantlings are referred to as girders and transverses. Members along the longitudinal direction, i.e. along the length of a ship, are named as girders, whereas those along the breadth of a ship are known as transverses. Girders provide for longitudinal strength and transverses contribute to transverse strength as well as provide support to longitudinals.

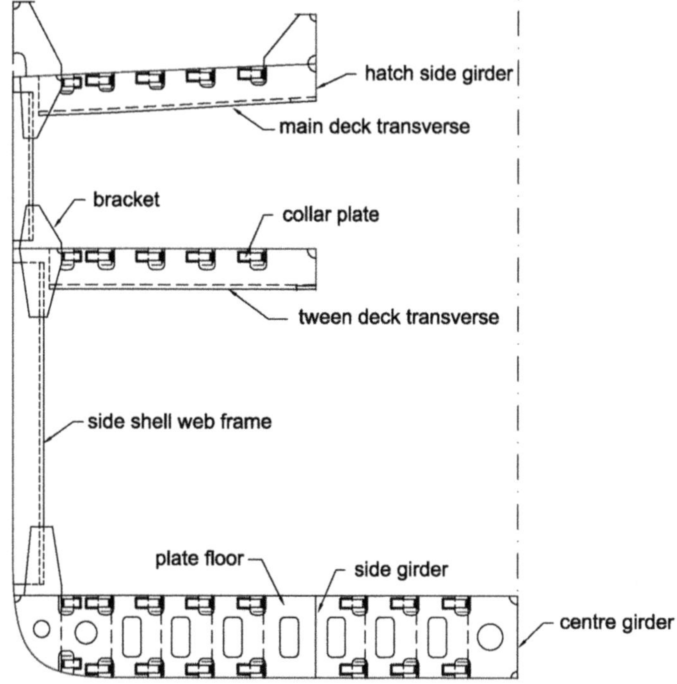

Fig. 4.2 Deck transverse, side shell web frame and floor forming a ring structure

4.2.1 Centre Girder

Centre girder also often referred to as centre keelson is a vertical flat plate in the double bottom, running along the centre line of the ship as shown in Fig. 4.2. It extends from the aft collision bulkhead to the forward collision bulkhead. Centre girder is welded to the keel plate at the bottom and to the inner bottom plating at the top, thereby divides the double bottom space in independent water tight port and starboard tanks. The general practice is to have centre girder running continuously and the port and starboard plate and bracket floors are welded to it.

4.2.2 Side Girder

Side girders are vertical flat plates within the double bottom running along the length of the ship from aft collision to forward collision bulkhead as shown in Fig. 4.2. However unlike centre girder, side girder is intercoastal, i.e. it runs from one plate floor to the next plate floor. It is terminated on the plate floor and welded to it. Side girders are generally not water tight, lightening holes are there to make them lighter and also to allow liquid flow within the double bottom space. In between the plate floors, at the bracket floor locations, struts are welded to the side girders. It provides necessary strength against buckling under vertical loading due to the cargo load acting downwards and buoyancy forces acting upwards. Depending on the ship's breadth, there can be one or more side girders on port and starboard side as per the requirement of classification rules or otherwise from functional requirement.

4.2.3 Hatch Side Girder

As the name suggests, it is the girder running at the side of the hatch opening as shown in Fig. 4.1 on both port and starboard side. It runs continuously piercing through the subdivision bulkheads along the length of the ship from forward to aft collision bulkhead. In the event of engine casing opening breadth being less than the hatch opening, the girder is terminated at the engine room bulkhead and a separate girder is provided at the side of the engine casing opening. The primary function of these hatch side girders is to compensate for the loss in longitudinal strength due to large hatch opening in the deck. These girders are generally built up sections, designed depending on the section modulus requirement. As these are at the side of the hatch opening, the girders are fabricated in the form of angle sections, such that the face plate of the girder do not interfere with cargo loading and unloading.

4.2.4 Hatch End Beam

The hatch openings in the deck plating have deep transverses at its forward and aft ends as shown in Fig. 4.1. These are named as hatch end beams. The primary function of these transverse stiffening members is to compensate for the loss in transverse strength caused by the large hatch opening in the deck. The scantlings of these deep beams are generally made same as those of the hatch side girders. The hatch side girders are kept continuous and the hatch end beams are terminated and welded to the girder.

4.2.5 Deck Transverse

These are transverse stiffening members of higher scantlings compared to the deck longitudinals. They are generally positioned at a spacing of about 3–4 primary frame spaces as shown in Fig. 4.1. In longitudinal framing system, these transverses

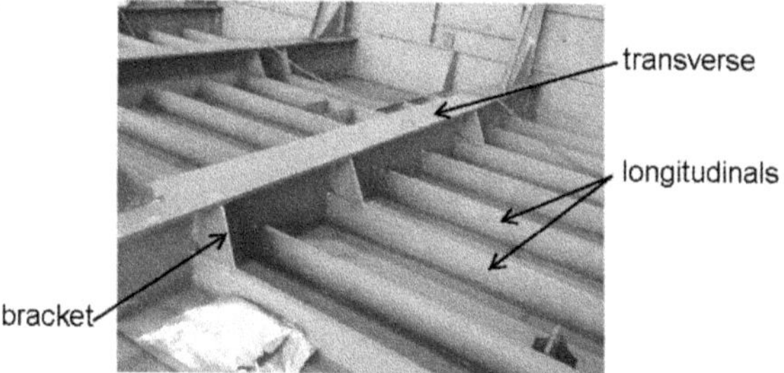

Fig. 4.3 Longitudinal framing system—longitudinals supported by the transverses

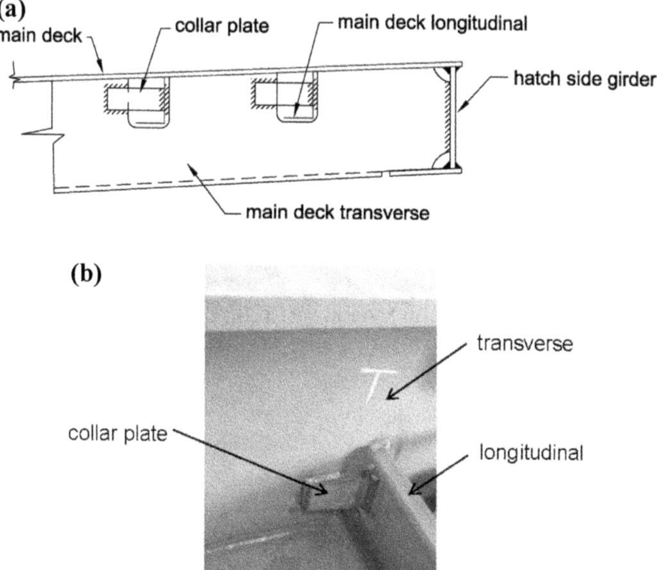

Fig. 4.4 **a** Intersection of deck longitudinals and deck transverse. **b** Intersection of longitudinal and transverse showing collar plate

provide for transverse strength as well as support the longitudinals as shown in Fig. 4.3. The longitudinal stiffeners run continuously piercing through the transverses.

At the longitudinal transverse intersection, the longitudinal is securely welded to the transverse web on one side and the on the other side, a collar plate is welded to make a rigid connection as shown in Fig. 4.4a, b. Thus the free span of a longitudinal becomes equal to the spacing between two adjacent transverses.

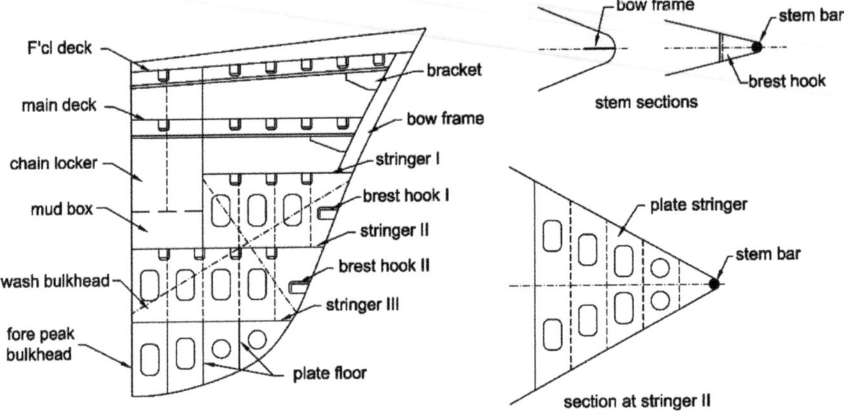

Fig. 4.5 Stringers and breast hooks in ford end construction

4.3 Stringers

Stringers are stiffening members of higher scantlings. Generally fabricated T sections or perforated stiffened plates are used as stringers. These are mainly used to additionally strengthen the side shell in fore peak structure of a ship against local loads coming from slamming or pounding of ship's bow. The bow plate or the so-called nose of the ship is generally made by sharp bending of plate. This bow plate is strengthened by welding horizontal pieces of brackets known as breast hooks. These breast hooks are placed in between the stringers. The stringers are horizontal members and run up to the ford collision or fore peak bulkhead as shown in Fig. 4.5. Vertical spacing between the stringers is kept as multiple of frame spaces as prescribed by classification societies.

Driving a ship at high speed in heavy weather may cause excessive slamming loads. Hence extreme care needs to be taken in the design, fabrication and maintenance of all structural connections and details used in the forward end of construction.

4.4 Floors

Floors are vertical transverse members in the double bottom of a ship. These floors are of three types: (i) Plate floor, (ii) Water tight floor and (iii) Bracket floor. Plate and water tight floors extend from centre line of the ship to the bilge and from bottom shell to the inner bottom plating of the ship. The bracket floors are essentially two pieces of brackets, at the bilge end and at the centre girder end.

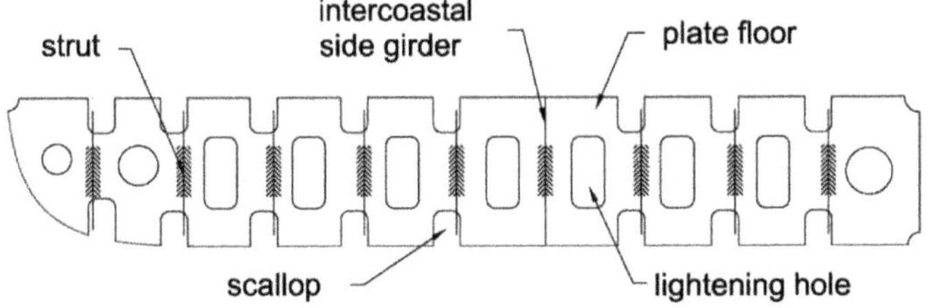

Fig. 4.6 A typical plate floor in longitudinal framing system

4.4.1 Plate Floor

Plate floors are generally laid out at spacing of 3–4 frame spaces. In between these floors, bracket floors are placed. However in places where there can be a situation of local loading like in engine room, for providing support to the main propulsion engine and other auxiliary machineries, it is recommended to have plate floors at all frame positions.

Plate floors often also referred to as solid floors are basically flat plate panels with struts stiffening them. The floors provide transverse strength to the hull structure. However, these floors need to be strengthened against compressive load due to buoyancy forces acting upwards and the cargo/machinery load acting downwards. This is done by using stiffeners welded to the floor plating. These stiffeners are known as struts. A typical plate floor in longitudinal framing system is shown in Fig. 4.6. In longitudinal framing system, the bottom and inner bottom longitudinals run along the length of the ship within the double bottom piercing the floors through the scallops. The web of the longitudinals are welded to the floor plate on one side and on the other side a collar plate is used to make a rigid connection between the longitudinal and the floor plate as shown in Fig. 4.4a. Thus the floors provide support to the longitudinals. Lightening holes of adequate dimensions are provided in the floors to lighten the structure as well as to provide for means of access inside the double bottom space.

4.4.2 Bracket Floor

A bracket floor is nothing but two pieces of plate brackets, connecting the centre girder to the immediate pair of bottom shell and inner bottom longitudinals and the other one at the bilge end connecting the pair of longitudinals in case of longitudinal framing system as shown in Fig. 4.7.

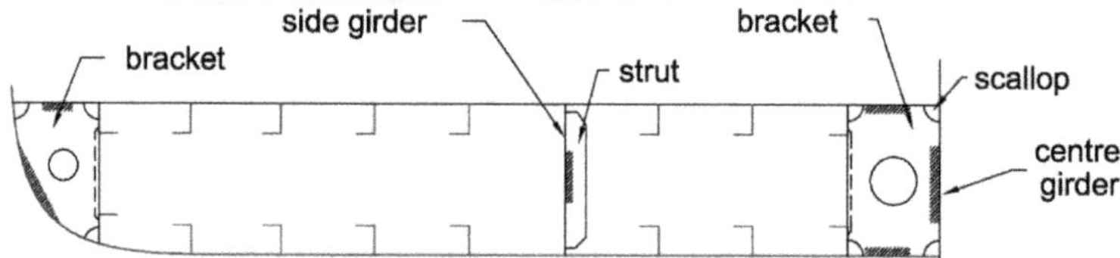

Fig. 4.7 A typical bracket floor arrangement in longitudinal framing system

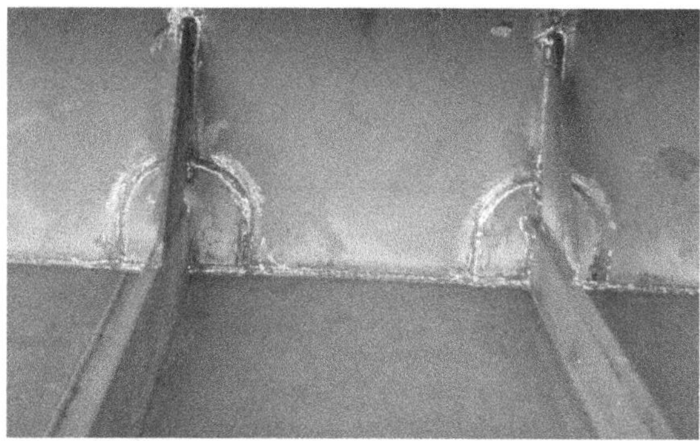

Fig. 4.8 Collar plate blanking off scallops in water tight floor

4.4.3 Water Tight Floor

Water tight floors are placed below the transverse subdivision bulkheads as well as in places where the double bottom space is used for tankage purposes. These are cut as plate floors without the lightening holes and subsequently, the scallops are blanked off using multiple collar plates as shown in Fig. 4.8.

4.5 Brackets

Brackets in a ship structure can be classified as a secondary structural member. It by itself may not directly contribute to structural strength, however helps in providing structural continuity as well as local strengthening, thus indirectly contributing to overall strength. Like structural continuity between the outermost deck longitudinal to the side shell frames as in general cargo ships is established through the brackets connecting them, Fig. 4.9a.

These corner brackets are often cut inside as shown in Fig. 4.9b. The predominant failure mode for these corner brackets is plate instability causing buckling. By this radiused cutting the straight flat plate brackets, it was fund to perform much better leading to lesser buckling failure. This may be attributed to improved stress flow and stiffness distribution of these radiused brackets [1]. Also curling, a flat bar, is often added to further enhance buckling strength of these brackets.

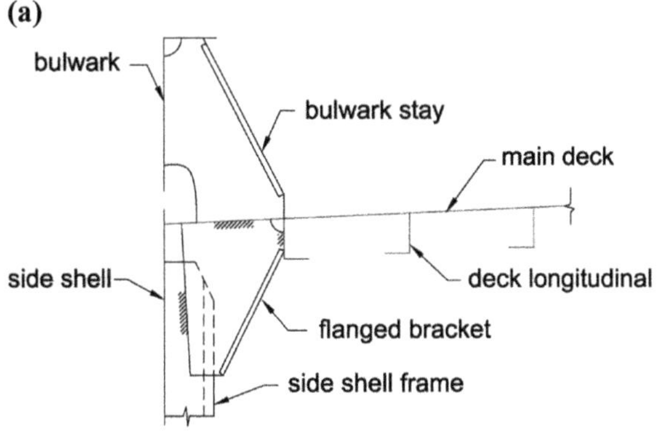

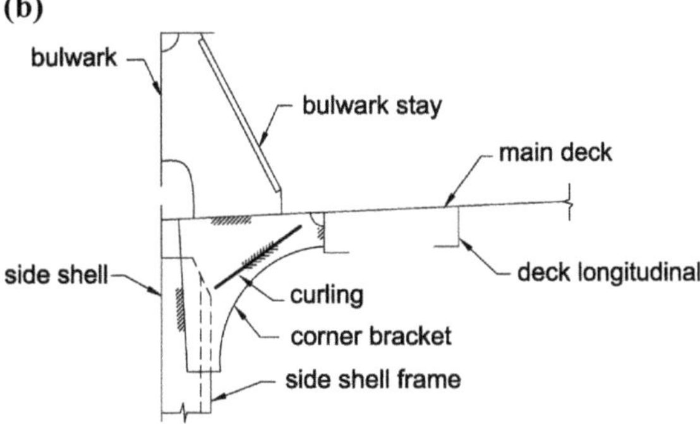

Fig. 4.9 **a** Bracket connecting side shell frame to deck longitudinal. **b** Rounded corner bracket connecting side shell frame to deck longitudinal

Similarly brackets are provided to connect the side shell hold frames to the sloping bulkhead of the bottom wing tank in bulk carriers, Fig. 4.10. This provides for effective load path continuity through bracketed end connection of the side shell frames.

It is to be noted that the bracket toe should always come on top of some stiffening member below as shown in Fig. 4.10. Buckling failures are mostly observed in such plate brackets. It is mostly eliminated by putting a flange at the free end or by putting a curling in the bracket.

The brackets that provide lateral support to framing and stiffening members are known as Tripping brackets as shown in Fig. 4.11. Support may be provided to either the web or the flange, or to both.

The primary failure cause of these brackets is panel instability, i.e. buckling of the brackets. The possible measures that can be taken to prevent such failures are use of flanged brackets or using curling, i.e. stiffener on the bracket, to provide for higher corrosion margin using higher thickness bracket and reduction or elimination of lighting holes and various cutouts in highly loaded areas [2].

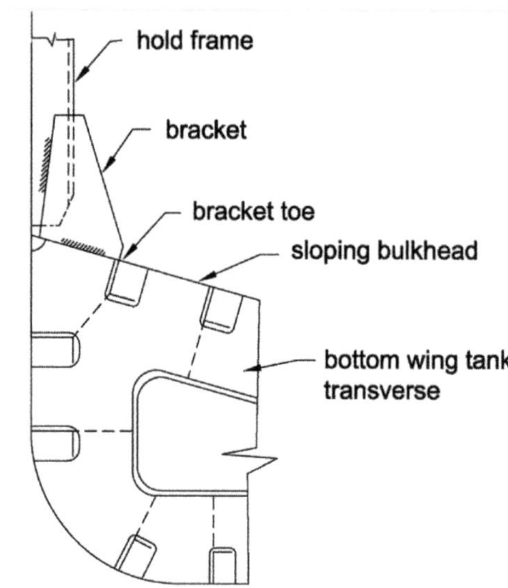

Fig. 4.10 Bracket connecting the hold frame to sloping bulkhead

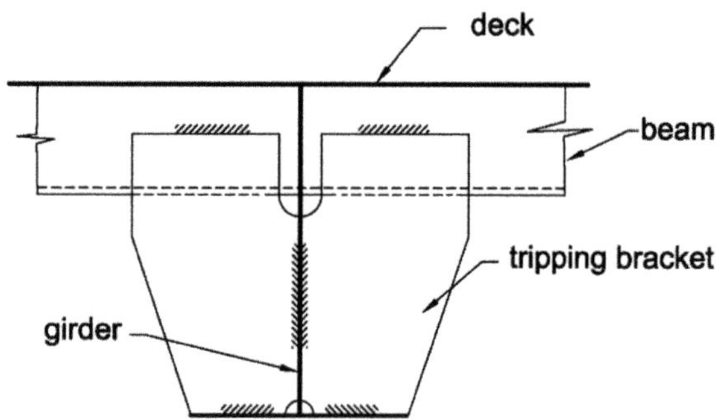

Fig. 4.11 A typical tripping bracket providing support to both web and flange of a girder

References

1. Design Guide for Ship Structural Details, SSC-331, Ship Structure Committee, 1990.
2. Standard Guide for Selection of Structural Details for Ship Construction, Designation: F1455 – 92 (Reapproved 2011), ASTM F1455.

Chapter 5
Structural Subassemblies

Abstract Shipbuilding involves assembly of various pre-fabricated structural components along with installation of various machineries and fitting out items. The products of first stage of assembly are referred to as subassemblies. These subassemblies put together yields assemblies and subsequently units, blocks and finally the complete ship. In the subassembly stage either flat or curved structures are fabricated. These are essentially two dimensional or simple three dimensional structures, comprised of flat or curved stiffened panels. These flat or curved stiffened panels would be either longitudinally framed or transversely framed.

Shipbuilding is essentially an assembly industry. It involves assembly of various pre-fabricated structural components along with installation of various machineries and fitting out items. The products of first stage of assembly are referred to here as subassemblies. These subassemblies put together yields assemblies and subsequently units, blocks and finally the complete ship.

The sequence of production activities carried out from steel stockyard to the stage of fabrication of subassemblies is shown schematically in the Fig. 5.1

In Fig. 5.1, one can see that, after plate preparation, the steel material is taken to the flame burning nay, i.e. plate cutting station. The plates are cut to the required size and shape. These cut plates are then fed to the 1st stage of prefabrication for production of so-called subassemblies. As can be seen from Fig. 5.1 some of the plates are fed directly from the cutting bay as flat and straight plates and some are fed through bending operation, that means curved plates. Thus in the subassembly stage either flat or curved structures are fabricated. These are essentially two dimensional or simple three dimensional structures, comprised of flat or curved stiffened panels. Some of the subassemblies are as shown in Fig. 5.2.

These flat or curved stiffened panels would be either longitudinally framed or transversely framed. That means, either longitudinal framing system will be implemented or transverse framing system.

N.R. Mandal, *Ship Construction and Welding*, Springer Series
on Naval Architecture, Marine Engineering, Shipbuilding and Shipping 2,
DOI 10.1007/978-981-10-2955-4_5

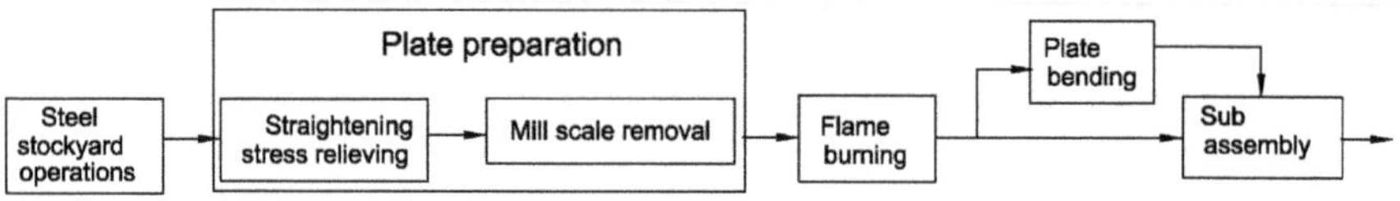

Fig. 5.1 Sequence of plate fabrication activities

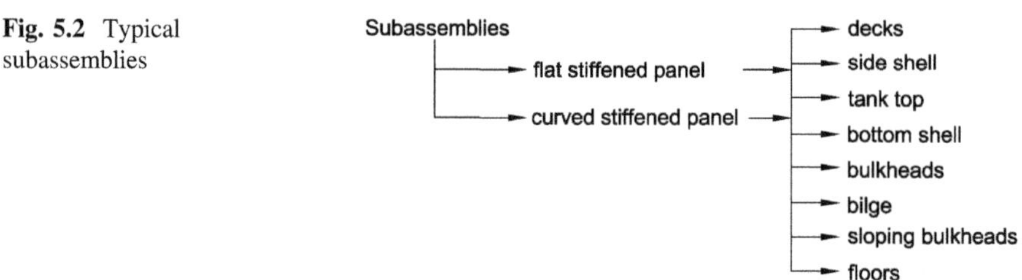

Fig. 5.2 Typical subassemblies

5.1 Flat Stiffened Panel

The very initial prefabricated structural items are those flat stiffened panels. The simplest ones are various types of brackets (Fig. 5.3) with curling followed by floors, transverses, girders and large stiffened panels.

Fig. 5.3 Typical flat subassembly

5.2 Curved Stiffened Panel

Subassemblies of curved stiffened panels unlike flat stiffened panels, are primarily fabricated as a part of a larger 3-D unit as shown in Fig. 5.4.

5.3 Bulkheads

Bulkheads can be classified under two heads: transverse subdivision water tight bulkheads and non water tight bulkheads as depicted in Fig. 5.5.

These transverse sub division water tight bulkheads subdivide the ship in several water tight compartments depending on floodable length and strength requirement. The concept of floodable length is the total length of a ship that can be flooded leading to heeling and trimming of the vessel however the deck will not get immersed. In fact the water surface, i.e. the water line will remain tangential to an imaginary line referred to as margin line that is 75 mm below the deck at side.

Longitudinal bulkheads in the cargo holds are widely used in case of oil tankers. The transverse sub division bulkheads divide the ship in several water tight compartments. Whereas the longitudinal bulkhead sub divides a vessel longitudinally along the transverse plane. The purpose is to divide the cargo space in longitudinal compartments. Depending on the breadth of the vessel, there can be one central line longitudinal bulkhead or there can be more number of longitudinal bulkheads port

Fig. 5.4 Typical curved subassemblies

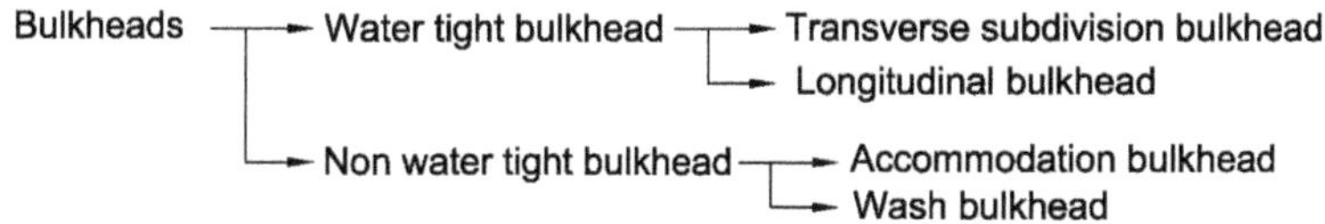

Fig. 5.5 Classification of bulkheads

and starboard. This longitudinal division is done to reduce the effect of free surface. In oil tankers and liquid cargo carriers, the free surface of the liquid in the cargo hold gives a negative effect on the vessel's stability. The reduction in the metacentric height is more if the free surface area is more. Hence longitudinal bulkheads are used to reduce this free surface area thus reducing the negative effect on stability.

The non-water tight bulkheads are generally the bulkheads in the accommodation region referred to as accommodation bulkheads. Also non water tight bulkhead is there in the fore end construction. It is referred to as wash bulkhead and provides additional strength to the fore end construction against slamming and pounding forces. This wash bulkhead is generally placed at the centre line of the ship, hence in essence it can be termed as longitudinal bulkhead.

5.3.1 Transverse Water Tight Bulkhead

Transverse subdivision bulkheads can be of flat stiffened plate construction or of corrugated construction. The basic functions of these bulkheads are:

(i) They divide the ship into several watertight compartments.
(ii) If by accident any compartment gets flooded, the flooding is kept confined within that compartment by these bulkheads. These are designed to take the hydrostatic load in case of flooding.
(iii) These bulkheads provide support to the longitudinals. These longitudinals run continuously piercing through the bulkheads. They are welded to bulkheads and the opening is thoroughly sealed to make them water tight. Thus it provides support to the longitudinals.
(iv) These bulkheads are one of the major members providing transverse strength to the hull structure. It prevents racking of the hull.
(v) Should any fire break out in any cargo hold, these bulkheads should also be able to confine the fire within the hold. This talks about the material of the bulkhead, it cannot be made of any easily combustible or low melting material.

5.3.1.1 Flat Stiffened Bulkhead

The bulkheads provide transverse strength. However in the event of some hull damage leading to water ingress in the compartment, the hold bulkheads will be subjected to hydrostatic loading as shown in Fig. 5.6. As it is well known that the hydrostatic loading is directly proportional to depth of water within the hold, the loading on the bulkheads will be maximum at the bottom of the bulkhead.

Because of this the plating arrangement of these bulkheads are horizontal streaks of plates with reducing thickness from bottom to the top of the bulkhead as shown

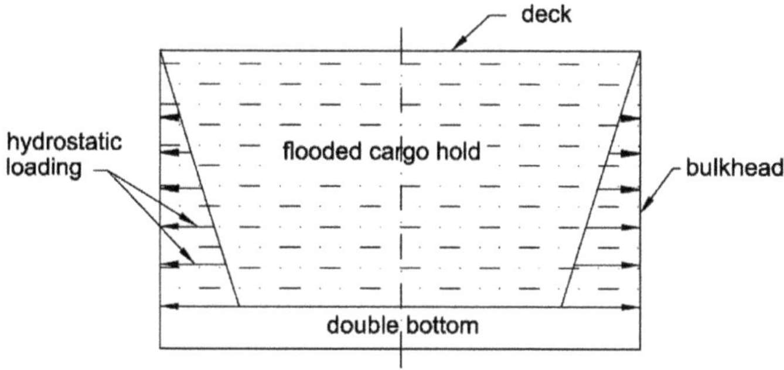

Fig. 5.6 Hydrostatic loading on the bulkheads of a flooded cargo hold

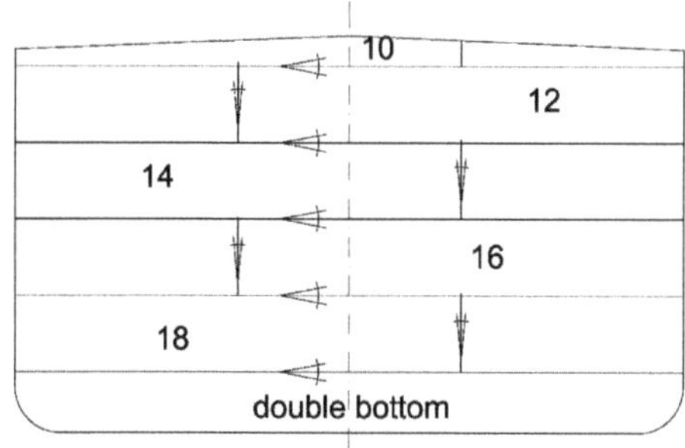

Fig. 5.7 Plating arrangement of a typical flat plate bulkhead

in Fig. 5.7. In this case the plates are arranged breadth wise along the height of the bulkhead. Thus with reducing thickness along the height, the net weight of the bulkhead is reduced.

Stiffening of flat plate bulkheads can be either vertical or horizontal. The stiffener orientation should be such that results in minimum free span of the stiffeners. That means minimum length between the support points. With increase in the span, for the same loading condition, the bending moment increases proportional to the length squared. This will require stiffener of higher section modulus to keep the stress level within the permissible limits. This will call for stiffener of higher scantlings. Hence it is important to have the stiffening arrangement such that it provides for minimum free span.

In case of flat plate subdivision bulkheads of general cargo carrier vertical stiffeners are used. Since such vessels will generally have at least one lower deck and tank top, therefore the vertical stiffeners of the bulkhead will get a natural support at these intermediate points. This will reduce their free span and thus the resulting bending moment for the same load will be highly reduced. Hence stiffeners of lower scantlings can be used to attain same amount of strength. A typical section is shown in Fig. 5.8.

In case of oil tankers or bulk liquid cargo carriers, there will not be any tween deck instead it is very likely that there will be longitudinal bulkhead for reducing

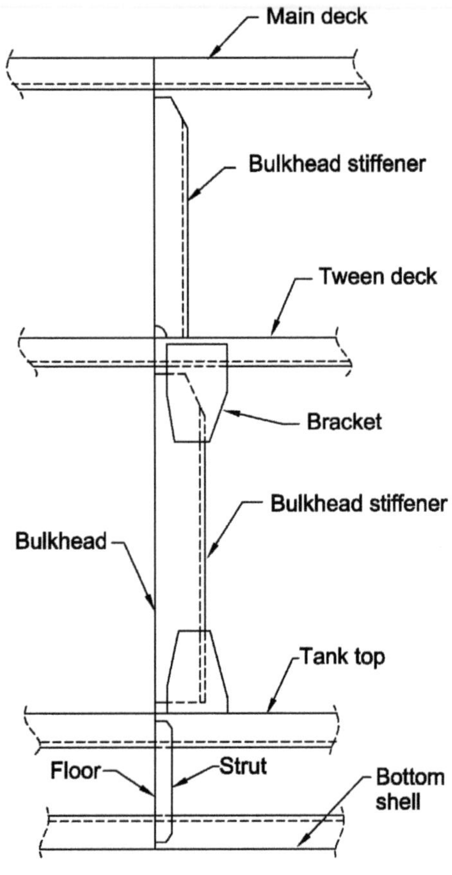

Fig. 5.8 Bulkhead stiffening arrangement in general cargo ship

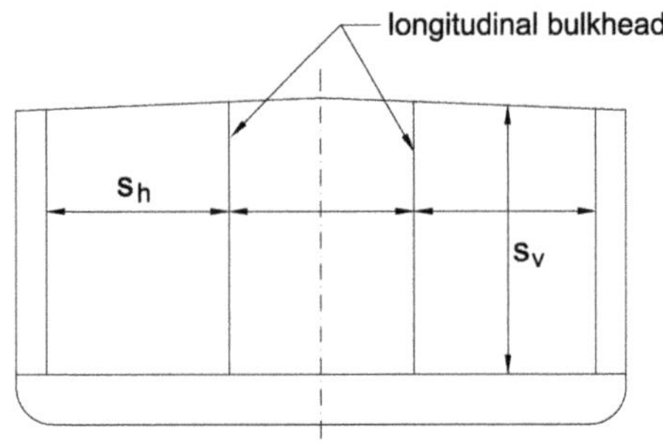

Fig. 5.9 Comparison of free span of stiffener orientation in oil tanker

free surface effect. Hence in such a condition the stiffeners of transverse subdivision bulkheads will be laid horizontally, having support at the side shell and the longitudinal bulkheads. Thus the free span S_h in this configuration will be less compared to having it in vertical orientation, S_v as can be seen in Fig. 5.9.

5.3.1.2 Corrugated Bulkhead

The stiffness is achieved by providing corrugations to the plate. Here the geometry of the corrugation is decided based on the section modulus requirement, as in case of stiffened bulkhead, the section modulus of stiffener is decided. The section modulus depends on the depth and width of the corrugations as shown in Fig. 5.10.

From fabrication point of view, it is advantageous to use corrugated bulkhead, provided, the shipyard has adequate facility in terms of hydraulic press and necessary die for fabricating the corrugated units. Once the corrugation parameters are worked out depending on the section modulus requirement as per the classification rules, these individual corrugated units are fabricated. These are produced by a single stroke in a hydraulic press using suitable male and female die. A typical V-type male female die is shown in Fig. 5.11.

Each plate is thus individually corrugated and the final bulkhead is constructed by butt welding these individual units. The involvement of welding is much less here compared to that of flat plate bulkheads with stiffeners. The corrugations are given along the plate length. In general where ever corrugated bulkheads are used, the corrugations are kept along the vertical. Therefore the advantage of reducing plate thickness as obtained in flat plate bulkheads is not there in case of corrugated bulkheads. These type of bulkheads are generally used in bulk carriers and oil tankers.

5.3.2 Non Water Tight Bulkheads

The partition bulkheads in the accommodation region and the wash bulkhead in the fore end construction are non water tight bulkheads. The wash bulkheads even have fairly large openings to make the structure lighter. These openings are referred to as lightening holes. These bulkheads can be of flat stiffened plate construction or of corrugated construction.

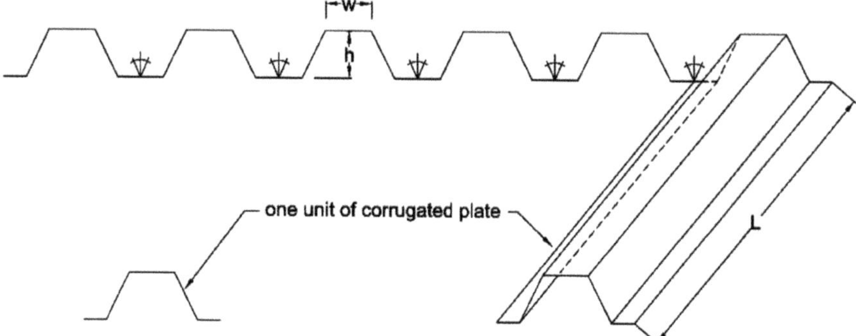

Fig. 5.10 A typical geometry of a corrugated bulkhead

Fig. 5.11 V-type male female matched die

5.4 Decks and Shells

The external skin of a ship comprises of weather deck, side and bottom shell. Depending on ship type and its functional requirement, a ship may have multiple decks as well as inner and outer side shells. As such now it is mandatory to have double bottom construction, i.e. outer bottom or bottom shell and inner bottom or tank top. All these decks and shells are essentially flat or curved stiffened plate structures.

5.4.1 Main Deck and Lower Deck

The uppermost continuous deck of a ship is generally referred to as the main deck. It is also called weather deck as it is exposed to the weather. The continuous deck to which the transverse subdivision bulkheads extend is referred to as bulkhead deck. Hence the main deck, the weather deck and the bulkhead deck can be one and the same deck and in most of the cases it is so.

Depending on functional requirement, a ship may not have any lower deck or can have a single lower deck or multiple lower decks. Bulk carriers, oil tankers, container ships do not have lower deck, however there will be at least one lower deck in general cargo ship. This is also referred to as tween deck. Whereas in RO RO vessels, there will be multiple lower decks. All these decks are continuous and run through the cargo holds along the ship's length and thereby they contribute towards longitudinal strength of the ship. Hence these decks are accordingly designed to withstand the local deck loading as well as the loading due to longitudinal bending of hull girder. As it has been already explained, longitudinal framing system is adopted in construction of these decks.

For increasing functional versatility, a ship can be fitted with folding tween deck, thus with the tween deck in lowered condition, the ship will function as a general cargo ship and in the folded or lifted condition the same vessel will function as a dry bulk carrier. In this case the tween deck in the cargo hold is hinged at the side shell. About those hinges, the deck can be made horizontal or lifted up and kept inclined as sloping bulkhead in a bulk carrier. As because these decks are not continuous along the ship's length, they do not contribute to the longitudinal strength of the ship. They are designed based on the cargo loading only that is likely to act on these decks.

The structural arrangement of main deck with hatch opening is essentially same for all types of vessels. Ships not having hatch opening, the structural arrangement remains the same without the portion of the cross deck structure as in case of decks with hatch opening.

Structural arrangement of main deck of a general cargo carrier

Main deck of a general cargo carrier has longitudinal framing system, i.e. primary stiffening members are longitudinal stiffeners. This provides for higher buckling strength and results in higher strength to weight ratio. At the same time these deck longitudinals also provide for longitudinal bending strength. The secondary stiffeners are the deck transverses of higher scantling which provide support to the deck longitudinal and also reduce their unsupported span. These members provide for transverse strength. The space between the two consecutive hatch openings, referred to as cross deck structure, is transversely stiffened. Since the length of this cross deck structure is much less than 15 % of the length of the ship, this part does not contribute to longitudinal strength. Hence the deck plate used in this place in line of opening is of much smaller thickness compared to the plates outside line of opening. Therefore this part becomes prone to buckling due the action of compressive transverse loads. Therefore transverse stiffening is done on the cross deck. Hatch side girders and hatch end beams are provided around the hatch openings to partially compensate for the loss in strength due to large hatch opening. Deck transverses are provided at about 3–4 frame space to provide support to the longitudinals as well as to provide for transverse strength. Because of narrowing of the sections, the fore and aft end of the deck is stiffened by stiffeners laid out somewhat radially, known as cant beams. A typical structural arrangement of a deck with hatch opening is shown in Fig. 5.12.

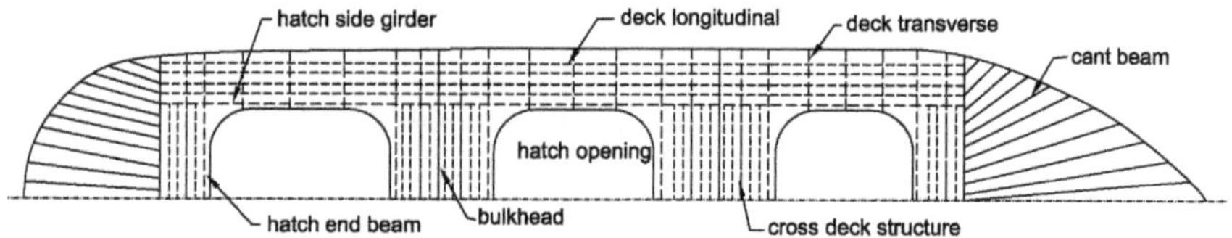

Fig. 5.12 Structural arrangement of main deck

5.4.2 Side Shell

The side shell of a ship normally forms the side boundary of the cargo hold and the engine room. Hence depending on the type of cargo as well as structural strength requirement, it may be of single shell construction or cellular, i.e. double wall construction. In oil tankers, it is mandatory to have double wall construction to provide for additional safety against oil spill in the event of any accident. Whereas in case of large container ships, it is preferable to have cellular construction to provide for adequate torsional strength to the hull girders. Rest of the ship types has single skin construction.

In case of cellular construction, longitudinal framing system is adopted for both the outer and inner side shells, as it provides for higher buckling as well as higher longitudinal strength. Whereas in case of other ships, depending on ship type or the type of cargo, either longitudinal or transverse framing system is adopted for stiffening of the side shells.

Transverse framing system is adopted for stiffening of side shell of a general cargo carrier. If longitudinal framing system is used then it would require transverse web frames having web depth at least double to that of the longitudinals. In that case it would encroach more in the cargo stowage space. Whereas in transverse farming system, the transverse web depth will be smaller than that in longitudinal framing system. Hence one will achieve higher clear cargo stowage space.

Similarly in case of dry bulk carrier, the side shell in the cargo hold region is transversely stiffened. However the wing tanks are longitudinally stiffened. In bulk carriers, loose cargo is carried in bulk. If the side shell is stiffened with longitudinal members, then they will act as shelves, thus even after unloading of cargo, some amount of cargo will always remain on them. Hence transverse framing system is adopted for the cargo hold side shell of bulk carriers.

5.4.3 Inner Bottom Plating

As per Safety of Life at Sea (SOLAS) requirement, all ocean going ships are required to have double bottom, i.e. outer bottom and inner bottom. The space in the double bottom cannot be used for carrying cargo. How the same is used for carrying fuel oil, lubricating oil, fresh water and for ballasting. Thus the space either remains empty or is used for carrying some or other liquid. Hence longitudinal framing system is adopted in the double bottom. The upper plating of the double bottom unit is referred to as inner bottom plating or tank top plating. This is longitudinally framed and the tank top longitudinals are supported by the floors in the double bottom. Thus the inner bottom plating provides for a smooth flat surface for stowage of cargo.

The inner bottom panel is prefabricated complete with the inner bottom longitudinals and then it is put over the double bottom unit on the skid. It is then welded to the bilge plate at the sides and to the floors at the bottom.

5.4.4 Bottom Shell

The bottom shell is the bottom most plating of the hull girder. The central plating of the bottom shell plate is the so-called keel plate. The adjacent plating is called garboard strake, and the plating at the round of bilge is called the bilge plate. The bottom shell is also longitudinally stiffened and the bottom shell longitudinals are supported by the floors in the double bottom.

The bottom shell plates are laid over the skid and butt welded to form the bottom shell panel. Then the bottom shell longitudinals are positioned, aligned and tack welded in position. The floors are then placed in position and tack welded. Finally all the longitudinals and the floors are welded to the bottom shell panel as per the prescribed welding sequence.

Chapter 6
Structural Assemblies

Abstract Shipbuilding is essentially an assembly industry. The steel material flows through different workstations and gradually grows in size. The structural components are assembled together through these stages of assembly. The structural assemblies dealt here are double bottom structure, wing tank, duct keel, fore and aft end structure, bulbous bow and rudder. The double bottom provides for adequate longitudinal strength to the hull girder. For convenience of cargo stowage and its unloading, wing tanks are provided in cargo holds of bulk carrier. Duct keel is a structural arrangement within the double bottom forming a tunnel all along the length of a ship. The hull girder comprises of the middle body and the two end units, namely fore end and aft end. A bulbous bow is an extension of the hull just below the load waterline. The basic purpose is to create a low pressure zone to reduce or eliminate the bow wave and reduce the resulting drag. Rudder provides the ability to steer the ship to its destination. Rudders are designed as balanced, semibalanced or unbalanced types.

Shipbuilding is essentially an assembly industry. The steel material flows through different workstations and gradually grows in size as shown in Fig. 6.1. The structural components are assembled together through these stages of assembly.

6.1 Double Bottom Construction

The double bottom space is not meant for carrying any cargo but may be used to carry some liquids like ballast water, fuel oil, lub oil, etc. Longitudinal framing system is adopted in double bottom structure. It provides for adequate longitudinal as well as buckling strength and thus it also gives maximum benefit of higher strength to weight ratio.

Generally a double bottom has a centre girder, also known as centre keelson and side girders port and starboard. In some cases, a duct keel arrangement is also used

N.R. Mandal, *Ship Construction and Welding*, Springer Series
on Naval Architecture, Marine Engineering, Shipbuilding and Shipping 2,
DOI 10.1007/978-981-10-2955-4_6

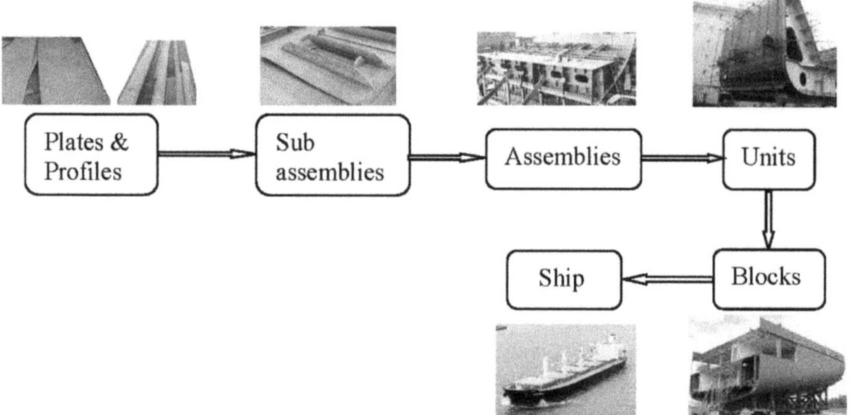

Fig. 6.1 Stages of assembly

in place of centre girder. The bottom shell and the inner bottom plating can be either longitudinally framed or transversely framed.

In smaller vessels like river launches, often transverse framing system is used. The length being less, the loading due to longitudinal bending is not very significant. From construction point of view, it becomes convenient to have transverse framing system.

However for bigger, i.e. longer vessels, it is preferable to have longitudinal framing system as shown in Fig. 6.2.

Apart from these longitudinal stiffeners, there are plate floors at intervals of about 3–4 frame spaces. The longitudinals are rigidly connected to the floor plating. The web of the longitudinal is welded to the floor and on the other side a collar plate is welded as shown in Fig. 6.2 to ensure rigid connection between the longitudinals and the floor plate. Thus the plate floors provide support to the longitudinals. These longitudinals provide for longitudinal strength, whereas the floors provide required transverse strength to the hull girder.

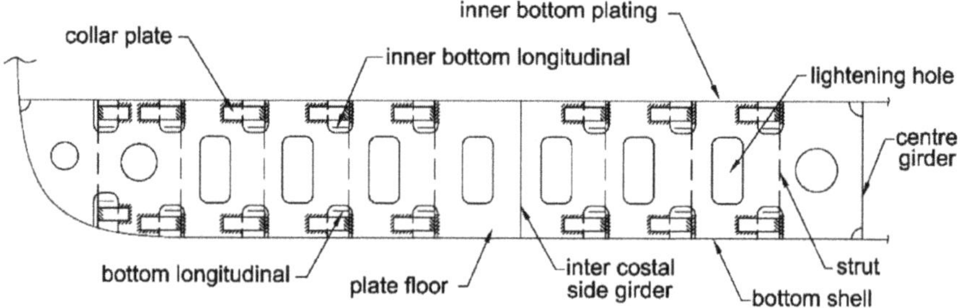

Fig. 6.2 A typical *double bottom* with longitudinal framing system

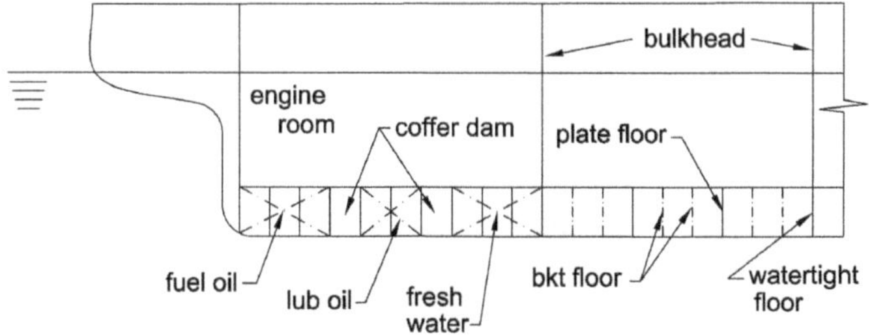

Fig. 6.3 *Double bottom* spaces

Depending on the breadth of a vessel, as per classification rule requirement additional intercostal side girders are also provided as longitudinal strength members. These girders are generally non water tight and are continuous between the plate floors. These are terminated and welded to the plate floors.

The double bottom under the engine room is additionally strengthened to support the concentrated loads coming from the main engine as well as other machineries and equipment. This is done by providing plate floors at every frame space. Whereas in the double bottom space under cargo hold region, bracket floors are there in between the plate floors. A typical bracket floor arrangement in longitudinal framing system is shown in Fig. 4.7. The plate floors below the transverse subdivision bulkheads do not have lightening holes. These are made water tight by completely sealing off the scallops for longitudinals using collar plates as can be seen in Fig. 4.8.

The double bottom space is used for carriage of different liquids, like, fuel oil, fresh water, lubricating oil, ballasting, etc. For service convenience, fuel oil, lub oil and fresh water is carried in the space below the engine room. Hence the double bottom space below the engine room is divided into required number of separate water/oil tight compartments. To prevent any contamination of liquids because of possible leakage between two adjacent compartments, an empty space of one frame space is kept. This is called coffer dam as shown in Fig. 6.3. Leakage, if any, the liquid will get collected in the coffer dam. It will not contaminate the liquid in the adjacent compartment.

6.2 Wing Tanks

For convenience of cargo stowage and its unloading, wing tanks are provided at the top and bottom side of the cargo holds of bulk carrier as shown in Fig. 6.4.

The wing tanks are not meant for carrying any cargo, either they remain empty or are used to carry some liquids like ballast water, fresh water, etc. Therefore longitudinal framing system is conveniently adopted to derive the maximum benefit of higher strength to weight ratio. The sloping bulkheads as well as the part of the side shell, bilge plating and upper deck are stiffened with longitudinals. These are

Fig. 6.4 *Bottom wing* tank in a bulk carrier

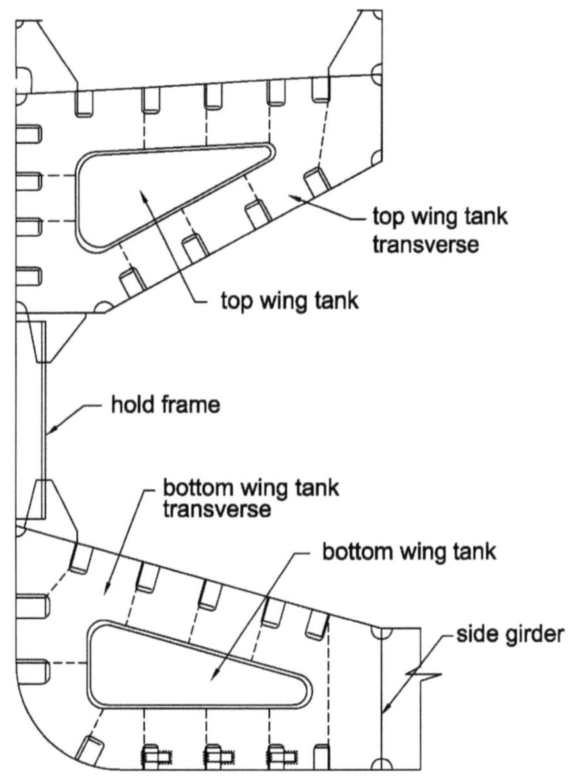

Fig. 6.5 Typical section through *top* and *bottom* wing tank of a bulk carrier

supported by transverse members referred to as wing tank transverses as shown in Fig. 6.5. All the longitudinals are rigidly connected to the web of the transverses as shown in some of the bottom longitudinals in Fig. 6.5.

6.3 Duct Keel

Duct keel is a structural arrangement within the double bottom in which centre girder is replaced by two side girders forming a tunnel all along the length of a ship. This arrangement of two adjacent side girders near the central line is referred to as duct keel as shown in Fig. 6.6.

This structural arrangement has several advantages:

(i) These two water tight side girders make a protected closed duct.
(ii) The duct can be used for laying pipes, electrical cables, etc. Otherwise, they would have run through the double bottom space.
(iii) The maintenance of these pipes and cables becomes easier.
(iv) Any leakage in a pipeline will not contaminate with any other liquid.
(v) It provides for a stronger support for the ship structure for putting the ship on keel blocks.
(vi) It provides for higher longitudinal strength.

Access to the duct is provided through engine room. For safety reasons, it is to be kept closed watertight at all times, unless some operation is required inside the duct.

The double bottom is longitudinally framed, however the duct keel is transversely framed as shown in Fig. 6.6. It comprises of two longitudinal side girders with a spacing of not more than about 2 m. The spacing should be such, that it should not exceed the width of keel blocks. In the docked condition, the girders need to be well supported by the keel blocks. The keel plate and the inner bottom plating between the two side girders are stiffened by transverse bottom frame and reverse frame respectively. The ends of these frames are well bracketed to the side girders as shown in Fig. 6.6.

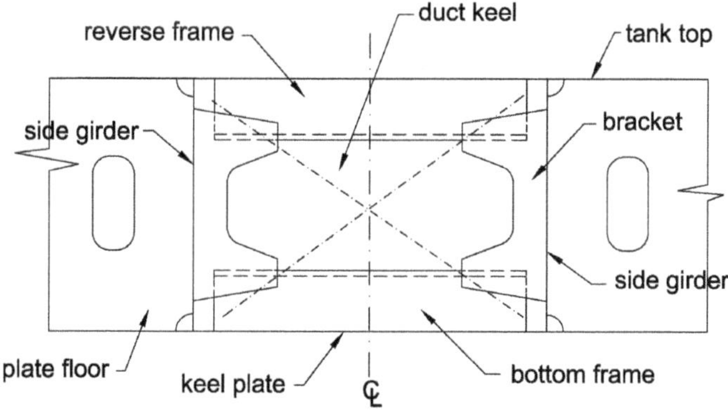

Fig. 6.6 Duct keel arrangement

6.4 End Construction

A ship hull can be structurally broken down primarily in four zones, i.e. aft end, middle body, fore end and superstructure. The hull girder comprises of the middle body and the two end units, namely fore end and aft end. The middle body undergoes maximum bending moment due to longitudinal hull girder bending. The load due to longitudinal bending plays the dominant role in structural design of this middle body, whereas the end structures are designed based on local loading. The fore end of the hull girder is subjected to extreme loading due to its pitching and heaving motion in rough weather condition. Also in sever bad weather condition, with huge waves and along with ship motions, huge mass of water flows onto the fore end deck of the ship. This is often referred to as 'green water' and the deck structure may suffer extreme loading due to this green water loading.

Similarly considerable attention is needed in designing the aft end structure of a ship in order to improve flow into and away from the propeller. Also the aft end construction assumes importance as both the propeller shaft and the rudder stock pierce the water tight hull and also supports them. The aft end should be so designed that it should be able to withstand the engine and propeller shaft excited vibration.

6.4.1 Fore End Construction

The fore end structure being the foremost part of a ship, it undergoes different types of loading which are essentially local in nature. Due consideration of these local loadings are to be given in designing the fore end structure of a ship. The basic components of a fore end construction are: main deck/fore castle deck, stem comprising of bow plate and stem bar, forward side shell panels, bulbous bow, chain locker, hawse pipe and fore peak bulkhead.

The loading that needs to be considered are due to, (i) green water loading on deck, (ii) local loading on deck due to the heavy anchor windlass, (iii) slamming load on the side shells due to pitching/heaving motions of the ship in rough weather, (iv) wear and tear of hawse pipe by the anchor chain caused by hoisting and lowering of anchor, (v) impact load due to accidental head on collision of the fore end structure.

In most cases, ocean going ships are provided with a forecastle deck. The purpose are primarily, (i) to have considerably higher freeboard to prevent shipping in of green waters, (ii) increased deck area to accommodate anchor windlass and any other deck equipment and (iii) a storage space for mooring ropes, other deck rigging items, etc. This storage space between main deck and forecastle deck traditionally referred to as bosun's store.

The forecastle deck structure does not experience loading from longitudinal bending of the hull girder. However it experiences loading from local loads,

(i) green water loading, (ii) heavy deck equipment, e.g. anchor windlass. Accordingly the higher thickness deck plating as well as additional local stiffening of the deck structure along with pillars for extra support are provided.

Often higher flare of the forward end is provided to increase deck area and to prevent water spray onto the deck also it adds to esthetic look of the vessel. However all these add to the slamming load experienced by the fore end side shell structure. In calm weather condition it does not affect but in rough weather causing the vessel to heavily heave and pitch, the fore end side shell panels and the bottom structure undergo extreme slamming load. Hence local stiffening of the stem is done by providing breast hooks, the side shell is strengthened by stringers and side shell frames put at reduced frame spacing, generally 610 mm. At the same time plate stringers and center line wash bulkhead are provided for additional stiffening. The plate floors are put at all frame position from forward of the fore peak bulkhead as shown in Fig. 6.7.

Bow stem

The upper part of the bow stem is made up of radiused plate. Whereas at the lower water lines, for medium to fine form vessels, the radius of curvature required becomes so small that bending a plate to that radius becomes difficult. Here a round bar is used and the side shell plates are terminated tangentially on this round bar and welded. This is called stem bar. Horizontal plate webs, known as breast hooks as shown in Fig. 6.7 are also often used to strengthen the radiused part of the bow. To further strengthen the radiused bow plate, flat bar stiffener is also used.

Ships specifically designed for ice breaking operation and having 'icebreaker' notation, the forward end construction requires members of higher scantlings.

Chain locker

The chain locker, as the name suggests, is storing place of the anchor chain cable. It is located forward of the collision bulkhead. The dimensions of the chain locker are

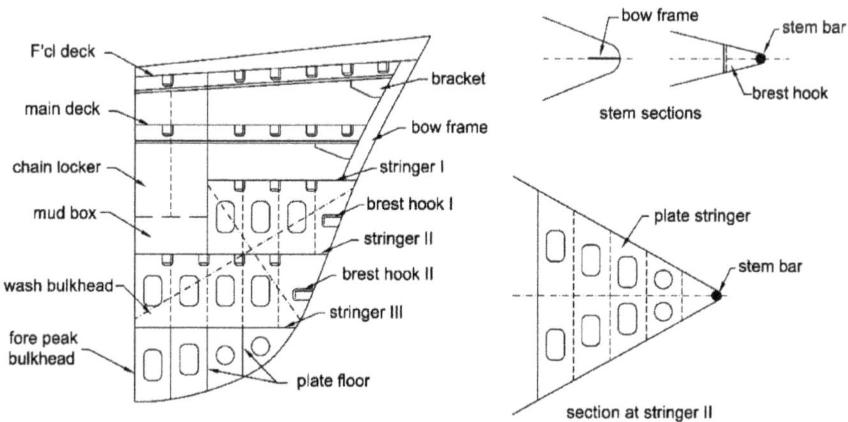

Fig. 6.7 A typical fore end construction showing additional stiffening arrangement

decided based on the size and length of the anchor chain cable. The depth of the locker should be such that the cable can be easily stowed. A centre line partition bulkhead is provided in the locker to segregate the stowage of the port and starboard cables for facilitating easy hauling and lowering of anchor. This bulkhead does not continue upto the top of the chain locker but allows a working space above the two compartments. The inboard end of the cable is securely connected to the bottom of the centre line bulkhead or at the bottom deck of the chain locker. It is preferable to provide a false bottom made of perforated plate above the bottom deck of the chain locker to facilitate drainage of mud and water from the anchor cable. The bottom deck of the chain locker is made sloping inboard for better drainage of muddy water, which is drained off through a suction line provided at the centre.

The stiffening of the locker bulkheads is done outside of the locker to prevent possible damage by the chain cable. Foot holes are cut at the forward part of the center line bulkhead of the chain locker to access the bottom of the locker. An access hatch is provided at the deck with an access ladder leading to a standing platform at the forward bulkhead. It is followed by the foot holes in the centerline bulkhead. The internal centre line bulkhead is stiffened with hat shaped rolled steel sections to avoid any entanglement with the anchor cable.

The chain cable is guided to the individual compartments through port and starboard pipes from the forecastle deck leading to the chain locker. These pipes are called hawse pipes. The lower half of these hawse pipes are of much higher thickness compared to the upper half of the pipes. The reason being the lower half of the hawse pipe undergoes heavy wear and tear due to chain cable friction.

Bulbous bow

A bulbous bow is an extension of the hull just below the waterline. They can be of various shapes and sizes. A typical bulbous bow is shown in Fig. 6.8. The basic

Fig. 6.8 A typical bulbous bow

purpose is to create a low pressure zone to reduce or eliminate the bow wave and reduce the resulting drag.

A bulbous bow works best under certain conditions and good design gives efficiency gains throughout the range of these factors. At low speeds a bulbous bow will trap water above the bulb without forming a low pressure zone to cancel the bow wave. This will actually cause increased drag and lead to loss of efficiency. It is often suggested that the usefulness of a bulb is limited in the interval $0.238 \leq Fn \leq 0.563$, F_n being the Froude Number. Generally at low speeds the effect of the bulb is negative. With increasing Froude Number, its effect becomes positive and increases up to a maximum value.

The bulbous bow offers several advantages apart from reducing bow wave, it works as a robust "bumper" in the event of a collision. It allows the installation of the bow thrusters at a foremost position, making it more efficient. It provides for a larger reserve of flotation or a larger ballast capacity forward and also reduces pitching of the vessel.

The bulb shapes can be classified as Addition Bulbs and Implicit Bulbs. The Addition bulbs are as if added to the hull as a separate entity. Here the bulb shape is completely independent from the hull shape. Typically there is a knuckle resulting from the intersection of the bulb and the hull as shown in Fig. 6.9a, b. Whereas in case of Implicit bulbs, they are continuation of the hull and therefore no knuckles are required as shown in Fig. 6.10a, b.

The bulbs generally have heavily curved surfaces making them expensive for production.

Some compromise in the hydrodynamic efficiency of the bulbs could result in reduction of their production costs. Following guidelines may be considered while designing a bulb to make it less cost intensive:

- Maximise use of single curvature plates.
- Use conic sections.

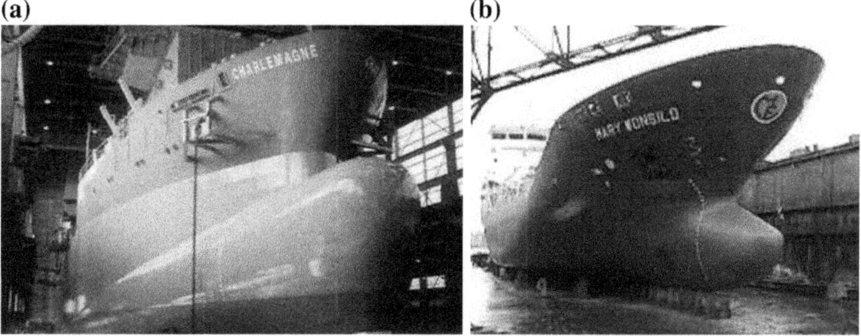

Fig. 6.9 Typical bulb of addition with a knuckle at the bulb-hull intersection

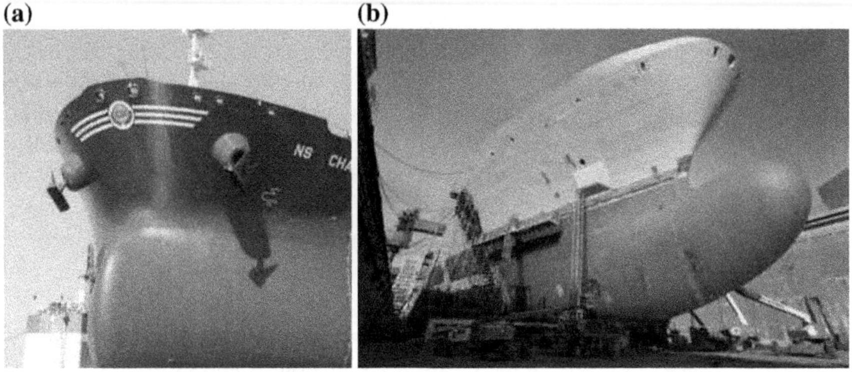

Fig. 6.10 Typical implicit bulbs integrated with the hull

- Decrease the curvature of the free-form curved plates, decomposing them into smaller ones.

Examples of simplified bulb form are shown in Figs. 6.11 and 6.12.

6.4.2 Aft End Construction

As in case of forward end construction, the aft end structure also assumes great importance from the point of view of flow past the propeller as well as the type of rudder that is going to be installed to have the best manoevring characteristics.

There are various types of stern forms. Most widely used forms for ocean going ships are cruiser stern and transom stern. Similarly there are several rudder forms as well. The shape of the stern and the type of the rudder along with the propeller size dictate form of the stern frame. If the propulsion is designed based on twin propeller then 'A' or 'P' brackets are used to hold the outboard overhang of the shaft. For larger vessels, 'A' bracket is preferred whereas for smaller vessels, 'P' bracket serve the purpose.

Fig. 6.11 A schematic of a simplified bulb form

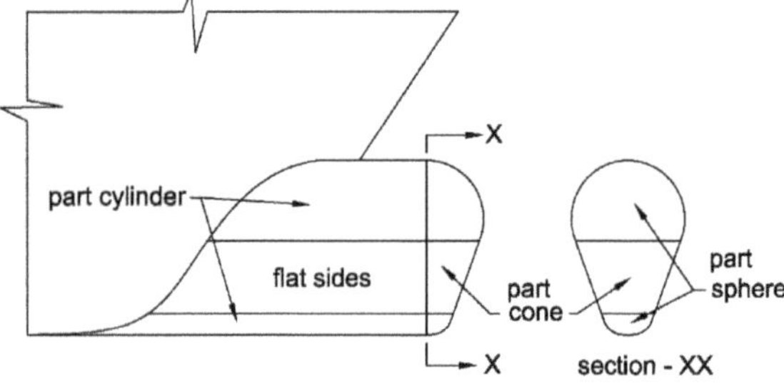

Fig. 6.12 An example of a simplified bulb form

Cruiser Stern

Traditionally, cruiser stern used to be the most favoured option for ocean going vessels. It presents esthetically more pleasant profile and also it is hydrodynamically efficient. The cruiser stern has an upward curved profile from the after perpendicular to the main deck or poop deck as shown in Fig. 6.13.

A cruiser stern generally has a large overhang making it vulnerable to large slamming forces. Therefore it is necessary to have adequate stiffening of the stern section. As can be seen in Fig. 6.13, plate floors are provided at every frame positions. The decks are stiffened by deck beams and cant beams along with centreline and side girders. The stern shell plating is stiffened by cant frames and webs.

Transom Stern

Unlike cruiser stern, transom stern is flat, thus it offers the advantage of construction as well as higher deck space. The transom stern gives the effect of apparent increase in length of water plane providing hydrodynamic advantage. In this case cant frames are not required and the transom plate is stiffened by vertical stiffeners. The aft peak houses the steering gear equipment and the rudder stock is connected to the steering gear through the rudder trunk. The aft end construction is adequately strengthened by deep floors at every frame space and centre girder to support the local loads of steering gear equipment, propeller and propeller shafting arrangement. In the aft end generally a poop deck is provided. The decks are transversely

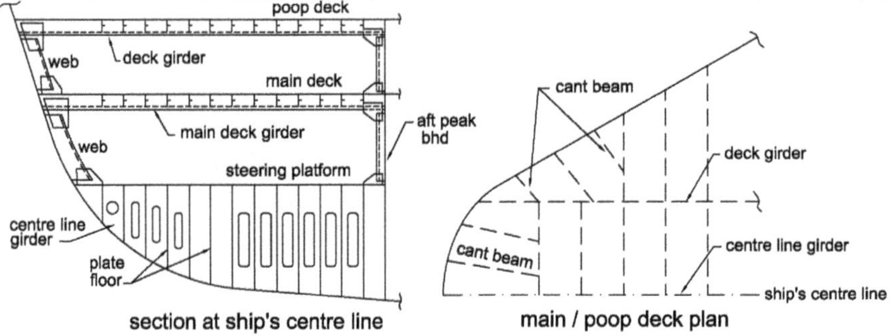

Fig. 6.13 A typical cruiser stern

stiffened by deck beams and additionally supported by centre line and side girders. Structural arrangement of a typical transom stern is shown in Fig. 6.14.

Stern Frame

One of the important components of aft end structure is the stern frame. It can be either a fabricated stern frame, a cast frame, or a forged one. A typical fabricated stern frame is shown in Fig. 6.15.

Steel plates and round steel bars are used to fabricate stern frames. Depending on the type of stern frame often cast components are also used in fabricated stern frames as shown in Fig. 6.15. Fabricated or cast frames are generally used in large ships. A typical cast stern frame is shown in Fig. 6.16. To overcome the difficulty of casting and transportation of very large stern frames, they can be cast in two three pieces and then welded together at the time of erection in the shipyard.

The shape of the stern frame depends on the type of rudder being used. As can be seen in Fig. 6.15, the stern frame has a sole piece to support the rudder, whereas in Fig. 6.16, it is meant for a spade rudder, which is hanging type, without any support at the bottom. The stern frame design depends primarily on the requirement

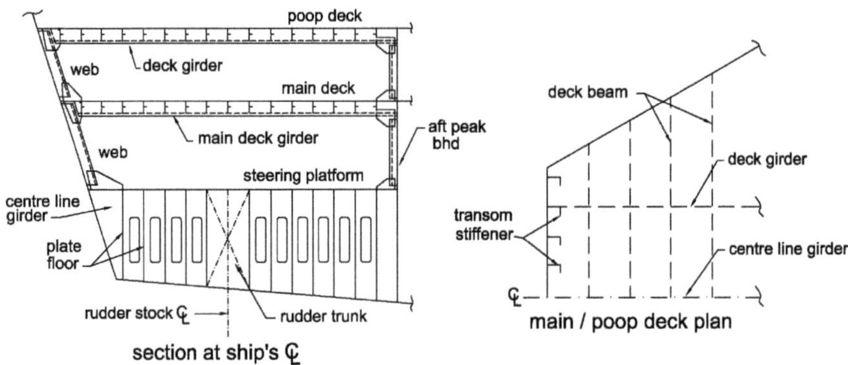

Fig. 6.14 A typical transom stern

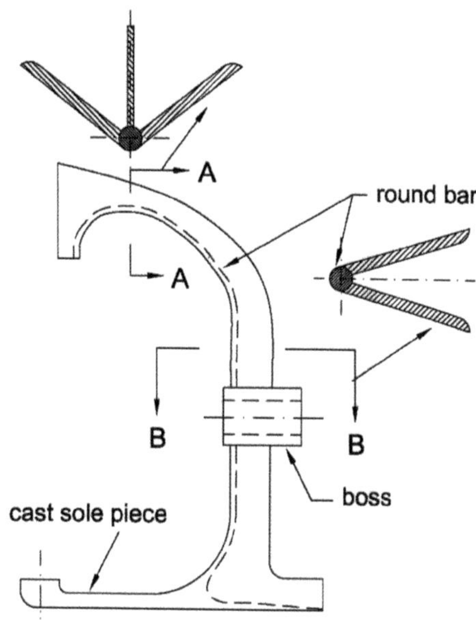

Fig. 6.15 A typical fabricated stern frame

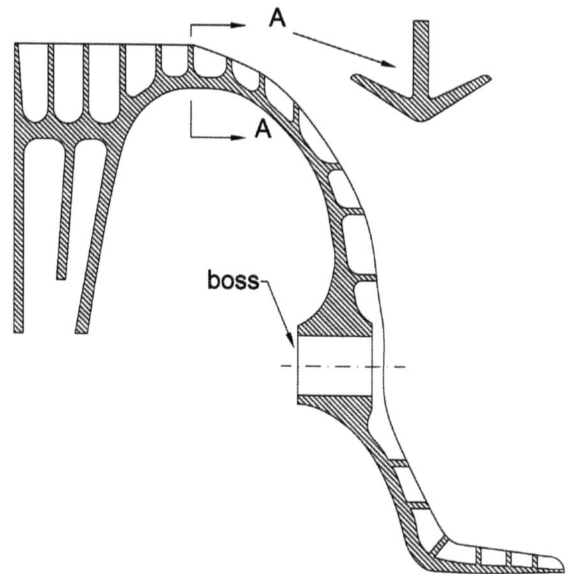

Fig. 6.16 A typical cast stern frame

of the propeller aperture to house the propeller. Adequate clearance between the stern frame and the propeller blade tip needs to be provided to avoid propeller excited hull vibration.

For strengthening the stern frames horizontal plates are welded at intervals to both the cast as well as fabricated frames. Since these frames arc of substantial thickness, necessary edge preparation and required preheating is done during welding of these frames to the hull structure.

6.5 Rudders

Rudder is one of the very important items in a ship. It provides the ability to steer the ship to its destination. Hence its design and construction is of equal importance. Rudders are designed as balanced, semibalanced or unbalanced types. Balanced rudders have larger proportion of about 25–30 % of their lateral area forward of the turning axis, semibalanced ones have a smaller proportion, less than 20 % of the area forward to the axis whereas unbalanced rudders have full area aft of the turning axis. In balanced rudders, the centre of lateral pressure is brought nearer to the turning axis to achieve reduction in torque required for rudder operation.

In smaller vessels simple stiffened flat plate rudder is used. However in large vessels, symmetrical aerofoil section is used for fabricating the rudders. These rudders are fabricated from shaped plates. The vertical and horizontal stiffeners are welded to one side of the plate and then the other plate, often referred to as closing plate is put in place and welded to the internal stiffening members by means of slot welding. Drain hole with plug is provided at the bottom of the rudder to facilitate drainage of water which may have accumulated through some leakage developed in the plates. In some designs, a lifting hole passing laterally through the plates is provided. Here a short tube is welded with doubler plates.

Rudder being a closed structure, it is not possible to see what is happening inside. That is why it becomes all the more important to protect the internal structure from corrosion. To achieve this, the internal surfaces are appropriately coated and also often the internal space is filled with some foam material. After fabrication rudders are tested under pressure against any form of leakage.

The rudders which are supported by sole piece, the rudder pintle is supported in the gudgeons fitted in the sole piece. The pintle and the gudgeons have suitable bronze, synthetic or composite bearing material. The lubrication medium is water only. However some of larger ships have used oil-lubricated metal bearings for the pintles.

The shaft which holds the rudder is called rudder stock. It is made of cast or forged steel. Its design depends on the torque and bending moment that it may have to withstand during its operation. The rudder stock is bolted to the rudder with bolts of adequate strength to sustain the torque and the load of the rudder. The rudder stock is taken inside the hull through a passage called rudder trunk and connected to the steering gear. The top of the trunk is fitted with a watertight gland through which the rudder stock enters the intact hull. The forward and aft ends of the trunk comprise of the floors of the aft end structure and flat plates are used on the port and

Fig. 6.17 A typical spade rudder with intermediate support

starboard sides to make the trunk. An access manhole is provided in any one side of the trunk to allow inspection of the stock from inside the hull when necessary.

The rudder bearing fitted inside the hull supports the weight of the rudder and partly it is supported by the pintles in case of intermediate or bottom supported rudders. Atypical spade rudder is shown in Fig. 6.17 having intermediate support. Here the weight of the rudder is partially supported by this intermediate pintle.

6.6 Stern Tube

Propeller shaft needs to pierce out of the hull keeping the hull safe against any kind of flooding through the shaft opening. Suitable water lubricated glands are used at the openings to prevent ingress of water. However because of possible wear and tear over a period of usage, some unavoidable leakage may develop. Therefore the propeller shaft is taken through a shaft tunnel, in which towards the propeller end the stern tube is located. This acts as a casing for the stern tube with the tail shaft (aft section of the propeller shaft), segregating it from the aft end or the engine room compartment, thus preventing leakage water to enter these compartments. When water lubricated bearings are used, the propeller end of the stern tube is open to the sea and the bearing is placed at the engine room end. These bearings used to be traditionally made from a type of wood called lignum vitae and currently composite material is being used. There are also patented oil lubricated metal bearings. In this case the bearings are put at both ends of the stern tube to retain the oil and prevent ingress of water.

6.7 Shaft Boss and 'A' Bracket

In multiscrew vessels the port and starboard propeller shafts pierce the hull somewhat forward of the aft end. The shaft is continued to the aft to get the necessary propeller aperture such that required propeller blade tip clearances are obtained. Thus in such cases there can be substantial overhang of the propeller shaft. Theses bossings or 'A' brackets are used to support this shaft overhang. Bossings have hydrodynamic advantage over the 'A' brackets. However from production point of view fitting a 'A' bracket is easier than providing bossing in the hull. To reduce hydrodynamic drag of these 'A' brackets, suitable aerofoil sections are used in the struts of the 'A' brackets. These struts are either fabricated from plates and sections or cast in the required shape. Atypical 'A' bracket is shown in Fig. 6.18.

Fig. 6.18 A typical 'A' bracket, stern tube is being installed

Chapter 7
Midship Sections

Abstract The longitudinal strength of hull girder depends on the section modulus of the midship section. This in turn depends on the scantlings and layout of the structural members in the midship region. The midship region extends one forth length of the ship forward and aft of midship. Over this midship region the scantlings of the structural members are kept the same. Maximum longitudinal bending moment is experienced by a hull girder within this midship zone. Therefore midship section plays an important role from longitudinal strength point of view, at the same time it depicts the structural layout depending on the type of cargo the ship is going to carry. Thus different types of ships have different midship sections. The structural arrangement and their scantlings are shown in these plans. These are statutory structural plans which are to be approved by the concerned classification society prior to actual construction of ship.

The overall structural design and its layout of a ship hull are given by the following structural plans:

(i) Decks and Profile
(ii) Midship Section
(iii) Aft end and Fore end construction
(iv) Bulkheads.

The structural arrangement and their scantlings are shown in these plans. These are statutory structural plans which are to be approved by the concerned classification society prior to actual construction of ship. The details of midship sections of various types of ship will be taken up in this chapter. The others have already been dealt with in previous chapters.

The longitudinal strength of hull girder primarily depends on the section modulus of the midship section. This in turn depends on the scantlings and layout of the structural members in the midship region. The midship region extends one forth length of the ship forward and aft of midship. Over this midship region the scantlings of the structural members are kept the same. Maximum longitudinal bending moment is experienced by a hull girder within this midship zone. Therefore midship section

N.R. Mandal, *Ship Construction and Welding*, Springer Series on Naval Architecture, Marine Engineering, Shipbuilding and Shipping 2,
DOI 10.1007/978-981-10-2955-4_7

plays an important role from longitudinal strength point of view, at the same time it depicts the structural layout depending on the type of cargo the ship is going to carry. Thus different types of ships have different midship sections. The basic features of various types of ships have already been discussed in Chap. 1. Here in this chapter only structural arrangements of midship region of these ships will be taken up.

7.1 General Cargo Carrier

General cargo carriers are expected to take and deliver cargo at several ports between port of origin and destination. Therefore proper stowage of cargo is essential, such that the required cargo can be accessed at a given port where it needs to be discharged. To facilitate this, some form of cargo segregation is done by providing one lower deck. Hence cargo can be stacked in the main hold or in the tween hold depending on the cargo discharge schedule. A typical midship section is shown in Fig. 7.1. It is to be noted that in general cargo ship longitudinal framing

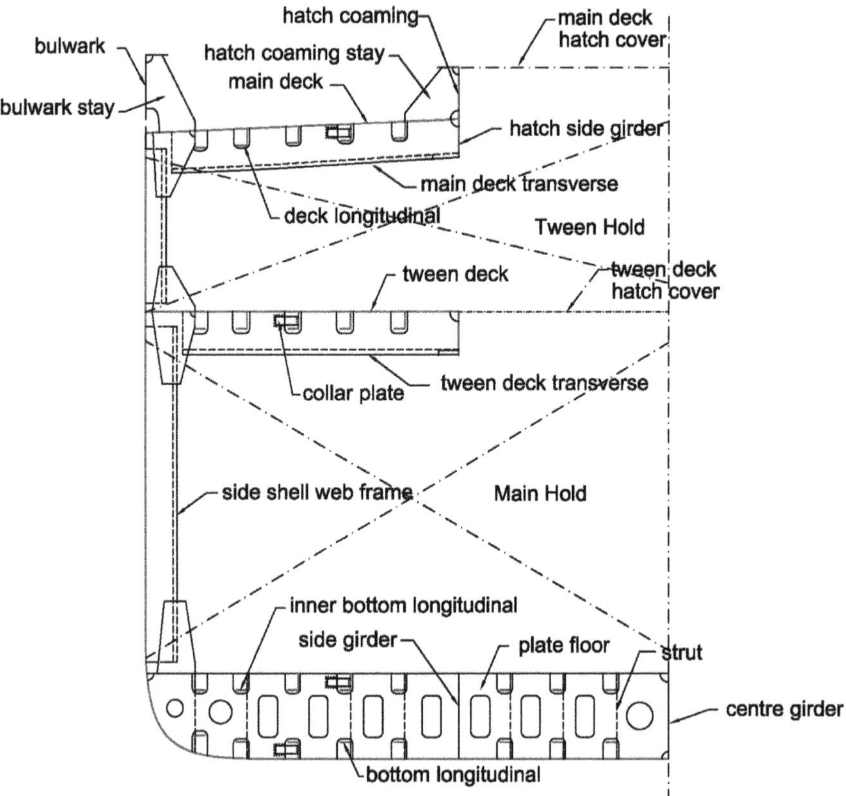

Fig. 7.1 Midship section of a general cargo carrier, section through deck transverse

system is adopted in the decks and in double bottom. However transverse framing system is adopted for stiffening of the side shells.

7.2 Bulk Carrier/OBO Carrier

Bulk carriers are meant for carrying cargo in bulk, i.e. in loose condition and not packed in bags. Hence the loading is done through belt conveyor and hopper system and unloading is done by pumping out the cargo or using grabs. Since cargo is loaded through hopper, it forms a heap. Hence the cargo hold upper corners remain empty, at the same time while unloading the cargo, again some cargo remains at the bottom corners. Hence to avoid both these situations, the corners are cut off from the hold region by providing what is known as sloping bulkheads. Thus in the midship section this arrangement of sloping bulkheads forming the upper wing and lower wing tanks are shown in case of bulk carriers. A typical midship section of a bulk carrier is shown in Fig. 7.2. Longitudinal framing system is adopted in the

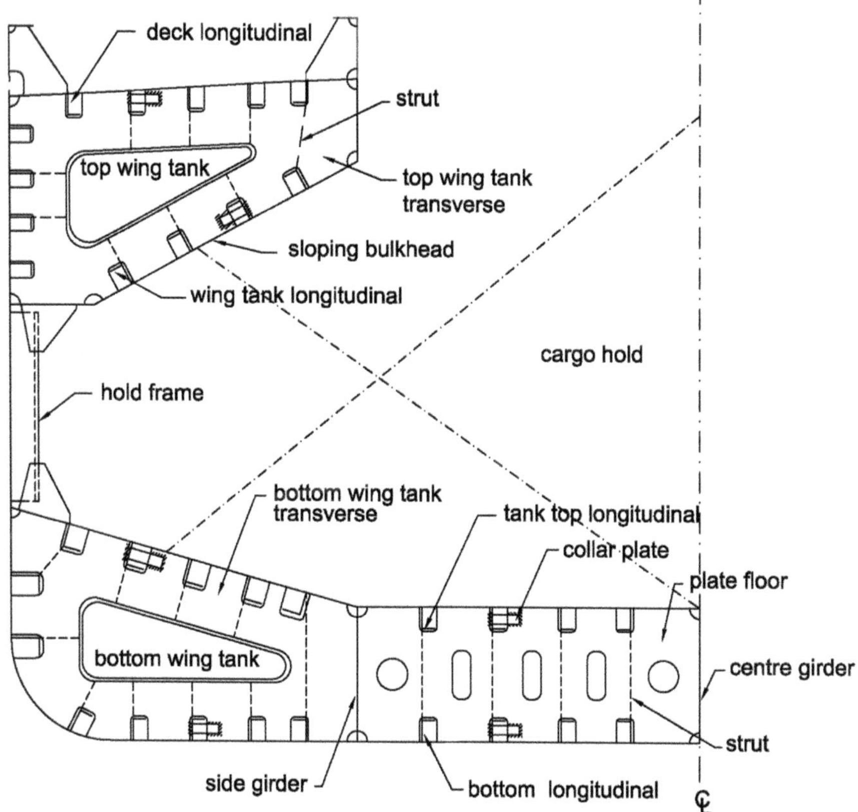

Fig. 7.2 Midship section of a bulk carrier

wing tanks and in the double bottom region, whereas the side shell in the hold region is transversely stiffened.

The concept of ore and bulk oil (OBO) carriers came for carrying high density cargo, like coal, iron ore, etc. In case of such high density cargo, major part of the cargo hold will remain empty in a conventional bulk carrier. In a turbulent seaway, it may cause shifting of cargo within the hold leading to listing of the vessel. This will cause loss of directional stability of the vessel. At the same time as the cargo is at the lower half of the cargo hold, the centre of gravity of the loaded vessel remains very low, causing the metacentric height to increase substantially. This makes the vessel stiff causing sever rolling of the vessel. To avoid this, the double bottom height is increased, causing the CG to rise. Also inner side shell is provided to reduce the cargo hold space just, such that the hold space gets fully filled up and shifting of cargo does not take place. A typical midship section of an OBO carrier is shown in Fig. 7.3. The central hold region is used for carrying the high density cargo and in return voyage, the space between the side shells is used for carrying oil preferably edible oil. In the event of any side shell damage, the oil cargo should not cause any environmental pollution.

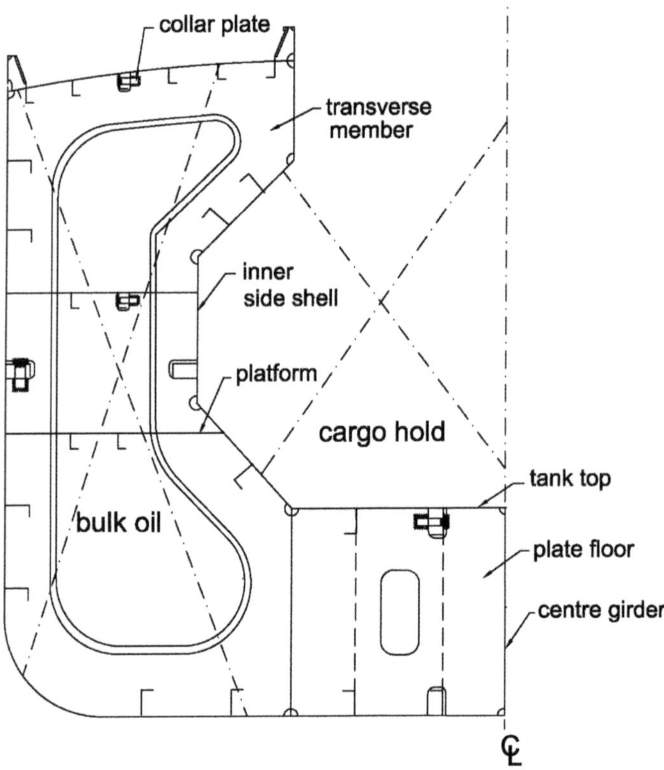

Fig. 7.3 Midship section of a OBO carrier

7.3 Container Ship

The most significant structural feature of container ships is it has very large hatch opening. These ships are also referred to as open deck ships. This feature is needed for easy and fast loading and unloading of containers. However this results in a structural challenge for providing adequate longitudinal and torsional strength to the hull girder. The large deck opening leads to a serious loss of longitudinal as well as torsional strength. New Panamax vessel of 12,500 TEU capacity has a length of 364 m, breadth 48.8 m and depth 29.8 m [1]. It is projected to build 22,000 TEU container ship by 2018 [2].

One of the challenges of designing ever increasing size of container ship is to provide for necessary longitudinal and torsional strength. To provide for necessary torsional strength, box girder and transverse framing of side shell is adopted in case of smaller container ships as shown in Fig. 7.4. To compensate for the loss in longitudinal strength due to large hatch opening, stringers are used in the side shell along with longitudinally stiffened box girder.

Whereas the larger container ships are of cellular construction having longitudinal framing all over. With increasing requirement of longitudinal and torsional strength, the box girder configuration is actually extended up to the double bottom giving it a cellular configuration as shown in Fig. 7.5. In these vessels generally duct keel is provided to facilitate laying of cables and pipes.

7.4 Oil Tanker

Oil tankers being bulk liquid cargo carrier, these vessels do not have any hatch opening. Contrary to the container ship configuration, it can be considered as closed form structure. Hence structural challenge is less compared to container ship, which is rather open form structure because of its very large hatch opening. However because of the extreme large size of the VLCCs and ULCCs, the following need to be considered for structural design [3]:

- Longitudinal Strength (Global bending/Shear)
- Transverse Strength (Local loads)
- Buckling (In combination with above)
- Fatigue (Long term effect)
- Ultimate Strength (Extreme load/Plasticity)
- Sloshing (Critical for: $L_{tank} > 0.1L$ and/or $B_{tank} > 0.5B$)
- Slamming (Critical at low draft Fwd and high speed)

Global bending and shear is of prime concern because of the size of these vessels. Sagging bending moment dominates in full load condition whereas hogging moment dominates in ballast condition.

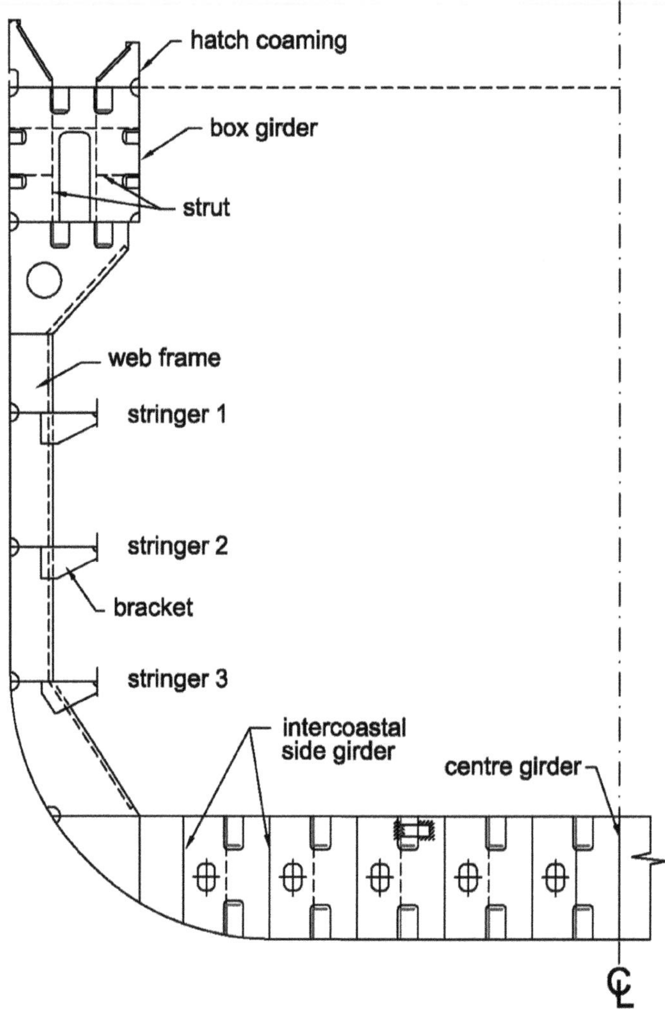

Fig. 7.4 A typical midship section of a container ship with box girder

From pollution prevention requirement as stipulated by MARPOL convention, double wall construction with double bottom is mandatory for all types of oil tankers. Depending on the breadth of the vessels, one or more longitudinal bulkheads are used in oil tankers to reduce the so-called free surface effect. All these contribute to longitudinal strength of the structure. At the same time longitudinal framing system is adopted for the entire structure, making it further advantageous from longitudinal strength point of view. The longitudinal stiffeners are supported by transverses which provide for adequate transverse strength to the structure. A typical midship section of a VLCC is shown in Fig. 7.6.

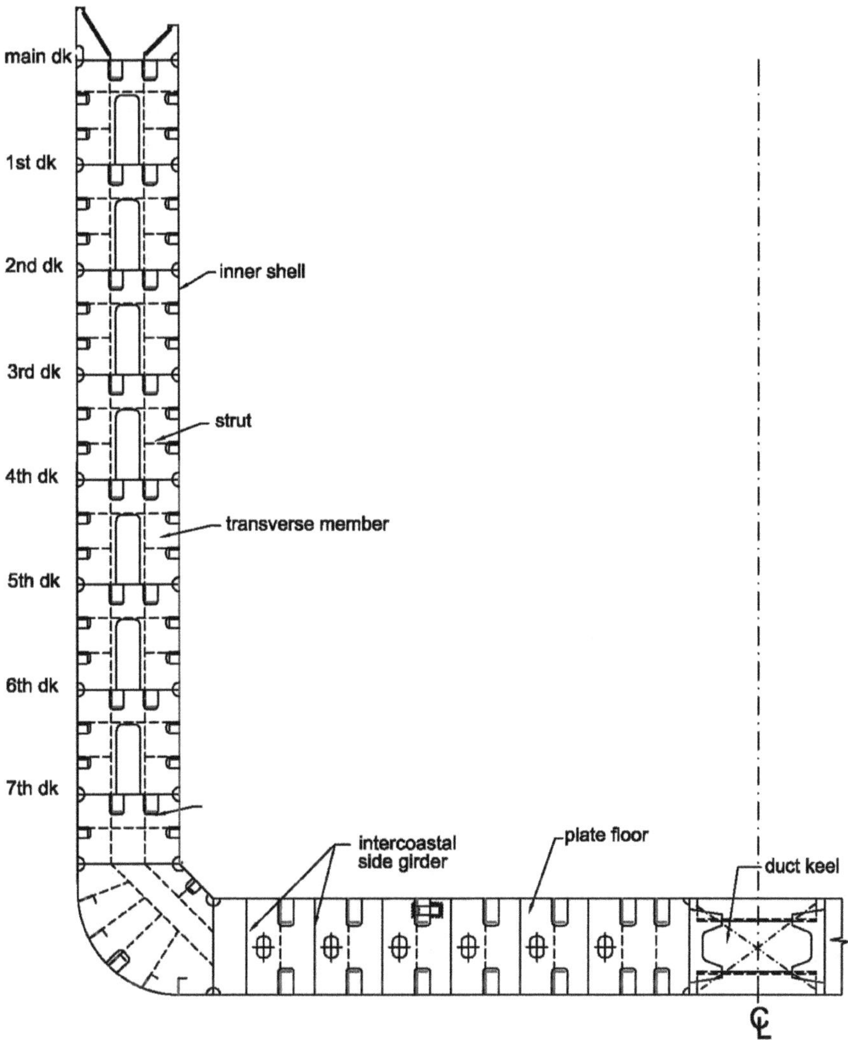

Fig. 7.5 A typical midship section of cellular container ship

7.5 RO-RO Ship

Roll On Roll Off ships are specially meant for exclusive transportation of automobiles. As the name suggests, the vehicles roll on the vessel and at the destination port they roll out of the vessel. Thus these ships do not have any hatch opening, instead they have ramp at a suitable location for loading and unloading of automobiles. These ramps are generally located at the port and starboard side of the forward end or bow loading ramp is provided.

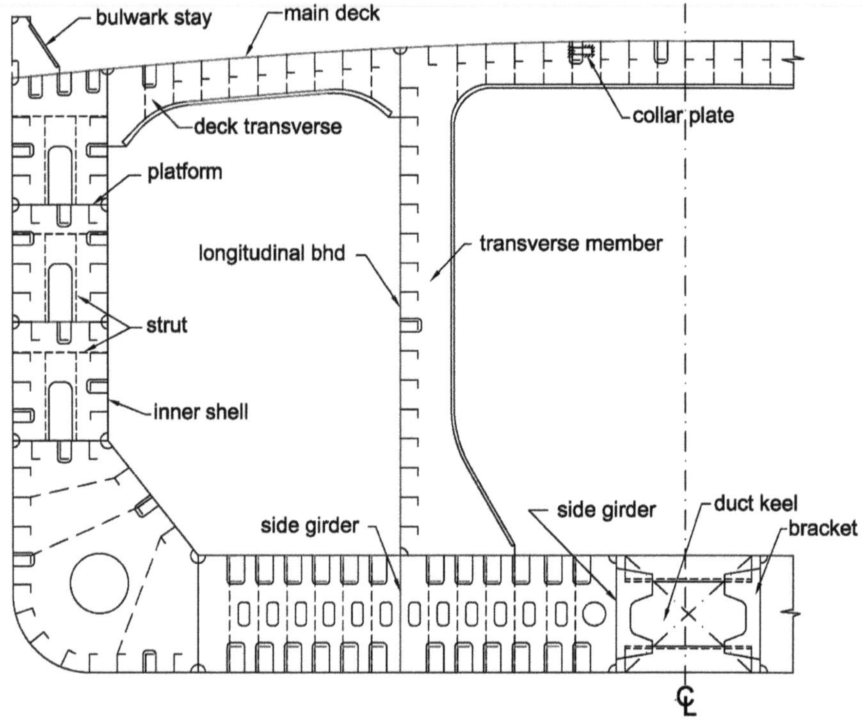

Fig. 7.6 Midship section of a VLCC

The current trend is to increase the cargo carried above the freeboard deck. Therefore it naturally calls for increasing the number of decks above the freeboard deck. The governing criterion for limiting the deck cargo, thereby the number of decks above freeboard deck is dictated by the damaged stability requirements of a RoRo ship.

Hence these ships are of multideck type. There are several decks to accommodate the vehicular cargo. The decks are strengthened suitably to support the wheel load of the vehicles. Complete longitudinal framing system is adopted for the stiffening the decks and shells. Thus these vessels naturally have adequate longitudinal, transverse as well as torsional strength.

References

1. Jefferies, H., & Probst, J.-F. (2007). *Hull design and limitations for container ships*. Singapore, Germanischer Lloyd.
2. Approximate ship capacity data. Container-transportation.com.
3. Germanischer Lloyd—Design of Double Hull Tankers. Presentation at National Technical University of Athens, May 2005.

Chapter 8
Structural Alignment and Continuity

Abstract A ship is assembled from various structural components. The question of structural alignment arises at the fit up stage of any production process. It can be alignment of two flat plates for butt welding, fitting of stiffeners of a stiffened panel, erection of the side shell panel above the double bottom unit or installation of the superstructure block above the main deck, etc. In all these cases, alignment of the mating edges and all the other associated stiffening members pose a serious challenge to the production team in the shop floor. This aspect of alignment is closely connected with structural continuity or discontinuity. Alignment is a production aspect whereas continuity needs to be implemented in the design stage. However they are interrelated, a structural misalignment during production will lead to a case of structural discontinuity. Thus to achieve structural continuity or to avoid serious discontinuities, precision structural alignment is a necessity.

A ship structure is essentially a product assembled from various structural components. Hence in the process of assembling these components, naturally the question of alignment arises. This aspect of alignment is closely connected with structural continuity or discontinuity. However as one can observe that, alignment is a production aspect whereas continuity needs to be implemented in the design stage. However they are interrelated, a structural misalignment during production will lead to a case of structural discontinuity. Thus to achieve structural continuity or to avoid serious discontinuities, precision structural alignment is a necessity.

8.1 Structural Alignment

The question of structural alignment arises at the fit up stage of any production process. It can be alignment of two flat plates for butt welding, fitting of stiffeners of a stiffened panel, erection of the side shell panel above the double bottom unit or installation of the superstructure block above the main deck, etc. In all these cases,

N.R. Mandal, *Ship Construction and Welding*, Springer Series on Naval Architecture, Marine Engineering, Shipbuilding and Shipping 2,
DOI 10.1007/978-981-10-2955-4_8

alignment of the mating edges and all the other associated stiffening members pose a serious challenge to the production team in the shop floor.

In fabrication of stiffened plate panels, there can be misalignment between the two mating edges of the plates resulting from deformation due to thermal cutting. It can be due to transverse shrinkage of the plates, buckling of the edges or even due to bad edge preparation as can be seen in Fig. 8.1. In such situation, depending on the extent of root gap, extra metal has to be deposited on the plate edge before commencing actual welding of the joint.

In any case it will give rise to the problem of fitup. Unless the plate edges are properly aligned, welding cannot be carried out or the will result in a poor weld quality. In fabrication of thin stiffened panels, it is extremely difficult to maintain alignment of the plate edges because of inevitable edge deformations due to the prior welding operations on the plates. The edge alignment is done by applying external force and is subsequently restrained from any possible misalignment by applying tack welds as well as holding the plates in position by magnetic clamps. In absence of magnetic clamps often flat bars as strong backs are welded at the back side of the aligned edges. However the use of strong backs or magnetic clamps is to be minimized to avoid development of locked-in stresses in the plate panels. Once these restraints are removed, it may lead to release of these stresses causing deformation of the welded panel. Similarly the stiffeners are positioned and need to be aligned to the plate before welding. There can be case of misalignment because of inherent deformation of the plate surface as shown in Fig. 8.2, i.e. lack of required flatness or inherent deformation of the stiffeners. By using suitable fixtures or pin jacks such misalignments are to be removed at the fitting up stage.

Fig. 8.1 A typical example of bad workmanship resulting in alignment problem

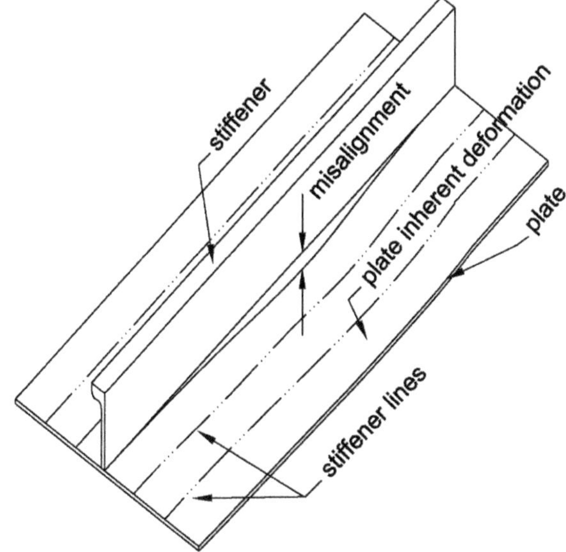

Fig. 8.2 Stiffener edge misalignment due to inherent deformation of plate

It is observed that the edges of the subassemblies are often heavily deformed particularly in case of thin stiffened panels as shown in Fig. 8.3. Therefore in subsequent assembly of these panels prior fairing is required for proper alignment of the panels. The Rules and Regulations for construction of steel hulls [1] prescribe the permissible extent of misalignment in butt joints as shown in Fig. 8.4.

The permissible deviation in alignment of flange of T-section stiffeners is shown in Fig. 8.5.

The gap between the bracket and the stiffener to which it is connected as well as the gap between the stiffener cut out in the web and the stiffener is to be less than 2 mm. In any case it should not exceed 3 mm.

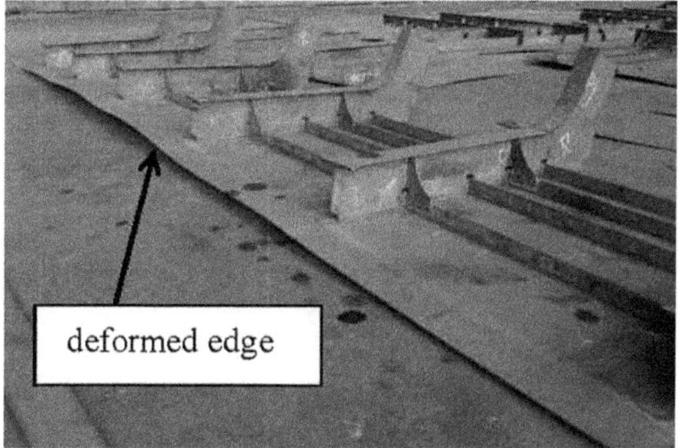

Fig. 8.3 Deformed edge leads to misalignment during *panel* assembly

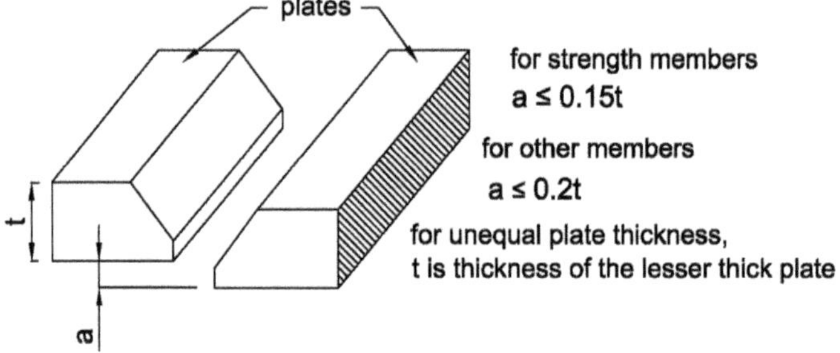

Fig. 8.4 Limit of misalignment in butt joints

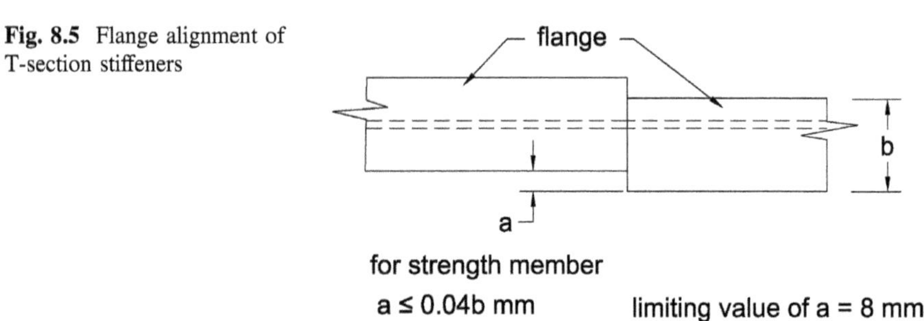

Fig. 8.5 Flange alignment of T-section stiffeners

8.2 Structural Continuity

Structural continuity provides for a continuous load path through the structure, e.g. the deck load is transmitted through the stiffened side shell to the bottom shell structure, which is eventually supported by buoyancy forces. Any discontinuity in this will result in stress concentration locally. In ship structure there are several unavoidable structural discontinuities for functional requirement. Hence adequate measures are to be taken to compensate for these structural discontinuities.

There are various unavoidable openings in a ship's structure for different functional requirements. Structural discontinuity caused by these openings may lead to local stress concentrations as well as loss of structural strength. Hence appropriate structural measures should be taken to compensate for these discontinuities.

(i) *Welding flat bar*
   In case of medium sized openings like doors, where the opening may cut through a stiffener, additional flat bar stiffeners are used to properly terminate the cut stiffener as well as a flat bar is welded all around the opening as shown in the Fig. 8.6.

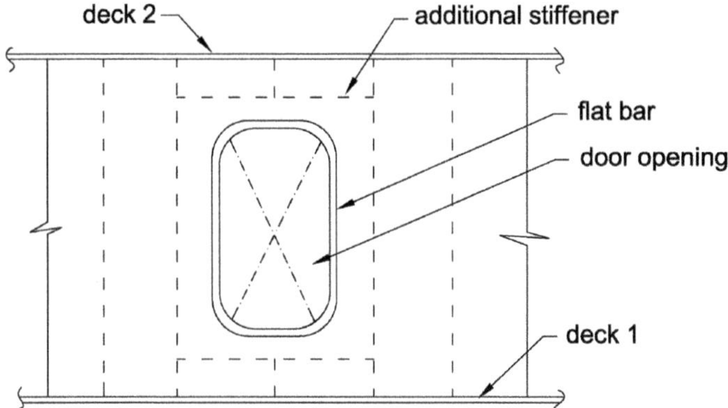

Fig. 8.6 Flat bar ring and additional stiffening around door opening

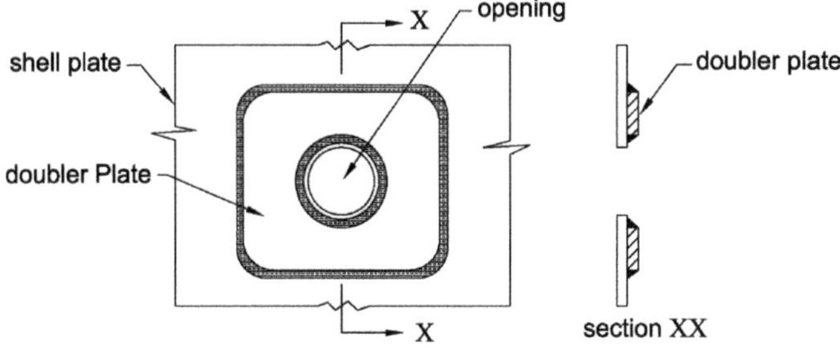

Fig. 8.7 Doubler plate in way of shell opening

(ii) *Doubler plate*
In case of small openings like various discharge openings on shell side, doubler plate is welded on the inner side of the shell plate as shown in Fig. 8.7.

(iii) *Insert plate*
In the areas of large openings as in case of cargo hatch opening, stress concentration is likely to occur at the corners of these openings because of sudden change in sectional area of the deck plate. Here plate of higher thickness compared to the adjacent plates is inserted at the corners. These plates are referred to as insert plates as shown in Fig. 8.8. This is done to enable the corner plates to withstand higher levels of stress because of inevitable stress concentrations.

(iv) *Beams and Girders of higher scantlings*
Stiffening members of higher scantlings in the form of hatch side girders and hatch end beams are used to make up for the lost strength due to large hatch opening as shown in Fig. 8.8.

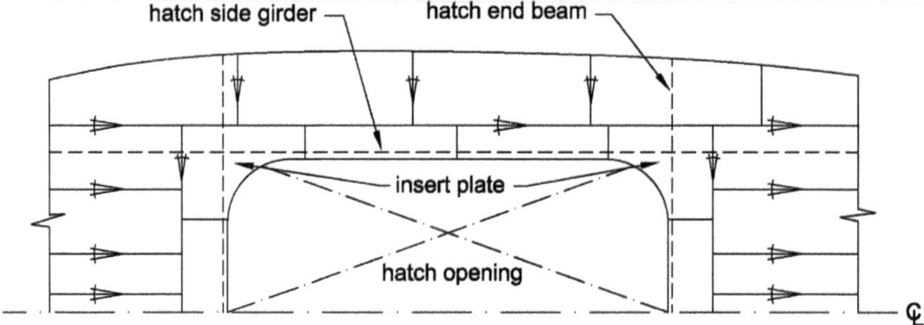

Fig. 8.8 Insert plate, hatch side girder and hatch end beam in way of cargo hold hatch opening

Reference

1. 47 Shipbuilding and Repair Quality Standard, (Rev. 7, June 2013), Part A Shipbuilding and Remedial Quality Standard for New Construction, IACS "Bulk Carriers - Guidelines for Surveys, Assessment and Repair of Hull Structure".

Chapter 9
Material of Construction

Abstract Wide variety of materials is used in ship construction. For ship hull construction several materials are available, however one needs to select the one which suits the intending purpose the best. The objective is to have a product that will be economical to manufacture and maintain, should also be durable and reliable. Various characteristics of steels, marine grade aluminum alloys and glass fibre reinforced plastics as materials for ship construction have been dealt with in this chapter.

In ship construction one can see use of wide variety of materials for wide ranging requirements. Here material for hull and superstructure construction will only be dealt with. For ship hull construction several materials are available, however one needs to select the one which suits the intending purpose the best. The objective is to have a product that will be economical to manufacture and maintain, should also be durable and reliable. The following aspects need to be checked while selecting a material for a given construction.

Design Aspects
From the designer's point of view, the material selection for hull fabrication depends primarily on the following three types of mechanical responses,

(i) Strength and stiffness of the parent material (σ_{ult}, σ_y and E).
(ii) Time-dependent material behavior such as,

- creep,
- stress corrosion cracking,
- fatigue and
- corrosion wastage.

(iii) Fatigue and corrosion fatigue of parent material.

Traditionally, the design of marine structures is based on strength and stability criteria to resist yielding or buckling, with the emphasis on (i) and (ii) above. The primary material properties used in this approach are ultimate strength, yield strength, modulus of elasticity, and fatigue endurance behavior (S-N curves).

N.R. Mandal, *Ship Construction and Welding*, Springer Series
on Naval Architecture, Marine Engineering, Shipbuilding and Shipping 2,
DOI 10.1007/978-981-10-2955-4_9

Resistance to fracture was usually taken into account by assessing the impact properties of the materials as given by Charpy-V notch impact test. For fracture dominated failure, following macroscopic defects are most important,

- cracks,
- inclusions,
- weld flaws,
- porosity.

The essential feature of the fracture process is connected with the interaction of the local stress-strain field with the defect.

Manufacturing Aspects

From manufacturing point of view, the material of construction should possess the following attributes:

- Ductility/Formability
- Machinability
- Weldability
- Resistance to corrosion
- Maintainability.

As ship's hull has compound curvature plates, the material should have adequate ductility, such that necessary plate bending can be achieved without any defects occurring in the plate. The material should be amenable to machining, like drilling, punching or cutting using guillotine shear. At the same it can be cut using oxy fuel/plasma cutting methods. The material should be easily welded using conventional welding techniques. In case of non-metallic material, like fibre reinforced plastics, conventional moulding techniques can be used for fabrication. Since ships operate in harsh corrosive environment, it will be an additional advantage to have good resistance to corrosion.

Economic Aspects

The economic aspect is equally important in the selection of the material for construction. There can be options where the economic aspect may over right the justification of material selection. Using a particular material should be economically viable. In the process some other aspect may have to be somewhat compromised.

Environmental Aspect

With ever growing concern on environmental issues related to all kinds of industrial products and production process, it is of utmost importance to select materials which have the least environmental impact. The environmental impact needs to be looked into during production, operation and decommissioning phase of the product.

9.1 Steels

Steels make a good choice as material of construction because of their,

- relatively low cost,
- ease of fabrication by welding, and
- moderately good mechanical properties.

The appropriate choice of structural steel should satisfy the requirements of classification societies.

The factors which affect the properties and quality of mill steels are,

- Chemical composition
- Melting practice
- Type and extent of deoxidation
- Special processing e.g. vacuum degassing-desulphurisation
- Grain refining
- Rolling procedure e.g. controlled rolling and quenching
- Heat treatment e.g. normalizing, quenching and tempering.

The various grades provide differing fracture toughness and weldability. Rules for inspection and testing of hull materials are detailed in the Classification society Rules and Regulations for the Construction and Classification of Steel Ships.

Improper selection of structural steels can lead to catastrophic consequences. The classic examples of this are the failures of the Liberty ships and T-2 tankers in World War II. Of 2500 Liberty ships built, failures were reported from 700 ships, and 145 of them broke in two. Several important facts were learned from these ship failures.

These failures are attributed to using alloy compositions with high ductile to brittle transition temperatures and to welding-induced flaws. It was observed thicker plates and sections are more liable to brittle fracture than thinner plates due to triaxial stress state.

Carbon-Manganese and Low-Alloy Steels

Carbon-manganese and low-alloy steels are the most widely used variety of steels for marine structures. The major advantages of steel as construction material are,

- high strength,
- high Young's modulus.

Significant disadvantages are,

- poor marine corrosion resistance,
- relative technical difficulty in welding,
- embrittlement by hydrogen or by low temperatures.

The range of available alloys is vast. The division is made by virtue of their composition and treatment and, in some cases, their intended application.

Carbon-Manganese Steels
The characteristics of these steels are,

- Carbon content 0.12–0.22 %
- Manganese content usually between 0.5 and 1.5 %
- Without any other deliberate alloying elements
- Usually not heat-treated
- Grain structure is ferrite-pearlite,
- Pearlitic phase increases with carbon content.

The grain structure has a number of implications as regards mechanical properties.

- Increasing carbon content increases the amount of pearlite (mixture of ferrite and Fe_3C) in the structure and thus the strength of the steel.
- Strength can be enhanced without changing the chemistry, particularly the carbon content, by reducing the grain size.

Traditional method of reducing grain size in metals.

- Cold work followed by raising the temperature to a point at which recrystallisation takes place,
- Soak to allow complete or partial recrystallisation
- Cool, 'freezing in' the recrystallised grain size.

In case of steels such a method can only produce a limited degree of grain refinement. The modern grain refining practice is based on dynamic recrystallisation.

Dynamic recrystallisation for reducing grain size,

- Hot working at a temperature where recrystallisation takes place
- Micro-alloying additions of aluminum, vanadium, titanium or niobium are added to impede the growth of austenite grains.

Substantial grain refinement and hence high yield strength and excellent low-temperature toughness with good weldability is obtained through controlled-rolled and on-line accelerated cooling process (CR + OLAC). This has led to the production of HSLA steels.

Impurities
Sulphur and phosphorus content are to be controlled, because they affect weldability and make the steel susceptible to some forms of corrosion. To ensure better fracture toughness, improved weldability and improved through-thickness properties, both Sulphur and phosphorus content are to be kept to a specified minimum. Rules and Regulations of Indian Register of Shipping [1] prescribe maximum permissible Sulphur and phosphorus content to be 0.035 % in ladle samples.

9.1.1 Normal Strength Steel

Marine grade structural steels are primarily classified under 4 grades, A, B, D and E based on Charpy V-notch impact strength. The mechanical properties of normal strength steels are summarized in Table 9.1.

The broad chemical composition of these steels is as follows:

Carbon content −0.21 % max. for plates and 0.23 % for rolled steel sections in grade A, B and D steels. For grade E steel carbon content is to be less that equal to 0.18 %.

Manganese −0.6 % min.

Silicon −0.50 % for Grade A, 0.35 % for Grade B and D and 0.10–0.35 for Grade E steels.

Steels having good weldability means, they can be welded successfully with any of the conventional simple welding technique without any additional precaution. Weldability of steels is defined by a parameter, known as carbon equivalent. Carbon and alloying element content greatly affects the metallurgy of weld deposit. In unfavourable conditions, cracks may develop in the weld deposit as it starts to cool down. This happens primarily due to formation of a brittle phase, commonly known as martensite in the solidified weld deposit. In such a situation the steel material is referred to as non weldable steel or steel having poor weldability. This metallurgical transformation depends directly on carbon content and partially on the other alloying elements. Hence to have good weldability, carbon and alloying element content should be as low as possible. However at the same time these alloying elements contribute towards mechanical properties of the steels. A parameter named as carbon equivalent (C_{eq}) is widely used to quantify the weldability of steels. Various formulations have been suggested for calculation of this parameter. For normal strength steel, it is defined as,

$$C_{eq} = C + \frac{Mn}{6}$$

A weldable steel should have $C_{eq} \leq 0.4$.

Table 9.1 Mechanical properties of normal strength steels [1]

Grade	Yield strength (MPa)	Tensile strength (MPa)	Elongation on $5.65\sqrt{S_o}$ (%)	Charpy V-notch impact test		
				Test temp (°C)	Thickness (mm)	Av energy (J)
A	235	400–520	22	–	–	–
B				0	≤50	27
D				−20	≤50	27
E				−40	≤50	27
					50 ≤ 70	34
					70 ≤ 100	41

9.1.2 High Strength Steel

In high strength steels, carbon content is limited to 0.18 %. These steels contain aluminum, niobium, vanadium, titanium or other suitable grain refining elements as micro alloying elements. These are also referred to as high strength low alloy (HSLA) steels. The total content of (Nb + V + Ti) should not exceed 0.12 %. The chemical composition of these steels is as follows,

Carbon content	−0.18 % max.
Manganese	−0.9–1.6 %
Grain refining elements	−Al, Nb, V, Ti
Total (Nb + V + Ti) %	−0.12 max.

Other alloying elements:

Cu (max)	−0.35 %
Cr (max)	−0.20 %
Ni (max)	−0.40 % and 0.08 % for FH grade
Mo (max)	−0.08 %

The carbon equivalent is given by, $C_{eq} = C + \frac{Mn}{6} + \frac{Cr+Mo+V}{5} + \frac{Ni+Cu}{15}$.

As per IRS classification rules, these are sub divided into 4 grades—AH, DH, EH, FH. Test temperatures are 0°, −20°, −40° and −60° centigrade for AH, DH, EH and FH respectively.

The ductility parameter is quantified by elongation over a specified gauge length of $5.65\sqrt{S_0}$, where S_0 is the cross sectional area of the tensile test specimen. The elongation percentage over the specified gage length varies from 22 to 14 over a yield strength variation of 315–690 N/mm^2.

With increasing yield strength and plate thickness, the requirement of absorbing impact energy as per the specified impact test increases from 31 to 55 J for test specimens taken in longitudinal direction and from 22 to 37 J for transverse specimens.

Steel for ships' hull

Substitution of higher-strength steel ($\sigma_1 > \sigma_2$) for a ship of identical dimensions and load distribution leads to a deflection (δ_1), which is increased in the ratio of the increase in allowable stress as shown below,

$$\delta_1 = \frac{E_2}{E_1}\frac{\sigma_1}{\sigma_2}\delta_2$$

Considering the modulus of the steels to be same. i.e. $E_1 = E_2$,

$$\delta_1 = \left(\frac{\sigma_1}{\sigma_2}\right)\delta_2$$

where δ_1 and δ_2 are the deflections of ship-1 and ship-2 respectively, E_1, σ_1 and E_2, σ_2 are the modulus of elasticity and yield strength of steel material for ship-1 and ship-2 respectively. One can observe that the weight savings are offset by the requirements of controlling the deflection. Thereby no benefit is obtained by using steels of higher yield stress than 340 MPa. The reduction in scantlings due to high strength steel leads to reduction in stiffness resulting in increased deflection.

For plates up to 10 mm in thickness, mild steel is usually adequate. For thicker sections (up to 25 mm) a lower inclusion count and reduced porosity are preferred to obviate the danger of brittle fracture. For this grade B steel is preferred.

Steels for submarines

Here the main load is hydrostatic external pressure having atmospheric internal pressure. In such a loading case yield strength plays an important role as a selection parameter. As can be seen, prior to 1940, steels of 220 MPa were used, whereas now steels, quenched and tempered to 550–690 MPa yield strength and even higher, are being used.

The development required for an increase in yield strength retaining weldability and toughness. This was achieved by the use of reduced grain sizes and subsequently by thermo-mechanical treatment coupled with accelerated cooling. The improvements in mechanical properties, weldability and low-temperature toughness have been achieved in the background of general decrease in carbon content and using micro-alloying elements with special thermo-mechanical treatments.

Steels for Offshore Structures

Offshore structures are subjected to repetitive cyclic loads due to the continuous impact of waves at sea. This makes the structure susceptible to fatigue failures. The initiation of such fatigue related cracks often takes place at the welded joints, particularly at the joints causing stress concentrations. Cast steels have been used as viable alternatives for structural steels in critical fatigue-prone areas of offshore structures. This is primarily because of their ability with which sections can be contoured, drastically reducing stress concentration factors. Also, through casting, more intricate shapes can be conveniently made, which are otherwise difficult to produce from plate or tube material.

The use of higher-strength structural steels in offshore structures may lead to 10–15 % saving in weight over conventional structural steels. Some higher-strength steels may be more fatigue-resistant compared to the lower-strength steels for the same level of stress intensity. However this applies only to the initiation stage in the base metal. The effect in welded joints of offshore structures is far more complicated by the fact that the structural life is propagation-dominated. Since propagation is only marginally affected by yield strength, both high and normal strength steels will show approximately the same fatigue life in offshore structural applications.

Therefore use of high strength steels is not much effective in case of offshore structures as they are not weight sensitive structures. At the same time, use of high strength steel will increase the working stress level making the structure more vulnerable to fatigue. It is therefore wiser to use normal strength steel which is economically favourable option.

9.1.3 Wear-Resistant Applications

Carbon-manganese steels with up to 0.7 % carbon may be used in wear-resistant applications where stresses are relatively low. The hardenability of a low-alloy steel is dependent upon both its carbon content and its alloy contents. The continuous cooling transformation (CCT) diagram shows the structures expected under various cooling conditions for a given type of steel. These diagrams have cooling rate as the horizontal axis, so that the transformation behaviour at a given cooling rate can be traced by following a vertical line downwards from the austenitising temperature. The carbon content determines the final hardness of the untempered martensite, whereas the ease with which this martensite can be produced (i.e. hardenability) is dependent on the carbon and alloy contents.

9.1.4 Stainless Steels

The difference between the stainless steels and the carbon-manganese and low-alloy steels is primarily in the use of chromium as a substantial alloying element. Stainless steels are primarily ferrous alloys, and the main alloying elements are chromium and nickel. Chromium normally forms a body-centred cubic (BCC) crystal structure whereas nickel forms a face-centred cubic (FCC) structure.

The chromium enhances the atmospheric corrosion resistance of stainless steels. As chromium content is increased, the transition to 'stainless' behaviour is a gradual one. At 13 % chromium, the steel can definitely be regarded as stainless, and most commercial alloys contain at least this amount of chromium. The chromium forms a passive layer of a chromium-rich oxide on the surface of the steel. This layer makes the steel resistant to atmospheric corrosion. This layer has a self-healing capability if disrupted, provided that the environment is conducive to this. However, the oxide film breaks down in case of complete and sustained immersion of stainless steel in sea water, causing pitting, crevice and inter-granular corrosion.

In ocean environment for above water line application, the 300 series stainless steel (18 % Cr) is best suited. Whereas for under water application, type 304 and 316 are recommended but for immersed condition of only 4–6 months. The 300 series are nonmagnetic, whereas the 400 series are magnetic. Since stainless steels are susceptible to crevice corrosion, welded joints are preferred to mechanical fastener. Special grades stainless steels (e.g. type 304L and 316L) resist inter-granular corrosion.

The five main forms of stainless steels are,

- Austenitic
- Martensitic
- Ferritic
- Precipitation hardening
- Duplex.

Austenitic Stainless Steels

Austenitic stainless steels form the major proportion of those produced and, although relatively weak, have the advantage of cryogenic toughness and a high degree of homogeneity, which improves their general corrosion resistance. Austenitic stainless steels are based around two compositions, the so-called '18/8' and '25/20' steels. The 18/8 composition contains 18 % chromium, and about 8 % nickel. Likewise, 25 % chromium requires the presence of about 20 % nickel to stabilise the austenite.

The most widely used designation system for stainless steels is the AISI system, and austenitic stainless steels are specified by three-digit numbers in the 200 and 300 series.

Carbon is usually considered to be a nuisance in austenitic steels. It has an affinity for chromium and will preferentially form chromium carbides at grain boundaries when exposed to a temperature range usually about 500–800 °C.

Because of the relatively slow diffusion rate of chromium in austenite compared with that of carbon, such precipitation results in chromium-depleted zones, with consequent loss of corrosion resistance, usually along grain boundaries, To avoid this problem during welding or other processing, a low-carbon version of 304 is available for steels which are likely to be exposed to the above-mentioned temperature range.

An alternative method of resolving the difficulty is to add an element which has a greater affinity for carbon than has chromium. Niobium (columbium), tantalum and titanium have this property. Additions of these at about 5–10 times the amount of carbon form the basis of the stabilised grades such as AISI 321.

Molybdenum (to about 4 %) enhances corrosion resistance, especially that to pitting and crevice corrosion. SS316 contains molybdenum.

Ferritic and Martensitic Stainless Steels

These steels are essentially iron-chromium-carbon alloys, and are therefore of low nickel equivalent. At lower carbon and higher chromium contents (e.g. 410) the formation of austenite is completely inhibited, leading to the ferritic stainless steels. The ferritic stainless steels have improved weldability and corrosion resistance. At higher carbon and lower chromium contents (e.g. 440), it becomes possible to render the steel austenitic or into a state of austenite plus carbides, thus producing the martensitic stainless steels. High carbon content severely reduces corrosion resistance.

In brief, the martensitic stainless steels provide a high-strength, corrosion-resistant option. The low carbon grades are usable where strength coupled with adequate toughness is required. Whereas the higher carbon grades are used where increased wear resistance (i.e. hardness) is required.

Duplex and Precipitation Hardening Stainless Steels

Most duplex stainless steels contain about 70 % iron and are likely to have austenite-ferrite structures. Most commercial cast duplex alloys have a composition of around 19 % Cr and 9 % Ni, which with a wrought alloy would be associated with an austenitic structure.

Properties of Stainless Steels

The mechanical properties of stainless steels are primarily dependent upon the matrix than its composition. In broad terms, the austenitics are weakest but toughest, while the martensitics exhibit highest strength.

There are two major ways to compromise between these opposing sets of properties; the first is to use a mixture of the two phases (duplex stainless), the second is to employ precipitation strengthening.

The corrosion of stainless steels is a complex phenomenon. The major points to be considered are the possibility of local effects such as pitting and crevice corrosion and the dangers of stress corrosion.

The general corrosion rates of stainless steels in seawater and marine atmospheres are of course lower than for non-stainless steels. In case of stainless steels, weight loss due to material wastage as a measure of corrosion is not useful, since pitting is the predominant mode of corrosion.

In general, atmospheric performance is rather good, but pitting may be a problem in sub-merged service. The higher grades (higher chromium, austenitics and molybdenum-containing) show superior performance.

As far as pitting and crevice corrosion are concerned, the stainless steels are generally more susceptible in those marine environments where the supply of oxygen is low. Higher resistance to pitting or crevice corrosion is obtained with higher chromium contents (>20 %) along with some percentage of molybdenum.

The principal interest in stress corrosion in stainless steels centres around chloride Stress Corrosion Cracking (SCC). Ferritic stainless steels are rather less susceptible to chloride SCC. Dissolved oxygen is known to aggravate chloride SCC, and a critical applied stress is required. Chloride stress corrosion in non-austentitic grades is much less likely. In ferritic and martensitic grades, particularly those of high strength, the stress corrosion problem is more one of hydrogen embrittlement. Duplex stainless steels for marine service show better chloride SCC resistance compared to the austenitic stainless steels. Martensitic grade stainless steels have good corrosion-erosion resistance. Austenitic and precipitation hardened stainless steels have higher resistance to cavitation damage.

Stainless steels are not widely used in marine environments primarily because of the presence of chloride ions which causes chloride SCC. Stainless steels are preferably to be used for above water applications rather than in submerged conditions. However castings can be used in condensers, marine exhausts, propellers and pump impellers. They usually show good performance where flow is maintained, however corrosion problem may arise if they are left idle submerged in seawater.

9.2 Aluminum Alloys

The basic characteristics of aluminum alloy are as follows:

- Light in weight (2660 kg/m^3 while steel weighs 7850 kg/m^3)
- Some of its alloys have strengths exceeding that of mild steel

- High strength to weight ratio
- Low rigidity (elastic modulus 70 GPa as compared to steel 210 GPa)
- Retains good ductility at subzero temperatures
- High resistance to atmospheric corrosion
- Possibility of SCC in sea water environment
- Highly electro-negative
- Coupling with noble metals (e.g. copper) to be avoided
- Not toxic
- Melting range −482 to 660 °C
- No color change when heated to the welding temperature range
- High thermal conductivity (as compared to steel)
- Necessitates a high rate of heat input for fusion welding
- Thick sections may require preheating
- For resistance spot welding, aluminum's high thermal and electrical conductivity require higher current, shorter weld time, and more precise control of the welding variables than when welding steel
- Aluminum and its alloys rapidly develop a tenacious, refractory oxide film when exposed to air.

As the temperature decreases, aluminum alloys and their weldments gain strength. The aluminum alloys do not become brittle but maintain or increase in ductility as the temperatures decrease below zero. Pure aluminum is a relatively weak material. However its alloys with various strengthening mechanisms form a strong material of construction. Aluminum alloys are strengthened by,

- Cold work
- Precipitation hardening
- Solid solution strengthening
- Dispersion strengthening.

Alloy Designation System

This system involves a four-digit composition designation along with temper designation. This gives information about the alloy composition and condition of the alloy.

- First digit—Principal alloying element
- Second digit—Variations of initial alloy
- Third and fourth digits—Individual alloy variations (number has no significance but is unique).

1XXX—Pure Al (99.00 % or greater)
2XXX—Al-Cu Alloys
3XXX—Al–Mn Alloys
4XXX—Al–Si Alloys
5XXX—Al–Mg Alloys

6XXX—Al–Mg-Si Alloys
7XXX—Al-Zn Alloys
8XXX—Al + Other Elements
9XXX—Unused series

Temper Designation System
F—as fabricated
Applies to products of forming processes in which no special control over thermal or strain-hardening conditions is employed.

O—annealed
Applies to wrought and cast products which have been heated to produce the lowest strength condition and to improve ductility and dimensional stability.

H—strain-hardened
Applies to wrought products which are strengthened by strain-hardening through cold-working. The strain-hardening may be followed by supplementary thermal treatment, which produces some reduction in strength. This is followed by two digits:

The first digit indicates basic operations:

H1 = strain hardened only.
H2 = strain hardened and partially annealed.
H3 = strain hardened and stabilized.

The second digit indicates degree of strain hardening:

HX2 = quarter hard.
HX4 = half hard.
HX6 = three-quarters hard.
HX8 = full hard.
HX9 = extra hard.

T—thermally treated to produce stable tempers other than F, O or H
Applies to products which have been heat-treated, sometimes with supplementary strain-hardening, to produce a stable temper. The T is always followed by one or more digits indicating specific sequence of treatments.

9.2.1 Aluminum Alloy in Shipbuilding Applications

Aluminum alloys are classified into two categories: heat treatable and non-heat treatable.

Non-heat treatable aluminum alloys
The non-heat treatable alloys are mainly found in the 1XXX, 3XXX, 4XXX and 5XXX alloys series depending upon their major alloying elements. The strength of

the non-heat treatable aluminum alloys depends primarily upon the hardening effect of alloying elements such as silicon, iron, manganese and magnesium. These elements contribute towards strength either as dispersed phases or by solid-solution strengthening.

Magnesium is the most effective solution-strengthening element in the non- heat treatable alloys. The 5XXX series aluminum alloys are based on magnesium. They have the highest strengths amongst the non-heat treatable alloys. They can be easily fusion welded with a minimum loss of strength.

Heat treatable aluminum alloys

The heat-treatable alloys are found primarily in the 2XXX, 6XXX and 7XXX alloy series. Heat treatable aluminum alloys develop their properties by solution heat treating and quenching, followed by either natural or artificial aging. Cold working may add additional strength. The heat-treatable alloys may also be annealed to attain maximum ductility. This treatment involves holding at an elevated temperature and controlled cooling to achieve maximum softening. In 6XXX series alloy (e.g. AA6061), main alloying elements are Si ($\sim$0.6 %) and Mg ($\sim$1.0 %).

4XXX—Al–Si Alloys

High silicon content (4.5–6.0 % in AA4043 and 11.0–13.0 % in AA4047) provides good flow characteristic, which in the case of forgings ensures the filling of complex dies and in the case of welding ensures complete filling of crevices and grooves in the members to be joined.

- Good flow characteristics.
- Medium strength.
- Typical ultimate tensile strength: 186 MPa. [2]
- Typical tensile yield strength: 124 MPa.

Alloy 4043 is one of the most widely used filler alloys for gas-metal arc (GMA) and gas-tungsten arc (GTA) welding of 6XXX alloys for structural and automotive applications.

5XXX—Al–Mg Alloys

Magnesium is the main alloying element in the 5000 series. The higher the Mg content the greater is the as-welded strength or annealed strength of the base metal. 5086, 5083 and 5456 alloys contain 4–4.9 % magnesium [3].

- Excellent corrosion resistance even in salt water.
- Very high toughness even at cryogenic temperatures to near absolute zero.
- Readily welded by a variety of techniques.
- Typical ultimate tensile strength: 317 MPa.
- Typical tensile yield strength: 228 MPa.

Typical applications

Weight sensitive vessels employ AA5083 plate for hulls, hull stiffeners, deck and superstructure. Alloy 5083 is also used for carriage of cryogenic liquefied natural gas in LNG carriers. AA5454, 5086, and 5083 are also used for various welded constructions exposed to high humidity and water in offshore oil rigs.

Care must be taken to avoid use of 5XXX alloys with more than 3 % Mg content in applications where they receive continuous exposure to temperatures above 100 °C. Such alloys may become sensitized and susceptible to stress corrosion cracking. For this reason, alloys such as 5454 and 5754 are recommended for applications where high temperature exposure is likely.

6XXX—Al–Mg-Si Alloys

These alloys have Mg content of about 0.8–1.2 % and silicon content 0.4–0.8 % [3]. 6000 series alloys possess good weldability provided adequate filler metal is fed into the joint. Using 4000 series filler metal, it is best to allow about 50 % dilution of the filler into the weld metal composition. With a 5000 series filler metal, one needs about 70 % filler metal dilution. If a 3.18 mm thick section of 6061 aluminum is welded with insufficient 4043 filler metal, it will result in cracking.

Typical characteristics of 6000 series alloys are:

- High corrosion resistance,
- Excellent extrudability,
- Moderately high strength.
- Good weldability.
- Typical ultimate tensile strength: 310 MPa.
- Typical tensile yield strength: 276 MPa.

Typical applications

6061 alloys are used for stiffeners in ship structures. Roof structures for arenas and gymnasiums are usually of 6063 or 6061 extruded tube, covered with 5XXX alloy sheet. The new magnetically levitated trains in Europe and Japan employ bodies with 6061 and 6063 structural members.

Filler Metal

The end use of the weldment and desired performance are important considerations in selecting an aluminum alloy filler metal. Many base-metal alloys and alloy combinations can be joined using any one of several filler metals, but only one may be the optimum for a specific application.

The primary factors commonly considered when selecting an aluminum alloy filler metal are,

- Freedom from cracks
- Tensile or shear strength of the weld metal
- Weld ductility
- Service temperature
- Corrosion resistance
- Color match after anodizing.

Cracking

In general the non-heat treatable aluminum alloys (5XXX series) can be welded with a filler metal of the same basic compositions as the base alloy.

Generally, a dissimilar filler metal having a lower melting temperature and similar or lower strength than the base metal is used for the heat treatable alloys (6XXX). By allowing the low-melting constituents of the base metal adjacent to the weld to solidify before the weld metal, stresses are minimized in the base metal during cooling and intergranular cracking tendencies are minimized. High silicon content and the high magnesium content aluminum alloys are easy to weld due to low sensitivity to cracking.

6XXX series alloys are very sensitive to cracking if the weld metal composition remains close to the base metal composition. These can be welded easily, provided the edge preparation permits an excess of filler metal admixture with the base metal, e.g. V-groove of 90°. For alloy 6061, the weld metal should possess at least 50 % of 4043 filler metal or 70 % of 5356 filler metal.

Filler metals with high silicon content (4XXX series) should not be used to weld high magnesium content 5XXX series alloys. Excessive magnesium-silicide eutectics developed in the weld structure decrease ductility and increase crack sensitivity. Mixing the high-magnesium content and high-copper content alloys results in high sensitivity to weld cracking and low weld ductility.

Strength

Several filler metals are available that meet the minimum as-welded mechanical properties. The diffusion of alloying elements from base metal may increase the as-welded mechanical properties. Filler metal alloys 5356, 5183 and 5556 provide high shear strength for structural fillet welds. The 1XXX and 5XXX alloy series filler metals produce very ductile welds and are preferred when the weldment is subjected to forming operations or post-weld straightening operations.

Corrosion Resistance

Aluminum-magnesium filler metals are highly resistant to general corrosion when used with base alloys having similar magnesium content. However, the 5XXX alloy series filler metals can be anodic to the 1XXX, 3XXX and 6XXX alloy series base metals. In immersed service condition, the weld metal will pit and corrode. Thus an aluminum-silicon filler metal, such as alloy 4043 or alloy 4047, would be preferred for improved corrosion resistance over alloy 5356 filler metal when welding alloy 6061 base metal for an immersed-service application.

Which Filler Metal to Use?

The role of the filler metal in the welding of aluminum alloy is a critical one. The high magnesium content alloys (5356, 5183 and 5556) as filler metal for welding AA5083 provide the highest as-welded strength, permit the smallest fillet welds, and often result in financial advantages. However for welding of AA6061, high silicon content filler either 4043 or 4047 is to be used.

Selection of Filler Metal

Criteria for selecting a filler metal are,

Table 9.2 Guide to the selection of filler metal for general purpose welding

Base metal	6005, 6061, 6063, 6101, 6151, 6201, 6351, 6951	5086	5083
5083	ER5356	ER5356	ER5183

- ease of welding,
- strength,
- ductility,
- corrosion resistance of the filler metal/base metal combination,
- color match with the base metal after anodizing,
- service at elevated temperature.

The selection of the correct filler metal greatly influences the service life of an aluminum weldment. A guide to selection of filler metals for various aluminum alloy combinations is given in Table 9.2.

Storage and Use of Aluminum Filler Metal

High level of cleanliness is very essential in producing high quality, defect free welds in aluminum alloys. To avoid dirt accumulation and subsequent contamination of weldment, filler-metal spools or rods are to be kept covered and stored in a dry place at a relatively uniform temperature. Electrode spools temporarily left unused on the welding machine, as between work shifts, should be covered with a clean cloth or plastic bag if the feed unit does not have its own cover. Unfinished spools of wire which will not be used overnight are to be returned to its carton and tightly sealed to avoid ingress of any moisture. The 5XXX series electrodes are most likely to develop a hydrated oxide. They need to be stored in appropriate cabinets maintaining a relatively low humidity (less than 35 % RH).

9.3 Fibre Reinforced Composites

The ever-increasing use of composites in various industrial applications points to the several advantages it provides. Weight reduction, corrosion resistance, design flexibility, low maintenance costs and easy reproduction of complicated designs are some of the primary reasons for using composites as a preferred material of construction. However this material more commonly known as GFRP or CFRP, i.e. glass fibre reinforced plastic or carbon fibre reinforced plastic. The plastic or the polymer material forms the matrix and the glass or the carbon fibres act as the reinforcements. Both the polymer and the fibres remain as separate entities and therefore the material is termed as composite. The polymer matrix binds the reinforcements, whereas the reinforcements provide for the strength. In boat building, where weight is a constraint, GFRP is widely used. Where the requirement is that of very high strength to weight ratio to be achieved, CFRP is used. In some of the naval ships, CFRP superstructure is being used, to achieve a substantial weight reduction and also to provide for stealth properties of the superstructure.

The prime advantage of this material is, it can be engineered to the specific requirement. Depending on the strength requirement, the laminate thickness can be worked out and accordingly fabricated. The basic raw materials are resin and glass fibre reinforcements in the form of mats or cloths. The resin is the monomer which at the time of laying up of the laminates is mixed with necessary chemical ingredients referred to as accelerator and catalyst to start the process of curing (irreversible chemical reaction). The curing process sets in within 15–30 min of mixing with catalyst. This process being exothermic, the pace of reaction depends very much on the ambient temperature. With the curing process the monomer is transformed to hard polymer. Thus the rigid laminates are obtained. The laminate strength depend on its glass content. At the time of fabrication a humid environment will deteriorate the mechanical properties of the laminates by absorbing moisture. Hence it is necessary to have a temperature and humidity controlled manufacturing environment to obtain the desired mechanical properties.

The glass fibre reinforcements come in various forms—unidirectional fibre, chopped strand mat (CSM) and woven rovings (WR). Generally for laminate fabrication a combination of CSM and WR is used to produce the laminate of required strength. The required laminate thickness is achieved by laying up several layers of these CSM and WR.

There is a wide variety of resin available. However for marine construction most widely used resin is isopthalic resin. This resin has a denser molecular structure compared to orthopthalic resin. It exhibits higher strength, greater flexibility and is more waterproof compared to orthopthalic resin. It also provides for good resistance to water, acids, weak bases, and hydrocarbons.

In steel, failure takes place mostly through propagation of a crack, that gets initiated under load. In FRP laminates, high strength fibres are bonded together by resin. Under load, this bonding degrades and the fibres start breaking leading to a complete failure. A comparison of mechanical properties of GFRP, marine grade aluminum alloy and shipbuilding steel is shown in Table 9.3.

Table 9.3 Comparison of mechanical properties

Property	GFRP	Marine grade aluminum alloy	Shipbuilding steel
Density (g/ml)	1.8	2.7	7.8
Tensile yield stress (MPa)	–	–	235–690
0.2 % proof stress (MPa)	150–300	124–276	–
Strength/Weight	83–166	45–102	30–88
Tensile modulus (GPa)	6–8	70	210
% elongation at rupture	1.8–2.5	12–16	14–22

The main advantage of GFRP is its high strength to weight ratio and its weakness is its very low tensile modulus making the material prone to high deflection as length increases.

Advantages of FRP Composites

(a) **High Specific Strength and Stiffness**—Fiber reinforced composites exhibits substantially high strength-to-weight ratios. However its stiffness is rather poor, which limits its applicability as construction material to boats of length generally not exceeding about 15 m.

(b) **Enhanced Fatigue Life**—Most composites are considered to be resistant to fatigue to the extent that fatigue may be neglected at the materials level in a number of structures, leading to design flexibility.
To characterize the fatigue behavior of structural materials an S-N (stress amplitude or maximum stress versus number of cycles) diagram is typically used. Typically the number of load cycles to failure increases continually as the nominal stress level is reduced. The limiting value of stress, below which failure due to fatigue is not observed, is called fatigue or endurance limit.
The endurance limit for normal strength and some of the higher strength steels is observed at 10^5–10^6 cycles. Whereas in case of GFRP and CFRP composites, an well defined endurance limit may not be obtained, however at low nominal stress levels, the slope of the S-N curve is significantly reduced.
Therefore an endurance limit of 10^6–10^7 cycles is taken as a common design practice for FRP material. FRP composites, under cyclic loading, exhibit gradual softening or loss in strength and stiffness due to formation of microscopic damage before occurrence of any visible crack in the laminates [4].

(c) **Corrosion Resistance**—Composites do not rust and corrode. Composite material components can be used gainfully where standard metallic components incur high maintenance costs due to corrosion.

(d) **Controllable Thermal Properties**—FRP composites exhibit different coefficients of thermal expansion in the fiber direction and perpendicular to the fiber direction, i.e. along the laminate thickness. FRP composites have low thermal conductivity, making them typically good thermal insulators.

(e) **Parts Integration**—Large and complex structural parts can be fabricated in one operation by adopting suitable fabrication technique like vacuum resin transfer moulding (VRTM).

(f) **Tailored Properties**—Composite material can be engineered to obtain mechanical properties as specifically required in different directions thereby improving efficiency and economy.

(g) **Non-Magnetic Properties**—GFRP and CFRP composites being non-magnetic, they are effectively used as hull structural material where non-magnetic property is required, e.g. mine sweeper.

(h) **Lower Life-Cycle Costs**—Composites having high resistance to marine corrosion requires less maintenance resulting in lower overall life-cycle costs.

References

1. Rules and Regulations for the Construction and Classification of Steel Ships, Part 1, Regulations, Indian Register of Shipping, July 2015.
2. MatWeb, Material Property Data.
3. Metals Handbook. (1990). Properties and selection: nonferrous alloys and special-purpose materials (Vol. 2, 10th edn.). In *ASM International*.
4. Lopez-Anido, R. A., & Naik, T. R. (2000). Emerging materials for civil infrastructure. State of the Art. ASCE Publications.

Chapter 10
Steel Material Preparation

Abstract Any fabrication activity is preceded by the steel material preparation activity. Steel plates and sections are received from the steel mills in as rolled condition. The very process of production in steel mills leads to formation of residual stress and a hard layer of oxides over the steel surface. Both this locked-in stress and the oxide layer, known as mill scale, needs to be removed before taking the plates and sections for further production operation. The handling of materials from steel mill to shipyard steel stockyard may also lead to surface deformation, hence straightening is required. Thus steel material preparation involves, straightening, stress relieving and mill scale removal. The plates are fed from the steel stockyard to the straightening machine and from there to the surface dressing station. Surface dressing is necessary to remove mill scale present on the plate surfaces. Mill scale is a layer of ferric and ferrous oxides formed on the plate surface during the hot rolling operation of the steel plates in steel rolling mills. There are various methods of surface dressing, however shot blasting and chemical pickling processes are the most efficient ones.

The sequence of production activities carried out from steel stockyard to the stage of fabrication of subassemblies is shown schematically in Fig. 10.1.

Here one can see that any fabrication activity is preceded by the steel material preparation activity. Steel plates and sections are received from the steel mills in as rolled condition. The very process of production of these steel materials leads to formation of residual stress and a hard layer of oxides over the steel surface. Both this locked-in stress and the oxide layer, known as mill scale, needs to be removed before taking the plates and sections for further production operation. The handling of materials from steel mill to shipyard steel stockyard may also lead to surface deformation, hence straightening is required. Thus steel material preparation involves, straightening, stress relieving and mill scale removal. The plates are fed from the steel stockyard through roller conveyor to the straightening machine and from there again through roller conveyor to the surface dressing station.

N.R. Mandal, *Ship Construction and Welding*, Springer Series
on Naval Architecture, Marine Engineering, Shipbuilding and Shipping 2,
DOI 10.1007/978-981-10-2955-4_10

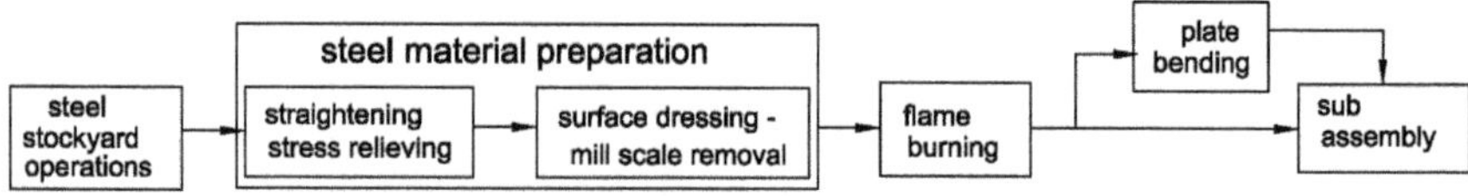

Fig. 10.1 Sequence of plate fabrication activities

10.1 Straightening and Stress Relieving

The straightening and stress relieving operations are done by passing the steel plates through a multiple roller machine subjecting the plate to several reversal of plastic bending as schematically shown in Fig. 10.2. For thicker plates (thickness ≥ 12 mm) generally a 5 roller machine is used. However for lesser thickness plates, several bending are to be imparted to relieve the residual stresses and to straighten the plates. For plates of 4–6 mm thick, often 21 roller machine becomes necessary. Another compromise could be, 2 or 3 plates can be stacked together and passed through a 5 roller machine for straightening and stress relieving.

10.2 Surface Dressing—Mill Scale Removal

Surface dressing is necessary to remove mill scale present on the plate surfaces. Mill scale is a layer of ferric and ferrous oxides formed on the plate surface during the hot rolling operation of the steel plates in steel rolling mills. It consists primarily of Fe_3O_4, of characteristic blue-gray colour covered by an extremely thin outer film of Fe_2O_3. The innermost layer contains fine metal grains and sometimes, residual black FeO which contribute to the roughness of descaled metal.

Initially the bondage of this layer with the steel surface remains very strong. However with passage of time it becomes weaker. On exposure to normal

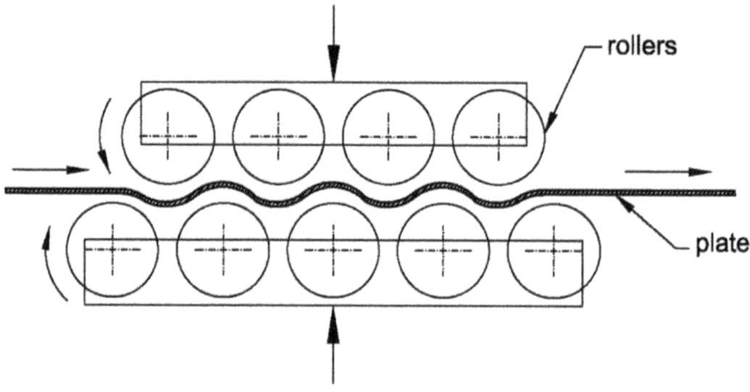

Fig. 10.2 Schematic diagram of plate straightening and stress relieving operation

atmospheric conditions in about 6 months time the layer of mill scale start peeling off of its own, thus exposing the bare steel surface. Hence if steel plates are used without removing the mill scale and the hull surface is painted for corrosion protection, over a period of time the mill scale along with the paint film peels off and the plate surface gets exposed to corrosive environment.

If mill scale is not removed prior to subsequent fabrication processes, it may interfere with cutting and welding operations. The mill scale may contain hydrated oxides, which may lead to release of hydrogen during welding. Thus defects in weld deposit may occur due to contamination from mill scale as well as it may cause hydrogen diffusion in the microstructure leading to hydrogen embrittlement.

There are various methods of surface dressing, e.g. natural, flame treatment, manual wire brushing, sand blasting, shot blasting and chemical pickling. However shot blasting and chemical pickling processes are the most efficient ones.

10.2.1 Shot Blasting

Shot blasting is an efficient method of surface dressing. It effectively and completely removes mill scale from steel plates and profiles. The process involves removal of mill scale using tiny metal shots which are fired onto the metal surface from both sides as schematically shown in Fig. 10.3. The shots impinge on the metal surface and are rebounded chipping off the mill scale with it exposing the bare steel plate. Next to the shot blasting chamber is the priming chamber. As the plate comes out of the shot blasting chamber, it is given a coat of primer paint, generally rich in zinc, to immediately protect it from atmospheric corrosion forming rust. To speed up the process of drying of the primer paint, the plate is moved to a drying chamber, where the plate is subjected to hot air jet. At one end of the shot blasting operation the straightened plate enters and at the other end a clean primed plate comes out, ready for subsequent fabrication activities.

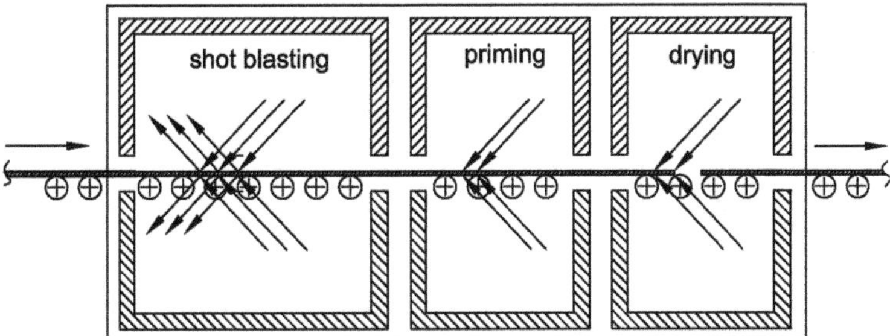

Fig. 10.3 Schematic view of shot blasting operation

The process is fully automated, plate feeding takes place using roller conveyor, schematically shown in Fig. 10.3. The shots are recycled. The various control parameters are,

- Velocity of the shots
- Angle of shot impingement
- Size and mass of Shots
- Speed of feeding the plate

These parameters are to be set appropriately to get the desired surface completely free from mill-scale achieving a surface finish of near white metal, conforming to some accepted standard, e.g. Swedish standard (SIS 05 5900) or International ISO-8501-1, Sa 2.5. Proper combination of these parameters will ensure complete removal of mill scale and at the same time will not cause any damage to the plate surface. Improper selection of parameters may lead to chipping off the steel material from the plate surface and also may deform the plate. The efficiency of this process is given by m^2/h. The merits and demerits of shot blasting operation can be summarised as follows:

Merits

(i) By setting the process parameters suitably the mill scale can be completely removed (100 % removal).
(ii) The complete process is carried out in a noise and dust insulated chamber and hence there will be no heavy dust and noise pollution which is produced as a result of the shots striking the plate.
(iii) The efficiency of this process is generally very high about 100–200 m^2/h.

Demerits

(i) Lesser thickness plates up to 6 mm may get deformed.
(ii) When the shots strike the plate, locally the stress may exceed the yield stress at that point of the plate leading to work hardening. As a result the reserve plasticity decreases and may lead to formation of cracks.
(iii) Depending on the process parameters the shots may chip off part of the metal along with the mill scale (or) may not remove the mill scale completely. So to avoid these problems, the parameters have to be adjusted suitably.

10.2.2 Acid Pickling

In this method of surface dressing the mill scale is removed through chemical reaction. Hence the prime advantage of the process is surface hardening, crack formation or plate deformation does not take place.

H_2SO_4 pickling

As the name suggests, an acid is used to dissolve the mill scale off the steel surface. The acid generally used is either HCL or H_2SO_4. Both have merits and demerits. The pickling time decreases with increase in temperature as well as acid concentration as shown in Fig. 10.4 for sulfuric acid pickling. The pickling rate gets nearly doubled with only 15 °C increase in temperature from 70 to 85 °C. With continuous pickling, the iron in the pickle bath gradually increases causing decrease in pickling rate. Typical pickling condition of hot rolled low carbon steel would be for acid concentration of 10 % by volume at 70 °C for 15 min. The acid bath is replenished to maintain a concentration level of about 10 % by volume. As iron build up approaches about 5 % and acid concentration drops to about 5 % increasing pickling time, the acid bath is then discarded. Before discharging the waste liquid, it is necessary to neutralize the residual acid in the liquid waste.

HCl pickling

Chloride salts have better solubility in water compared to sulphate salts, therefore salt removal is easier. However HCl is more volatile and more so at elevated temperature compared to H_2SO_4, therefore acid consumption becomes more in case of HCl at elevated temperature pickling. Hence it is preferable to use HCl at room temperature with a dilution of industrial grade concentrated acid with 1–3 parts of water. Hydrochloric acid has less pickling time compared to that of sulfuric acid. Some benefits of HCl is also obtained by adding about 2–10 % rock salt to the sulfuric acid bath. HCl can be more effectively used compared to H_2SO_4, as it can be used down to concentrations of 2 % with iron content of about 8 %.

For effective pickling operation with both hydrochloric and sulfuric acid, adequate agitation of the acid bath is done.

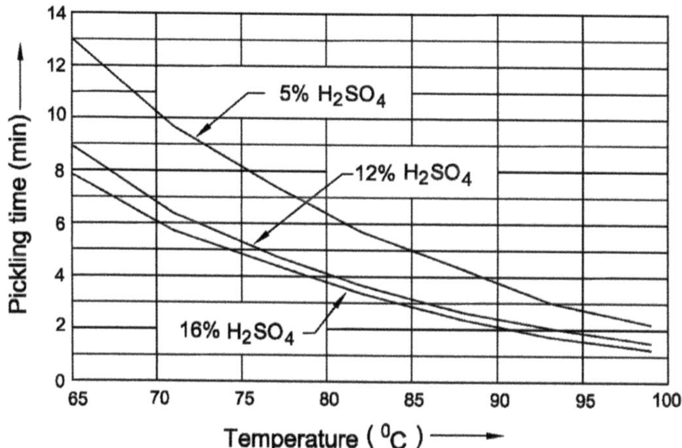

Fig. 10.4 Variation of pickling time with temperature and concentration (by weight) of H_2SO_4 for hot rolled low carbon steel

After the acid bath, a water bath follows which removes the acid from the plates. It is then given an alkali wash, generally $Ca(OH)_2$, which removes all traces of acid from the plate surface. This operation yields a shining steel surface devoid of any mill scale. The plates are then passed through a passivising bath, composed of 23–25 % Phosphoric acid with 19–20 % Ethyl Alcohol. It reacts with the plate to form a uniform layer of insoluble inert salt. This passive layer prevents the plate from further corrosion or rust formation due to exposure to weather. This process can be easily automated to give higher productivity.

The various control parameters for this process are:

- Acid bath concentration
- Pickling time
- The temperature of the acid bath

The reaction rate increases with increasing temperature reducing pickling time. Pickling time increases with decreasing acid concentration. Therefore these parameters are to be monitored and controlled to achieve proper pickling of plates. Improper parameters may lead to inadequate pickling or corrosion of the steel plates.

The merits of this process are:

- No work hardening takes place (as in shot blasting)
- Thin plates are not deformed.
- The process is noise free.

The demerits are:

- Operations involving acids and alkalis are hazardous.
- It has a high water requirement.
- The acidic fumes which develop in the pickling hall can corrode the cranes and other equipment.

Chapter 11
Plate Cutting

Abstract In shipbuilding, cutting of various plate parts of varying thicknesses with accepted dimensional accuracy is required. Plate cutting is done either by mechanical means or by thermal processes. In mechanical process, cutting is done by applying a shearing force either using guillotine shear or high pressure water jet. Whereas in thermal processes, cutting is done either through oxidation, i.e. oxy-flame cutting or through fusion, i.e. plasma arc cutting or through sublimation i.e. laser cutting.

In shipbuilding, one of the major activities is plate cutting. It necessitates cutting of various plate parts of thickness varying from few millimetres to several tens of millimetres with accepted dimensional accuracy consistently. Plate cutting is done using manual hand operated gas torch to numerically controlled plasma/laser cutting systems. At the same time for straight cuts mechanical guillotine shearing machines are also used. A comparatively new method involving water jet is also being used as a cutting head with numerical control.

The various methods of plate cutting that may be used in shipbuilding can be classified as shown in Fig. 11.1.

11.1 Mechanical Process

In mechanical process, cutting is done by applying a shearing force either using guillotine shear or high pressure water jet. When the stress exceeds the ultimate shearing stress, parting of the plate takes place.

N.R. Mandal, *Ship Construction and Welding*, Springer Series on Naval Architecture, Marine Engineering, Shipbuilding and Shipping 2,
DOI 10.1007/978-981-10-2955-4_11

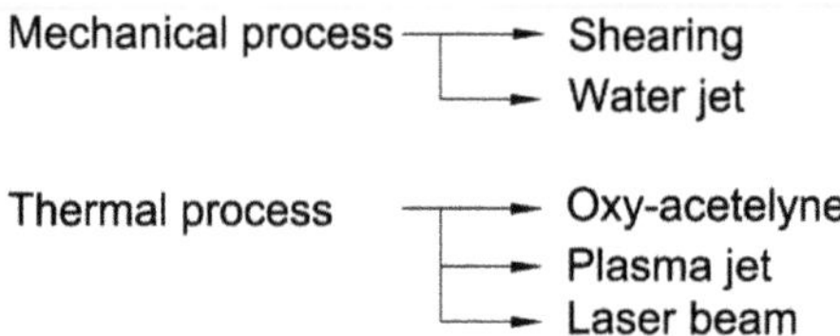

Fig. 11.1 Various methods of plate cutting

11.1.1 Mechanical Shearing

Mechanical cutting by shearing is done by the action of two blades, one fixed at the bottom and one moving vertically above. The moving blade has an inclined edge. The angle of inclination is generally ½ to 2½°. It is called the rake of the blade. The upper blade is usually angled. As it comes down, it progressively meets the plate placed on the lower blade from one end to the other. The cutting pressure thus remains concentrated exactly at the junction of the two blades achieving a cut parallel to the blades. The cut progresses from one end to the other, thus reducing the required force.

A gap of approximately 5–10 % of plate thickness is maintained between the two blades. This clearance and the rake are dependent on the thickness of the material to be cut. The fixed bed holding the bottom blade also has a series of hold-down pins to hold the plate in position at the time of cutting. The blades typically have a square edge and are made of high-carbon steel. The shearing machines are often referred to as guillotine shear. A typical guillotine shear is shown schematically in Fig. 11.2.

11.1.2 Water Jet

In this method of plate cutting, water is forced through a small orifice, under extremely high pressure. Thus it creates a water jet which has a tremendous concentration of energy. The flow through the tiny orifice creates high pressure and a high velocity jet. The water is pressurized between 1300 and 6200 bar. This is forced through the orifice, which is typically 0.18–0.4 mm in diameter [1]. This creates a very high velocity, very thin beam of water jet having a speed almost equal to that of sound. As the thin stream of water leaves the orifice, fine abrasive material is added to the water jet. The high velocity water exiting the orifice creates a vacuum which sucks in abrasive particles from a line feeding abrasive particles, Fig. 11.3. It then mixes with the water in the mixing tube. The abrasive particles are then accelerated through the orifice by the water jet and cuts through the metals as shown in Fig. 11.4.

The abrasive particles under the tremendous force of water erode the material. At the starting position the jet is kept stationary for a short while, till the jet pierces through the metal and then the abrasive jet stream is moved along the cutting path.

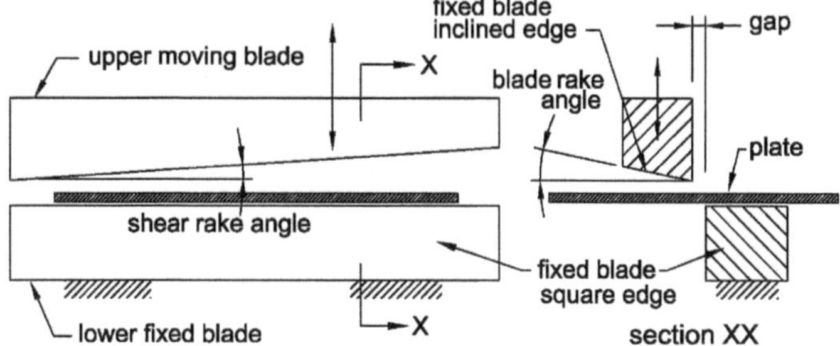

Fig. 11.2 Schematic view of a typical guillotine shear

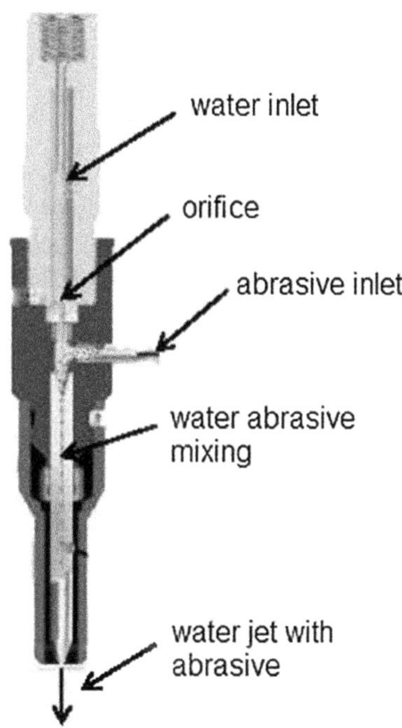

Fig. 11.3 A typical waterjet nozzle

The speed of cutting depends naturally depends on the type of the metal its shearing strength and thickness and water pressure and the type of abrasive.

General merits of waterjet cutting

Since it uses pressurised water with abrasive material, it can cut a wide variety of materials, e.g. copper, brass, aluminium, mild steel, 304 stainless steel, etc. It can cut steel plates as thick as 100 mm. However thicker the material, the longer it will take to cut.

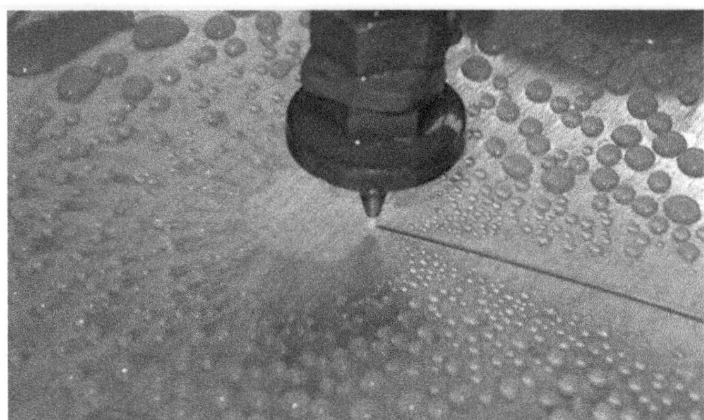

Fig. 11.4 Waterjet cutting of steel plate

Fast setup
The movement of water jet nozzle can be easily computer controlled as in case of oxy-fuel or plasma cutting head. As there are very low sideway forces with waterjet cutting, therefore no additional fixturing is required. The low side forces also allow for close nesting of parts, and maximum material usage.

Almost no heat generation
Whatever little heat is generated by the waterjet, it gets dissipated by the water. Therefore material itself experiences almost no temperature change during cutting. The result is, there is no heat induced deformation or residual stress developing in the plate as well as no adverse metallurgical transformation takes place. Thus heat treated material can be cut without affecting the property of the material.

No mechanical stresses
Waterjet cutting does not introduce any stresses into the material.

Safe Process
This cutting method is essentially very safe. It is non-explosive and the abrasive materials that are used are also inert and non-toxic. Any leak in a high pressure water system leads to rapid drop in pressure to safe levels. One of the hazards of water jet cutting is, it produces very sharp edges on the metal plates. Also one needs to keep in mind that the high pressurised jet, which is capable of easily cutting through a 50 mm thick steel plate, should never come in contact with any body part, it will simply cut through it.

Environment friendly
The spent abrasive and waste material from the cutting process is nontoxic and is suitable for land fill. The abrasive material is inert and can be easily disposed of. In most areas, excess water can be simply drained to the sewer or preferable would be to have a closed loop system where the water can be recycled. For achieving the

required level of pressurisation, the pumps consume a considerable amount of electrical energy, which has an indirect environmental impact.

Narrow kerf

Waterjet cutting produces a very narrow kerf. The kerf is typically about 0.5 mm wide. Thus a very little material is removed.

Comparison of waterjet with laser cutting

In laser cutting, it is focused on the material to melt, burn, or vaporize the material. To make the laser move over the material being cut, additional optics are required as the distance from the emitting end of the laser changes. Here the cutting speed being very fast, it is highly productive and also leaves behind very small heat affected zone.

Advantages of waterjet

Waterjet cutting has a number of advantages over laser cutting. No heat-affected zone (HAZ) with waterjets. These are more environmentally friendly and comparatively safer. Uniformity of material is not important. Laser cutting requires fairly even surface, otherwise the laser will become out of focus reducing the cutting power. A waterjet will retain much of its cutting power over uneven material. Lower capital equipment costs. The cost of a waterjet machine is substantially lower than that of a laser cutting machine. Better tolerances on thicker parts. Waterjets offer better tolerances on parts thicker than 12 mm. For thinner parts, both waterjets and lasers offer comparable tolerances. Waterjet can easily cut steel plates of 50 mm thickness, whereas laser has a practical limitation of about 12–19 mm thickness. Maintenance on a waterjet is simpler than that of a laser machine. Waterjet operations are much simpler compared to a laser cutting device. Waterjet provides better edge finish. Material cut by waterjets have a fine, sand-blasted surface because of the way the material was abraded, which makes it easier to make a high-quality weld. Material cut by laser tends to have a rougher, scaly edge, which may require additional machining operations to clean up.

Comparison of waterjet with plasma cutting

In plasma cutting, an extremely hot (about 15,000 °C) stream of plasma is created which melts the material and the molten material is blown away by high seed plasma jet.

Advantages of waterjets

The prime advantage of waterjet cutting compared to plasma cutting is that waterjets operate at much lower temperatures. During piercing, the temperature of the material may rise as high as about 50 °C, but the cutting operation typically happens at ambient temperature. The water in the cutting process takes away the heat. Hence there is no question of heat affected zone or heat induced distortion. Materials that can not be melted easily or get destroyed while melting can be cut by water jet. However plasma jet can cut steel plates faster than water jet particularly thicker sections.

Comparison of waterjets with flame cutting
In oxy-acetylene cutting of steel, the cutting location is heated up to about 1000 °C and then additional oxygen is supplied to start the oxidation of the material. At that temperature the oxide remains in molten state and is thrown off the plate under the action of the oxy-acetylene jet. Oxy-acetylene cutting can be used only for iron and steel.

Advantages of waterjets
Flame cutting can only be used for iron and steel, whereas water jet can be used for a wide variety of materials. Flame cutting involves heat, hence the plates being cut may suffer heat induced shrinkage and deformation, which is totally absent in water jet cutting. The plate edges in flame cutting need to be grinded for removing oxide layer as well as carbon impregnated metal. The edge finish in water jet is smooth and does not require any post cutting operation. The precision level in water jet cutting is higher compared to that of flame cutting. It also leaves much smaller kerf that in flame cutting. Flame cutting can be faster than waterjets, but it also requires post cutting operation, which is not required in case of water jet cutting.

11.2 Thermal Process

The Thermal Process of metal cutting can be further classified based on the physical mechanism of cutting as given below [2]:

Cutting through oxidation
The material is oxidised (burnt) and the oxide in its molten state is blown off the cutting location by the oxygen jet. Here the melting temperature of the material must be higher compared to that of its oxide. Example: Oxy-acetylene flame cutting.

Cutting through fusion
The intense heat melts the material and is blown out by the high energy gas jet. Here oxidation of the material does not take place, however if there is a oxide layer already existing as in case of aluminium alloys, the oxide layer along with the metal gets melted instantly and is blown off the cutting location. Hence in this method an intense heat source is required. Example: Plasma arc cutting.

Sublimation cutting
In this method of thermal cutting, the heat source is so intense that the material evaporates and the cutting is done. The metal vapour expands under its own vapour pressure and also additional gas jet helps to transport out the metallic vapour from the cutting zone.

In extreme cold conditions, it is necessary to bring the steel plates to a temperature level of about +20 °C before cutting. The warm-up time depends on the plate thickness. An indication for the required time is given for three different plate thickness as shown below [3]:

The warm-up time from −20 to +17 °C are as follows:

- about 8 h for 12 mm thick steel plate,
- about 12 h for 21 mm thick steel plate,
- about 17 h for 40 mm thick steel plate.

For stack of thick and large plates the warm-up time can be substantially larger. As a rule of thumb, the warm-up time from −20 to +20 °C for a steel plate of width 2 m and length 6 m, is about 24 h. Therefore it is preferable in cold countries, to bring the cold plates inside the shop floor a day before starting of plate processing.

As such thermal cutting of high strength steels up to about 500 MPa does not require any special measure. However with increasing strength beyond 500 MPa, some preheating is required. Preheating depends on the plate thickness, its chemical composition, thermal cutting process and cutting speed.

11.2.1 Oxy-Fuel Flame Cutting

It is a process of cutting through oxidation (burning). A fuel gas is burnt in presence of oxygen to generate the required heat. For industrial applications and in shipyards in particular, acetylene gas is used as the fuel. Acetylene has higher calorific value of 18,890 kJ/m^3 compared to other fuel gases, e.g. it is 10,433 kJ/m^3 for propane. Therefore it produces the highest flame temperature in oxy-fuel process. The maximum flame temperature with acetylene in oxygen environment is about 3,160 °C, whereas with propane it is about 2,828 °C. The hotter flame leads to rapid piercing at the start of a cutting process. The piercing time is typically about one third that of with propane [4]. This intense flame at the metal surface also reduces the width of the Heat Affected Zone (HAZ) and heat induced distortion.

In this process, the material is brought to the ignition temperature, about 1000 °C for steel, by an oxy-acetylene flame and is then burnt by supplying additional oxygen (cutting oxygen) to the heated zone. The oxide in the molten state is blown off by the gas jet.

To achieve successful oxy-acetylene cutting, the following conditions need to be satisfied:

- The ignition temperature (oxidation) of the material has to be lower than its melting temperature, otherwise the metal will melt and cutting cannot be done.
- The melting temperature of the oxides has to be lower than the melting temperature of the material itself such that the molten oxide can be blown off by the gas jet.
- As the cutting progresses, the ignition temperature has to be continuously maintained. This implies that a positive heat balance has to be maintained balancing the heat supplied by the gas flame and the heat losses by way of conduction and convection.

- The oxidation reaction between the oxygen jet and the metal must be an exothermic one to maintain a positive heat balance.

Only steel and titanium satisfy these conditions and therefore oxy-acetylene cutting is possible for these metals.

Cutting speed can be increased by increasing cutting-oxygen pressure, e.g. a heavy duty cutting nozzle can have cutting oxygen pressure of 11 bar. The mixing of oxygen and acetylene is done either in the torch handle or in the nozzle depending on the torch design. The selection of torch and nozzle depends mainly on requirement of cutting thickness. In Oxy-fuel cutting multiple torches can be used simultaneously to cut identical multiple components, thus enhancing productivity.

Adverse effects of oxy-acetylene flame cutting
In oxy-acetylene flame cutting, carbonisation (increase in carbon content) may occur at the cut edge. This results in hardening of the flame cut surface.

Carbonisation occurs due to the selective oxidation of the steel. The iron oxide layer that forms on the cut surface prevents the formation of carbon monoxide, thereby increasing the carbon content in the molten layer. In structural steels, the depth of this carbonised layer is less than 0.1 mm [3]. The width of the heat affected zone in flame cutting is generally less than 5 mm. With increase in carbon content, it increases the hardenability of the cut edge.

Flame cutting being a thermal process, the cooling rate depends on plate thickness. It increases with increase in plate thickness. The increase in cooling rate leads to increase in grain size. This along with carbonisation of the cut surface results in increase in hardness of the flame cut surface. The hardness of the surface increases with the increase in plate thickness. Therefore to prevent higher cooling rate in thicker plates, preheating is done. Preheating decreases the hardness of the flame cut edge and improves its formability. It also makes the machining of the plate edge easier and speeds up the flame cutting procedure. The need for preheating depends on the plate thickness, its chemical composition as well as cutting process (gas) and cutting speed. The preheating temperature T°C for different plate thicknesses can be calculated as follows [3]:

$$T = 500\sqrt{C_{eq} - 0.45} \quad \textit{for plate thickness } t = 5 \textit{ to } 100\, mm$$

$$T = 500\sqrt{C_{eq}(1 + 0.0002t) - 0.45} \quad \textit{for plate thickness } t > 100\, mm$$

$$C_{eq} = C + 0.155(Cr + Mo) + 0.14(Mn + V) + 0.115\, Si + 0.045(Ni + Cu)$$

The flame cutting operation may lead to heat induced shrinkage of the cut edges causing alignment problems in structural fabrication. Also the thermal stresses may cause buckling of the edges as shown in Fig. 11.5.

Fig. 11.5 Edge buckling due to oxy-acetylene cutting

11.2.2 Plasma Arc Cutting

In this thermal process plate cutting is done through fusion. In this process a high temperature, high velocity ionised gas jet (plasma) is formed which causes through thickness melting of the plate and blows off the molten metal achieving the cut. This plasma is often referred to as the 4th state of matter. The three states of matter are solid, liquid and gas. The difference between these states relates to their energy levels. Matter changes from one state to the other through the introduction of energy, such as heat. For example, when energy in the form of heat is added to ice (solid), it melts and forms water (liquid). When more energy is added, it vaporizes in the form of steam (gas). Further adding energy to steam, the gases that make up the steam get ionised. In this process, some of the electrons break away from their atoms in the steam. These free electrons make the resulting ionised gas (steam) electrically conductive. This state of the electrically conductive ionized gas is referred to as plasma.

Actually, in any electric arc welding, the welding arc is nothing but a column of ionized gas, i.e. plasma. However, in the plasma arc cutting process, the plasma is made to pass through a constricting nozzle to concentrate and direct the high velocity plasma to the workpiece. A constricting nozzle is a part of the plasma torch. The temperature of the plasma exiting the nozzle can be in excess of 20,000 °C and the velocity can approach the speed of sound [5].

The merits of plasma arc cutting can be summarised as:

- It can cut all metals.
- Higher cutting speed compared to oxy-acetylene cutting, especially on steels less than 25 mm thick.
- Heat affected zone and heat induced deformation are minimised.
- Carbonisation of cut edge as in oxy-acetylene cutting does not take place.
- Hazardous or explosive gases are not used.

Some of the demerits are as follows:

- Electric power is required unlike oxy-acetylene process.
- Metal fumes and UV radiation can pose a health hazard.
- Capital cost of equipment is substantially higher compared to oxy-acetylene process.

The process

A pilot arc is struck within the body of the torch between an electrode (cathode) and the constricting nozzle. The purpose of this pilot arc is to initiate the ionization of the gas. High frequency current is used to initiate and maintain the pilot arc. The pilot arc is non transferred, i.e. not between the electrode and the work piece. It is between the electrode and the nozzle.

For the actual cutting to begin, the arc must be transferred to the workpiece in the so-called 'transferred' arc mode. The electrode is kept negative and the workpiece positive, so that the majority of the arc energy (approximately two thirds) is transferred to the workpiece for cutting. When starting the cutting process, the nozzle must be close enough to the job such that the arc gets transferred to the job. For manual cutting, the distance from the nozzle to the work piece should be about 1.6–3.2 mm while the angle of the torch to the work piece is kept between 70° and 90° [6]. Whereas in case of automatic cutting, the torch is kept perpendicular to the work piece. The high open-circuit voltage and the ionized gas produced by the pilot arc will allow the arc to jump from the electrode to the workpiece. When the arc is between the electrode and the job, it is called a *transferred arc*. Once the transferred arc is established, the pilot arc is switched off. The metal below the plasma jet melts almost instantaneously and the high velocity of the jet coming from the constricted nozzle blow away the molten metal making a smooth kerf. The cutting continues as the torch is moved over the plate surface maintaining the necessary distance between the plate and the nozzle tip.

Plasma cutting process can have the following variations:

- *Plasma cutting with transferred arc*

In this process, the arc is between the plasma torch electrode (cathode) and the transferred to the workpiece (anode) as shown in Fig. 11.6. Thus only electrically conductive materials can be cut by this mechanism. Additional gas is supplied to the plasma arc torch for shielding the plasma jet as well as to cool the torch. This secondary gas flow also helps in constricting the plasma jet.

- *Plasma cutting with non-transferred arc*

Here the arc is established inside the nozzle between the electrode and the nozzle body (Fig. 11.7). The hot stream of plasma jet comes out of the nozzle orifice in the form of flame. It can be used to cut electrically non-conducting materials.

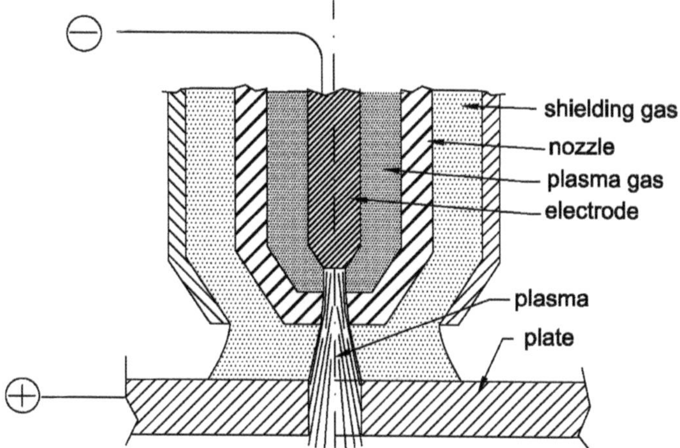

Fig. 11.6 A schematic of a transferred arc plasma cutting

Fig. 11.7 A schematic of non-transferred arc plasma cutting

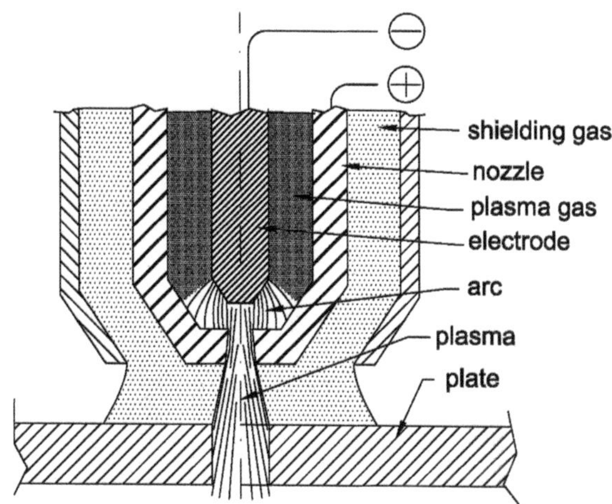

- *Plasma cutting with transferred arc and water injection*

In this method of plasma arc cutting, water injection is done as schematically shown in Fig. 11.8. It induces greater degree of constriction of plasma jet leading to increased temperature as high as 30,000 °C [5]. It has several advantages as shown below over conventional plasma cutting process:

- Improvement in cut quality (improved squareness of cut edge)
- Increased cutting speed
- Less chances of 'double arcing'
- Enhanced nozzle life (reduced nozzle erosion)
- *Plasma cutting with water shroud*

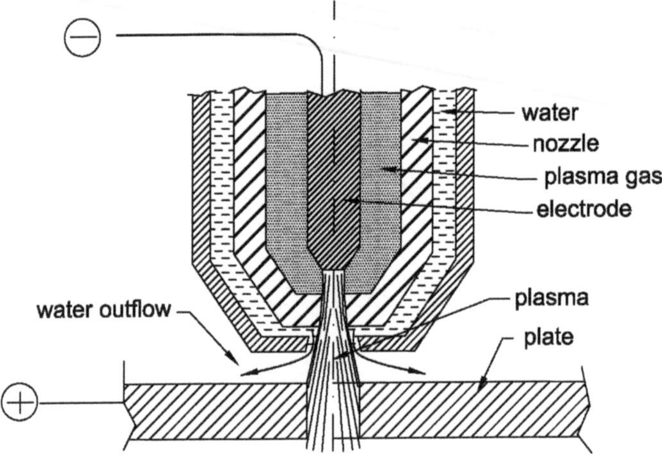

Fig. 11.8 A schematic of plasma torch with water injection

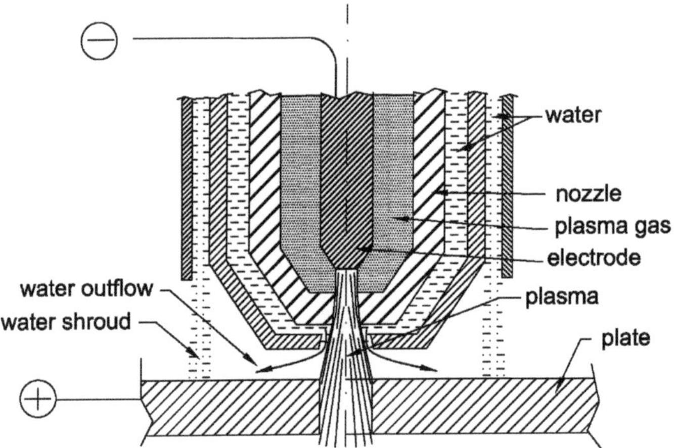

Fig. 11.9 A schematic of plasma torch with water shroud

In this variation of plasma arc cutting, the arc is shrouded with water as shown in Fig. 11.9. Along with providing water shroud to the plasma arc, often cutting is carried out keeping the job under 50–75 mm immersed in water. This water immersion adds to the following advantages [5]:

- Reduction in fume emission
- Reduction in noise levels (from about 115 dB to about 52–85 dB)
- Enhanced nozzle life

However this water shroud, unlike water injection does not increase the degree of plasma constriction and therefore does not appreciably lead to improvements in the squareness of the cut edge and the cutting speed.

Cutting in immersed condition is generally done for automated cutting at higher current levels. Water here acts as a shielding medium and also helps in the cooling of the cutting surface reducing the possibility of heat induced deformation of the cut edges.

- *Plasma cutting with air as plasma gas*

In this method air is used as the plasma gas. However in this case the electrode material needs to be of hafnium or zirconium mounted in copper holder. Also air is used for cooling the torch replacing water. The main advantage of using air plasma torch is that air is used in place of expensive inert gases. However hafnium tipped electrodes are expensive compared to tungsten electrodes.

Power source

A direct current constant current power source with high open circuit voltage (OCV) is used for plasma arc cutting process. Generally the voltage required to sustain the plasma is about 50–60 V, however to initiate the arc the OCV can be as high as 400. The circuitry of a plasma arc machine is schematically shown in Fig. 11.10.

Gas composition

Plasma torches having tungsten electrode commonly uses either argon, nitrogen or argon with hydrogen as the plasma gas. However often air is also used as plasma gas. In such a situation, hafnium electrode is to be used. The gas flow rate of plasma gas is to be set based on the current level and the nozzle bore diameter. The arc may get disrupted if the gas flow is too low for the given current level, or it is too high for the nozzle bore diameter. It results in so-called double arcing, i.e. the arc will be in series, electrode to nozzle and nozzle to workpiece. This may even cause melting of the nozzle.

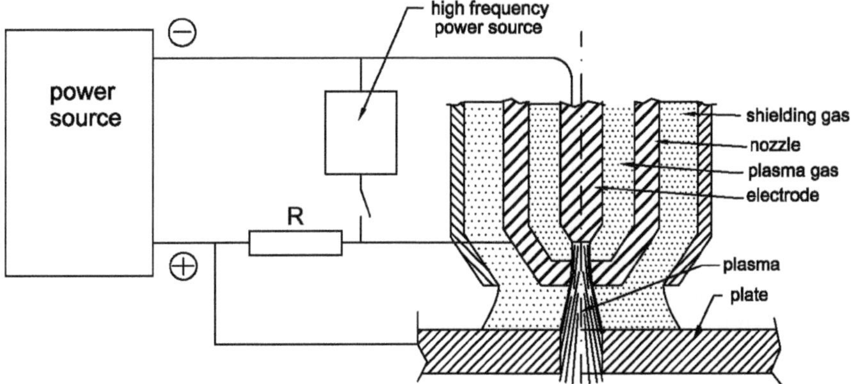

Fig. 11.10 A schematic of plasma arc circuitry

Cutting parameters

For mild steel plates of thickness from about 3 to 25 mm the current ranges from 40 to 80 A with cutting speed varying from 2300 to 250 mm/min respectively. For aluminium alloys of the same thickness range, the current and speed requirements are 40–80 A and 4060–300 mm/min respectively [6].

11.2.3 Laser Cutting

Laser as a heat source is also used to cut steel and aluminum plates. The basic mechanism of laser cutting can be summarised as follows [7]:

- A high intensity laser beam is generated.
- The beam is focused onto a very small area on the surface of the workpiece by means of a lens.
- The focused beam almost instantaneously heats up and melts the small (generally less than 0.5 mm diameter) area through and through along the plate thickness.
- The pressurised gas jet acting coaxially with the laser beam ejects the molten material from the plate.
- In some cases, the gas that is used reacts with the material exothermically and greatly enhances the cutting process, e.g. oxygen is used in cutting of carbon and/or mild steels. In presence of the laser heat the material burns in the oxygen jet and the molten oxide is blown off under the force of the oxygen jet. Thus it improves the efficiency of the process.
- The laser beam is moved over the plate surface or the plate is moved by a CNC X-Y table and the cut is achieved.

Merits of laser cutting [7]:

- Laser cutting is done at a substantially higher speed, e.g. A 5 kW CO_2 laser can cut a 10 mm thick mild steel plate at a speed of about 2.35 m/min and a 25 mm thick mild steel plate at 0.80 m/min.
- The actual heated area in laser cutting is very small and most of this heated material is removed during cutting. Also the speed of cutting being high, the overall heat input to the plates is very low. Thus the heat affected zones are minimized as well as the heat induced distortion.
- Laser cutting being a non-contact process, minimum of fixturing and clamping of plates is required. The plate can be just positioned under the beam.
- The kerf width is extremely narrow (typically 0.1–1.0 mm).
- The process can be fully CNC controlled.
- The kerf width being very small and the process being fully CNC operated, close nesting of piece parts can be done such that scrap or wastage of material is minimized.

- In most cases the laser cut components can be used for subsequent operation without any edge cleaning operation.
- The capital cost of a laser-cutting machine is quite high, however, the running costs are generally low. Hence break even can be faster.
- The laser cutting process is extremely quiet compared to other techniques, e.g. plasma cutting. It is important from the point of view of improving working environment and thus the efficiency of the shop floor workers.

The process

Industrial scale laser cutting is generally carried out by either CO_2 laser or Nd:YAG (neodymium-doped yttrium aluminum garnet) laser. The general principals of cutting are same for both. Both these methods generate high intensity beams of infrared light which can be focused and used for cutting. However CO_2 lasers being a comparatively cheaper option, it is favoured for general industrial applications.

Even if the wavelength of CO_2 laser is ten times that of Nd:YAG laser (10.6 microns and 1.06 microns respectively), still for shipyards' steel cutting requirements, it can be satisfactorily focused to sufficiently small area to provide kerf width less than 1 mm [7].

Laser cutting can be done by the following three mechanisms:

- laser beam combustion cutting
- laser beam fusion cutting
- laser beam sublimation cutting

The *laser beam combustion cutting* (oxidation cutting) is in principal similar to that of oxy-fuel cutting. Here the heat is derived from absorption of laser beam in the job and oxygen is used for oxidizing (burning) the metal. The schematic arrangement is shown in Fig. 11.11. It gives a very good bur-free cut edge for structural steels of thickness up to about 12 mm. The cutting speed being high along with extremely focused heat source, it gives near shrinkage/distortion free

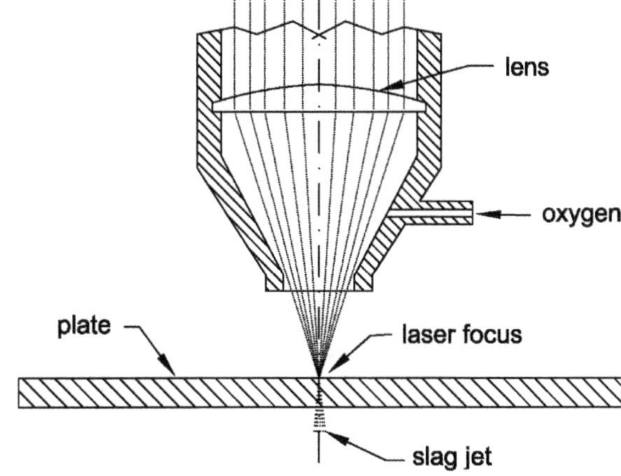

Fig. 11.11 A schematic of laser beam combustion cutting

parallel cut edges, unlike oxy-acetylene process where significant shrinkage/ distortion may occur.

In *laser beam fusion cutting* the focused laser beam melts the workpiece and the molten metal is ejected from the cut by the mechanical action of the cutting gas jet. For successful cut, suitable shielding gas and pressure are to be chosen, e.g. nitrogen is used for steels and argon is used for titanium. The pressure required is in the range of 8–14 bar. It increases with increase in thickness.

Laser beam sublimation cutting is not a suitable option for cutting of steels. Because it will require much more energy compared to combustion cutting or even fusion cutting. It is good for such materials where the melting and boiling points are very close, e.g. acrylic sheets are very effectively cut using CO_2 sublimation (evaporation) cutting method.

Application of laser cutting in shipyards

With all the high cutting quality, still the basic cost of laser cutting equipment remains as the prime deterrent for its wider application. However some of the shipyards have implemented CO_2 laser cutting systems of 4–6 kW power. The practical limit of these machines is 25 mm thick steel plates.

To increase the capacity to cut even thicker plates, so-called LASOX processes are being adopted. In this system, oxygen stream provides the cutting force while the laser only preheats the plate surface. Using such machines, it offers the advantage of a single machine that can cut a very wide range of plate thickness from few mm to thick plates of 50 mm and even more.

The limitation of thickness was overcome by a process which combined laser with a jet of oxygen. At the cutting location, the steel plate is heated up focusing laser on to it. As it attains the ignition (oxidation) temperature of about 1000 °C, a supersonic stream of oxygen is injected at the spot to initiate the process of oxidation and cutting. This is in essence same as that of oxy-fuel cutting, where the fuel gas or acetylene is replaced by laser. In principal it is also same as that of laser beam combustion cutting. The exothermic reaction adds to the heat energy in the cutting process. Unlike in any of the other oxygen assisted cutting process, here an extremely high speed jet of oxygen is used. This is a patented method named as LASOX [8, 9]. In this method steel plates of 50 mm thick can be cut with much less laser power of about 2 kW. Whereas any other laser cutting process can effectively cut only about 25 mm thick steel plates [9].

References

1. http://www.waterjets.org/.
2. Thermal Cutting. (2002). ISF - Welding and Joining Institute, Aachen University.
3. Hot-rolled Steel Plates, Sheets and Coils Thermal cutting and flame straightening. (2011). Ruukki Metals Oy, Suolakivenkatu 1, FI-00810 Helsinki, Finland, www.ruukki.com.
4. Oxyfuel cutting - process and fuel gases, Job Knowledge 49. http://www.twi-global.com/technical-knowledge/job-knowledge/oxyfuel-cutting-process-and-fuel-gases-049/.

5. Bill Lucas and Derrick Hilton, Job Knowledge 51, The Welding Institute. http://www.twi-global.com/technical-knowledge/job-knowledge/cutting-processes-plasma-arc-cutting-process-and-equipment-considerations-051/.
6. Althouse, A.D., et al. (2013). Modern welding. Illinois, USA: Goodheart-Willcox Publishers. ISBN: 978-1-60525-795-2.
7. Powell1, & Kaplan, A. (2004). Laser cutting: from first principles to the state of the art. In Brandt, M. & Harvey E (Eds.), Proceedings of the 1st Pacific International Conference on Application of Lasers and Optics.
8. Bender Pioneers Lasox Cutting In Production, National Shipbuilding Research Program, Advanced Shipbuilding Enterprise, 5300 International Boulevard, North Charleston, South Carolina-29418, December 10, 2002.
9. Oxygen-assisted beam cuts thick steel. *The world's first production laser-assisted oxygen system that can cut 50 mm-thick steel is installed at a US shipyard.* 24 Jan 2003, http://optics.org/article/16723.

Chapter 12
Plate and Section Forming

Abstract In a ship structure the percentage of curved plates is generally not more than 15 % of the total plates used in the hull structure. Here each individual plate has a different curvature. The stiffeners are also required to be bent matching with the curvature of the plates at the respective sections. The plates in the forward and the aft end have compound curvature. Mathematically these curvatures are referred to as Non-Gaussian curvature. These are non-developable curvatures. Plate and section forming is carried out either by mechanical means or through a thermal process using line heating technique. The concept of matched die has been used in devising universal press. Here the die is made flexible, such that it can easily take the required curvature. Controlled heating is applied in line heating method along predetermined line segments. This causes differential shrinkage along the plate thickness, resulting in angular bending of the plate. By applying this bending at the required location, the target shape of the plate can be achieved. This method of line heating can be gainfully utilized using a numerically controlled heating torch to apply the required thermal load to achieve the desired target shape.

In general the forward and the aft end of a ship hull are made of curved sections, whereas the middle part has a parallel middle body comprising of straight vertical side shell, flat bottom and rounded bilge. Overall the percentage of curved plates is generally not more than 15 % of the total plates used in the hull structure. However the production of these curved plates is quite cumbersome and time consuming. Here each individual plate has a different curvature and hence matched die cannot be used as is done in automobile industry. Similarly along with the plates the stiffeners are also required to be bent matching with the curvature of the plates at the respective sections. The plates in the forward and the aft end are of compound curvature, i.e. they have curvature in both the planes like that of a pillow cover or a football. Mathematically these curvatures are referred to as Non-Gaussian curvature. The basic characteristic is, these are non-developable curvatures. To achieve this, the plate needs to be stretched, i.e. stretching of the mid plane of the plate will take place.

These plate and section forming is carried out either by mechanical means or through a thermal process using line heating technique.

N.R. Mandal, *Ship Construction and Welding*, Springer Series
on Naval Architecture, Marine Engineering, Shipbuilding and Shipping 2,
DOI 10.1007/978-981-10-2955-4_12

12.1 Mechanical Methods

Plates and sections are bent by mechanical force using hydraulic machines either manually or through some automated process. The bending is carried out at ambient temperature, i.e. without application of any heat.

12.1.1 Roller Bending and Hydraulic Press

Plate Bending

Plate bending through mechanical means is carried out using 3 or 4 roller bending machines and hydraulic press. The bilge plates generally have cylindrical, i.e. single curvature. These are developable surfaces. This bending is done by roller bending machines. However for other complicated compound curvature plates in addition to roller bending they are further processed through the hydraulic press to achieve the required shape. Progressive bending of the plate is carried out. Prior to bending of the plates, the frame positions are marked on the plate. As the bending progresses, tame plates at the frame positions are used to check the correctness of the shape being obtained. This process of bending is fully manual and highly worker's skill dependent.

Stiffener Bending

The stiffener bending is carried out by 3-point frame bending machines. The stiffener is progressively fed to the machine from one end and through the strokes of the side rams of the frame bender the required curved shape is obtained. A schematic of a typical frame bending machine is shown in Fig. 12.1. Here also the above referred tame plates are used for the respective frames to check the curvature being attained progressively. This process of frame bending is also fully

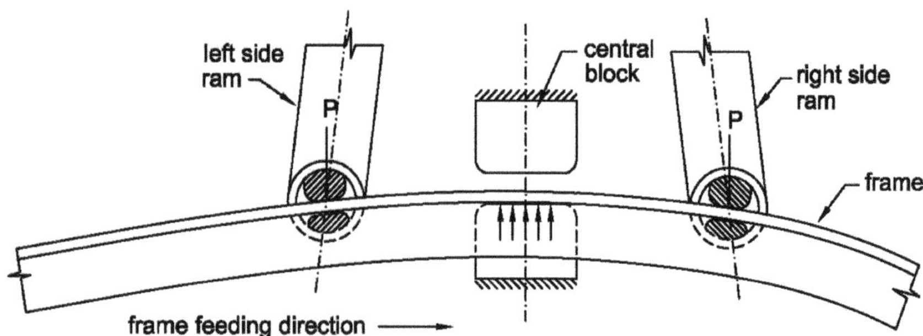

Fig. 12.1 Schematic of a typical frame bending machine

Fig. 12.2 A typical frame (concave) bending operation

manual and worker's skill dependent. A typical frame bending (concave, web in compression) operation is shown in Fig. 12.2.

These stiffeners are of rolled steel sections. Where a stiffener/girder of higher section modulus is required which is not available as a standard section, this is fabricated by welding face plate to a web. This face plate and web are cut in suitable dimensions from standard steel plates. These fabricated stiffeners/girders are required either as straight or curvilinear sections. There can be two ways of fabricating these curvilinear sections:

- by cutting the web to the required shape and welding the face place to it;
- by cold forming of the straight section in a frame bending machine.

For carrying out cold forming of the sections, the following needs to be checked beforehand to achieve successful bending of the frames [1, 2]:

- Fillet weld strength needs to be assessed to ensure that web-flange weld failure does not take place.
- Buckling of web due to compressive stresses in concave bending.
- Roofing of face plate due to secondary stresses in concave bending.
- Thinning of web edge caused by tensile stresses due to convex bending.

For assessing the fillet weld strength, an index called the Fillet Weld Loading Factor (FWLF) can be used. This index depends on the section dimensions, fillet throat thickness, radius of curvature and bending machine parameters [1]. Depending on the yield stress of the parent metal and the tensile strength of the weld deposit, limiting values of the FWLF are determined. By comparing the FWLF of the T sections with that of the limiting ones it can be predicted whether the fillet joining the flange to the web will be able to sustain the load during cold bending.

12.1.2 Universal Press for Plate Bending

The concept of matched die has been used in devising universal press. In shipbuilding, there is no scope of large scale mass production. Hence it is not feasible to keep large number of expensive dies for plate forming. However if the die can be made flexible, such that it can easily take the required curvature. Then plates having different curvatures can easily be formed using this pair of flexible die. The universal press essentially consists of a pair of flexible die, which are computer controlled. Using universal press, plate bending operation can be automated.

- In this method two arrays of hydraulic jacks are used, one at the ground and the other hanging from top. In this system of jacks, each jack can be operated independently. In this method the array of these jacks form the male and female dies as shown in the Fig. 12.3.

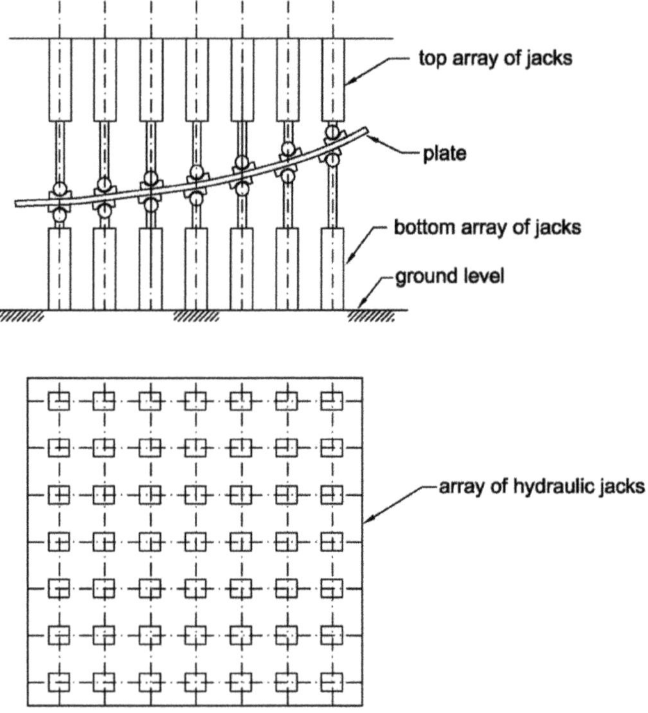

Fig. 12.3 Schematic presentation of universal press

- The curved shape of the plate is first generated on the bottom array of jacks.
- The plate to be bent is kept over this surface and pressed by the upper array of jacks.
- Under this pressure the stress level exceeds the yield stress of the plates leading to bending of the plates.
- Thus almost any curved shape as relevant in shipbuilding can be achieved through this method.

Difficulties in This Process of Automatic Plate Bending

When the jacks are retracted after bending, the bent shape deviates from the shape under pressure. It is known as spring back action. The extent of spring back that may take place depends on mechanical properties of the given plate, i.e. yield stress and modulus of elasticity. These values are not unique for a given plate. They generally lie within a range. Hence exact amount of spring back cannot be calculated. Hence in this process getting the exact bent shape will require monitoring of the plate shape after each stroke and evaluating the difference with that of the target shape and again apply the bending stroke. Thus it may require several strokes to finally get the desired bent shape.

12.2 Line Heating

Line heating is a method of applying thermal load for shaping a plate or section. Through this process, compound curved surfaces conforming to hull surface definition of a ship can be fabricated. As the name suggests, controlled heating is applied along predetermined line segments. A heat source is moved over a plate following these lines. This causes differential shrinkage along the plate thickness, resulting in angular bending of the plate. By applying this bending at the required location, the target shape of the plate can be achieved. Hence once the required heating pattern for a given target shape is generated, this method of line heating can be gainfully utilized using a numerically controlled heating torch to apply the required thermal load to achieve the desired target shape.

The heat source can be adjusted to derive the required effect of bending, shrinking, or a combination of both. The bending process depends on the intensity of the heat source, its speed of movement over the plate, and the pattern of lines where heat is applied.

Plate bending

When a plate is heated on one side, the temperature gradient across the thickness of the plate generates different expansion across the thickness of the metal plate, thereby causing the plate to bend. Line heating by oxy-acetylene flame is a commonly used method for thermo-mechanical forming, especially in shipyards. When

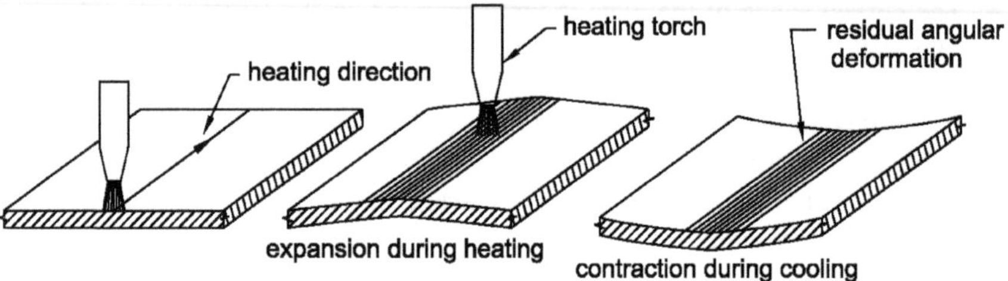

Fig. 12.4 Line heating kinematics

a plate is subjected to local heating, two things happen: the material becomes softer (lower yield limit) and at the same time it expands. The adjacent cold material which still has its original strength, restrains the expansion of the hotter zone. Thereby compressive stresses develop leading to material flow towards the heated zone resulting in thickening of the section. During the cooling phase, nonuniform contraction takes place causing nonuniform shrinkage forces to generate along the plate thickness leading to angular deformation. The sequence is illustrated in Fig. 12.4.

Transient thermal and structural analyses are done for determining the heating parameters and heating path for a given target shape.

In shipbuilding, one of the most tedious and human skill dependent operations is plate forming for producing compound curved shapes. Line heating technique can be gainfully utilized to develop necessary thermal strains in the plate to generate the desired target shapes. Thereby the process of plate forming becomes amenable to mechanization once the line heating pattern and heating parameters become available.

Frame bending

As plate bending is necessary conforming to hull shape, also the frames/stiffeners at the corresponding locations of the hull structure need to be bent conforming to the hull definition. This bending of frames can also be effected by line heating. Frame bending can be done by applying suitable bending moment to it. Therefore by properly designing the heating pattern, required moment can be generated to achieve the target bent shape. The heating zones should so located that the shrinkage forces generated by line heating will develop the required bending

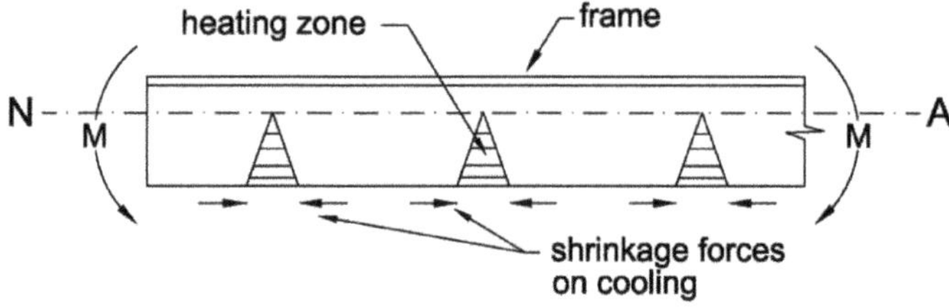

Fig. 12.5 Mechanism of frame bending by line heating

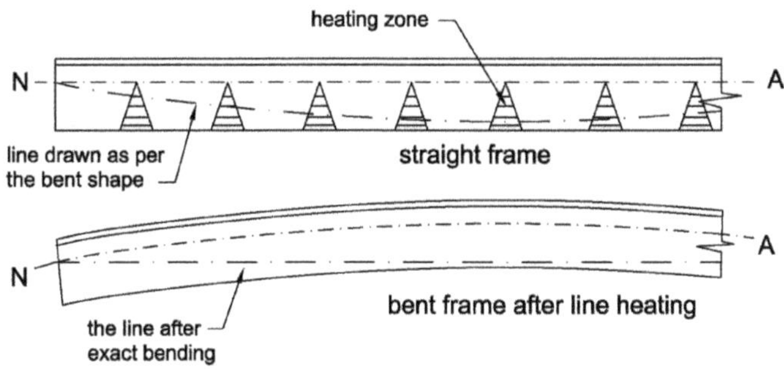

Fig. 12.6 Line heating pattern for frame bending

moment. Figure 12.5 shows the mechanism of frame bending by line heating. To generate a bending moment as shown in the Fig. 12.5, the shrinkage forces should be acting at the web edge gradually decreasing towards the neutral axis (NA). This is achieved by heating triangular zones on the web from its edge to the neutral axis.

To achieve a target concave bent shape of a frame shown in Fig. 12.2, the heating pattern should be as shown in Fig. 12.6. Triangular heating zones are located at suitable intervals from the web edge to the neutral axis. Prior to commencing line heating, the required bent shape is drawn on the web as shown by chain dot line in Fig. 12.6. On completion of line heating, this line will become straight. Therefore by checking the straightness of this chain dot line one can confirm the correctness of the bent shape.

To achieve a convex bending of a frame, similar triangular zones are to be heated on the face plate side of the neutral axis, because shrinkage forces are required on the face plate side to develop the required bending moment.

12.2.1 Compound Curved Surface Generation

Curved surfaces can be broadly divided into two categories (i) *developable* and (ii) *non-developable*. Mathematically the developable surfaces are the ones that have zero Gaussian curvature. This means that the developable surface can be developed without stretching or tearing the surface into a flat surface. The developable surfaces can be produced by rolling only on one direction. Therefore they are also referred to as rolled surfaces. Examples are, cylinder, cone, rounded bilge of a ship, etc.

On the other hand a non-developable surface is the one that has non-zero Gaussian curvature in at least some parts of the surface. This means stretching of the plate is required during the formation of the surface. Although developable surfaces are highly desirable from production point of view but sometimes hydrodynamic, structural or aesthetic requirements may require non-developable

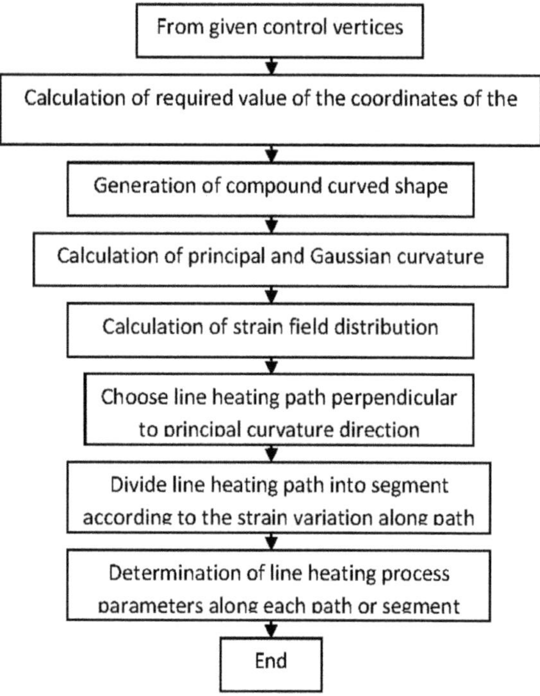

Fig. 12.7 Process flow chart for compound curved plate generation

surfaces. The flow diagram for generation of compound curved surfaces is shown in Fig. 12.7 [3].

In the context of line heating, the curved plates, that constitute the ship's hull are termed as patches. The developed form of a patch is called blank. The bending process takes B into P as shown in Fig. 12.8. In case of ship hull curved plates (patches), the surface definition of the 3-D patch is known and we are required to produce the 2-D blank that would on application of appropriate strains maps into the target 3-D surface. So, it is convenient to think the reverse process, i.e. from given *P*, what deformation would be required to reduce the curved surface to a flat surface? Then, this deformation can be reversed in order to create *P* from the blank *B*.

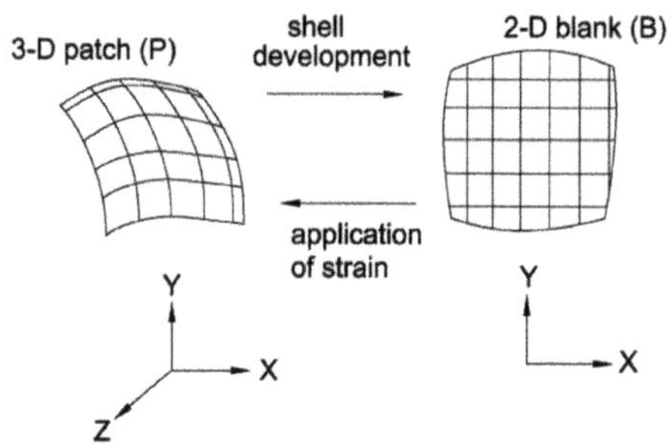

Fig. 12.8 Mapping of patch to blank and bending of blank to patch

12.2.2 Heating Path Generation

In line heating operation the most important component is generation of heating pattern and heating sequence. The heating line pattern is generated by drawing lines perpendicular to the direction of principal curvature. The distance between two adjacent heating paths is equal to or more than the width of the plastic zone caused by heating. A heating line can only produce a certain amount of bending. If higher bending is required the distance between two lines should be made smaller or multiple heating needs to be done on the same line segment. The heating path also depends upon the Gaussian curvature distributions. If the Gaussian curvature is positive, the heating path is placed only on one side of the plate whereas if it is negative, the heating paths are to be placed on both sides of the plate [3].

12.2.3 Types of Heat Sources Used in Line Heating

The line heating heat sources can be any one of the following:

(a) Oxy-acetylene gas flame.
(b) High-frequency induction heating.
(c) Laser beam.

The basic characteristics of the heat sources are given below.

(a) **Gas torch**
It is by far the cheapest and simplest method. However it is somewhat difficult to maintain uniform heat delivery to the plate and also produces comparatively a wider heated zone. A typical manual line heating operation for plate bending using oxy-acetylene gas torch is shown in Fig. 12.9.

Fig. 12.9 Manual line heating with oxy-acetylene gas flame

(b) **High-frequency induction heating**

It allows for much precise control of heat penetration, as it depends on the frequency of the induced electrical field. It is however unsuitable for heating at the edges of a plate as over heating is almost inevitable. Further the equipment is rather heavy, so it is not very convenient to use in manual line heating operations.

(c) **Laser beam**

It is a most well controlled process. A precise narrow zone heating is possible with this. Laser heating with a protective gas shielding reduces the risk of oxidation of the surface. However this is a very expensive option.

References

1. Mandal, N. R. (1992). Cold forming of welded steel T sections. *Thin Walled Structures, 13*, 319–335.
2. Mandal, N. R. (1991). Failure of web/flange welds during cold bending of T sections. *Thin Walled Structures, 12*, 213–227.
3. Mandal, N. R. (2009). Welding techniques, distortion control and line heating. New Delhi: Narosa Publishing House. ISBN:978-81-7319-971-4.

Chapter 13
Fusion Welding Power Source

Abstract In shipbuilding applications, electrical energy is most widely used in the form of arc for several welding processes and as resistance of molten slag in case of electro-slag welding. Arc welding requires a power source that can deliver electrical power at low voltage and high current such that it can establish and sustain an arc plasma column between the welding electrode and the work piece. The welding power sources can have output of alternating current or direct current. These power sources have volt-ampere characteristics typically of constant current or constant potential type. The output may also have a pulsing mode. The classification of the power sources are based on their static volt-ampere characteristics. Constant-potential power sources usually have near constant voltage output compared to constant-current sources having constant-current output. The fast response solid-state sources can provide power in pulses over a broad range of frequencies. These are known as pulsed power sources. The principal metal transfer modes in gas metal arc welding process are short circuiting transfer, globular transfer, spray transfer and Pulsed transfer.

Fusion welding is a thermal process. The joining takes place through melting of the mating edges. To achieve this, the required heat energy can be derived from various sources, e.g. oxy-acetylene flame, electrical energy in the form of arc, electrical resistance of molten slag, laser beam, solar energy, electron beam, etc. In shipbuilding applications, electrical energy is most widely and conveniently used in the form of arc for several welding processes and as resistance of molten slag in case of electro-slag welding.

Arc welding requires a power source that can deliver electrical power at low voltage and high current such that it can establish and sustain an arc plasma column between the welding electrode and the work piece. The line voltage in the mains supply is too high to be used directly in arc welding. Therefore either a transformer, solid-state inverter/rectifier or a motor-generator set is used to obtain the required open circuit voltage necessary for arc welding. It is generally in the range of 20–80 V. At the same time the same power source should be capable of delivering high current typically ranging from 50 to 1500 A. This welding power source can

N.R. Mandal, *Ship Construction and Welding*, Springer Series
on Naval Architecture, Marine Engineering, Shipbuilding and Shipping 2,
DOI 10.1007/978-981-10-2955-4_13

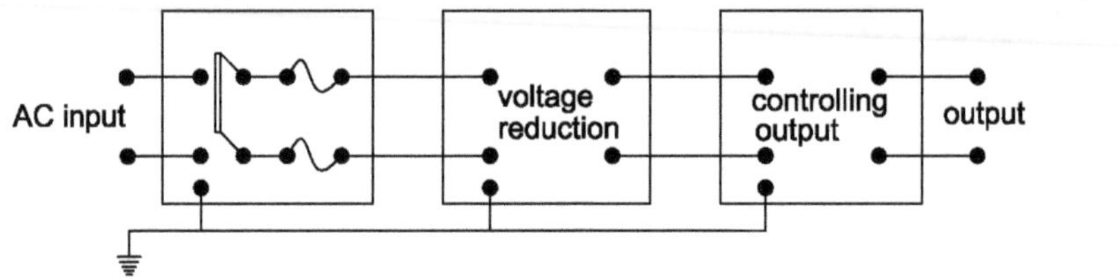

Fig. 13.1 Basic elements of an arc welding power source

have output of alternating current (AC) or direct current (DC). These power sources have volt-ampere characteristics typically of constant current (CC) or constant potential (CP) type. The output may also have a pulsing mode. A schematic of the basic elements of an arc welding power source is shown in Fig. 13.1.

13.1 Types and Characteristics

The classification of the power sources are based on their static volt-ampere characteristics. Constant-potential power sources usually have near constant voltage output compared to constant-current sources having constant-current output. However precision power sources are there which can hold the outputs truly constant. The fast response solid-state sources can provide power in pulses over a broad range of frequencies. These are known as pulsed power sources.

Volt-Ampere characteristics
The performance effectiveness of a welding power source is determined by its operating characteristics. These are defined as static and dynamic characteristics. They influence arc stability and thus the welding performance.

Static output characteristic can be readily measured under steady state conditions by conventional test procedure using resistive loads. A set of output volt-ampere curves obtained from this measurement are usually used to describe the static characteristic.

Whereas the dynamic characteristic of a welding power source is obtained by measuring the transient variations in the output current and voltage that appear in the arc. This parameter describes the instantaneous variations. Their occurrence is very short, typically of the order of 0.001 s [1].

In general welding arcs always operate in changing conditions. The transients occur:

- At the start of the welding, during the striking of an arc,
- during rapid changes in arc length,
- during the metal transfer across the arc,
- during arc extinction and re-ignition in each half-cycle in case of welding with AC power.

These transients in a welding arc can occur in milliseconds, during which significant change in ionization of arc column occurs. A good power source must respond rapidly to these demands. The static volt-ampere characteristic has little significance in determining dynamic characteristic of an arc welding system. The dynamic characteristic of an arc welding power source is controlled by the following design features:

- Local transient energy storage, such as DC series inductance or parallel capacitance circuits,
- feedback control in automatically regulated systems,
- modification of waveforms or circuit operating frequencies.

Controlling these design features results in improvement of arc stability which leads to:

- Improvement in uniformity of metal transfer,
- reduction in metal spatter,
- reduction in weld pool turbulence.

Manufacturers of welding power supplies give the static volt-ampere characteristic, however there is no universally accepted method and norm by which dynamic characteristics can be specified.

13.1.1 Constant Current Power Source

In this type of power source variation in arc voltage causes a small change in arc current. It implies that small variations in arc length will not affect welding current, it will remain fairly constant. Thus heat generation remains almost constant leading to uniform metal deposition. A typical volt ampere characteristic of a conventional CC power source is shown in Fig. 13.2.

This is also known as drooping power source because of the drooping nature of the volt-ampere curve. These power sources may have open circuit voltage control in addition to output current control. By changing either of these controls, the slope of the volt-ampere curve can be changed. Constant-current power sources are typically used for manual shielded metal arc welding (SMAW) and gas tungsten arc welding (GTAW). These power sources are either inverters, transformer rectifier or DC generators.

The open circuit voltage of a constant-current, rectifier type power source ranges from about 50 to 100 V. At the start of the weld, with striking the arc, sharp drop in voltage takes place from the OCV. A conducting column of ionized gases is formed along with heating up of the electrode. As there is a drop in voltage, simultaneous increase in welding current takes place. After a certain point the voltage/current variation becomes linear following Ohm's law. The static volt-ampere characteristics of a welding power source are shown in Fig. 13.2. The slope of the

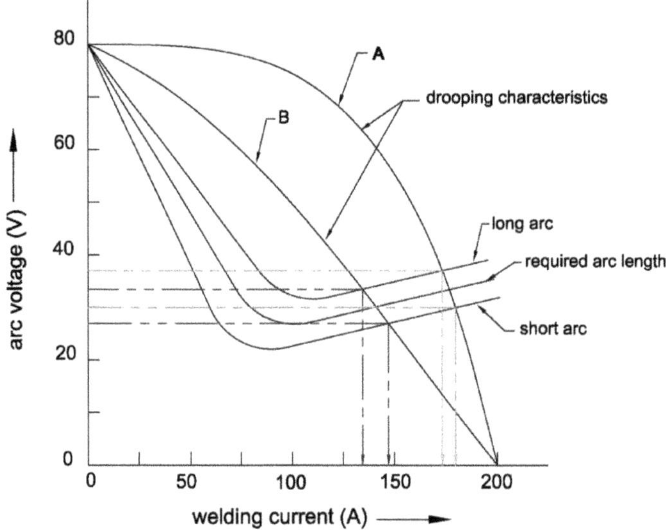

Fig. 13.2 Typical volt-ampere characteristics of a CC power source

volt-ampere curve determines the quality of a CC power supply. The steeper the curve better is the machine. In that case certain variation in arc voltage will produce much less variation in arc current, leading to a near situation of constant current. Welding current is the dominant parameter that determines electrode melting rate, deposition rate and depth of penetration.

The effect of the slope of the V-A curve is explained from Fig. 13.2. The curve-A indicates a power source having an open circuit voltage of 80 V. An increase in arc voltage from about 29 to 37 V i.e. for a 27 % change, it would result in a decrease in current from 180 to 173 A, i.e. 3.9 %. The relative change in current is much smaller. Whereas curve-B represents a different power source exhibits a flatter V-A characteristic. Here for an increase of arc voltage from about 27 to 33 V, i.e. for a change of 24.7 %, there is a decrease in current from 147 to 133 A, i.e. a change of 9.1 %. Thus power source having a stiffer V-A curve yields lesser variation of current for a larger variation of arc voltage. Therefore in case of SMAW, a small change in the arc length due to manual operation resulting in a change in arc voltage will lead to a much smaller change in welding current. The net effect is electrode melting rate and thereby deposition rate remains fairly constant resulting in a uniform weld bead.

Whereas if the open circuit voltage (OCV) is 50 V (curve-B), for the same change of arc voltage, i.e. from 20 to 25 V, the change in current would be from 123 to 100 A, i.e. 19 %, which is much higher than the previous case. Hence the machine having higher OCV is more closer to a constant current power source. At the same time having the same OCV but with a stiffer slope as in curve-C, the change in welding current for the same change in arc voltage is much less. Thus a power source having higher OCV and a stiffer V-A curve results in minimal variation of current for a substantial change in arc voltage, leading to a near true situation of constant current power source. Hence, a less skilled welder would prefer such a power source with much steeper V-A curve and higher OCV, so that the current remains more or less constant even if there are fluctuations in arc length.

However, a skilled welder might prefer a power source with much flatter V-A curve, because he can substantially vary the welding current hence the metal deposition by changing the arc length. This could be useful in case of out-of-position (i.e. other than down hand welding position) welding because a welder could control the electrode melting rate and molten pool size e.g. welding in the overhead position or when there is variation in root gap. Constant current power source is also used in submerged arc welding where the electrode feed rate and the welding current are matched with deposition rate of the melted electrode.

13.1.2 Constant Voltage Power Source

A typical volt-ampere (V-A) curve for a constant-voltage (constant potential) power source is shown in Fig. 13.3. Ideally the V-A curve should have been flat, i.e. parallel to current axis. However because of internal electrical impedance there is a minor drop in the output voltage with increasing current. This leads to a V-A curve with slight downward slope. This slope therefore can be changed by changing the internal impedance. A typical constant-voltage (constant-potential or Flat) power source has a negative slope of 1–2 V/100 A. Machines with V-A curves having slopes up to 8 V/100 A are still referred to as constant-voltage power supplies.

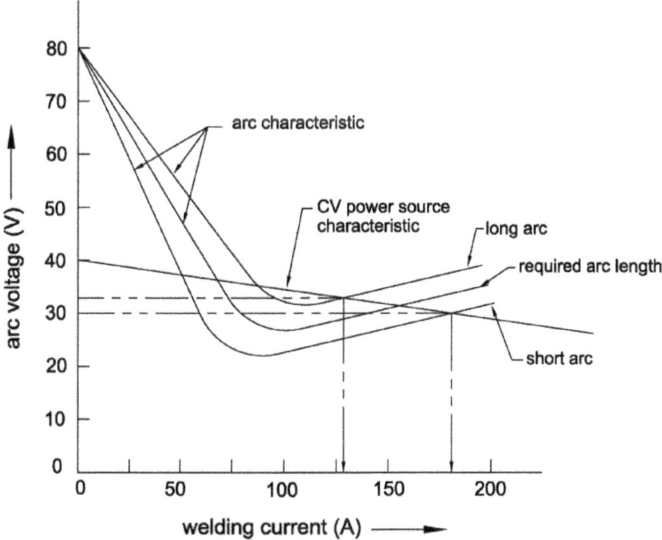

Fig. 13.3 Typical volt-ampere characteristics of a CV power source

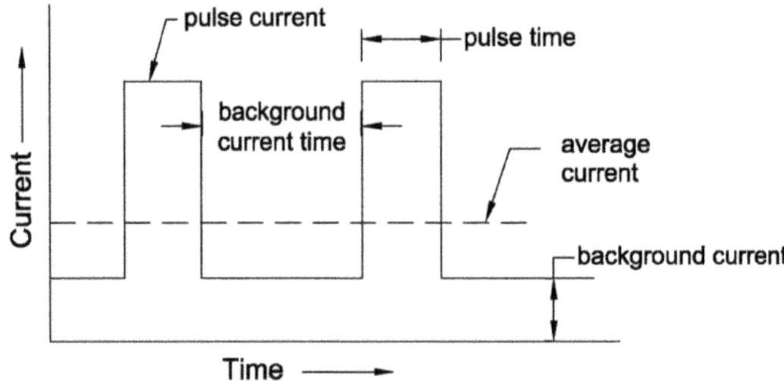

Fig. 13.4 Current wave form of a pulsed power source

It can be seen from Fig. 13.3 that with increase (or decrease) in arc voltage from 30 to 31.25 V i.e. 4.2 % change, will produce a much larger change in current from 180 to 130 A, i.e. 27.7 %. Hence welding processes having constant electrode feeding mechanism, such as gas metal arc (GMAW), submerged arc (SAW) or flux cored arc (FCAW) maintaining constant arc length are suitable for constant-voltage power sources. A slight change in arc length will cause a substantial change in welding current. This will automatically increase or decrease the electrode melting rate to regain the desired arc length. This effect is known as self-regulation.

The difference between static and dynamic characteristics of a power supply can be explained from Fig. 13.3. For example during short circuiting metal transfer in GMAW, the welding electrode tip touches the weld pool, causing a short circuit. At this point the arc voltage almost drops to zero and there is a rapid increase in current. If the power supply responded instantly, very high current would immediately flow through the welding circuit quickly melting the short-circuited electrode tip and freeing it with an explosive force. This would result in dispelling the weld metal as spatter. However this rapid increase in current gets limited by the circuit inductance. The inductance in the circuit opposes the sharp increase in the welding current. It slows down the rate at which the current increases during striking an arc or in short circuit mode of GMA welding. The inductance is made variable such that it can be adjusted to give a stable condition. By properly adjusting the inductance smooth bead shape can be achieved. This improves the dynamic characteristic of the power supply. Welding power sources designed with such dynamic characteristics control the rate of current change, thus decreasing the explosive force. This provides a better and less violent arc start.

13.1.3 Pulsed Mode Power Source

Generally in electric arc welding the power sources deliver a continuous power, whereas the metal transfer from the electrode takes place intermittently. Hence continuous high current is not required. Thus the arc power can be reduced by designing the power supply in such a way that it will provide higher current only when metal transfer takes place.

The current level during the metal transfer interval is kept sufficiently low so as to avoid any metal transfer however it is maintained at a level high enough to sustain ionization in the arc column. This current is known as 'background current' which helps in keeping the arc alive. Whereas in the metal transfer interval, the current is raised above the transition current for sufficient time, enough to allow transfer of one or two droplets. This current is known as peak current or pulse current as shown in Fig. 13.4.

By controlling the pulse variables i.e. peak current, background current, peak current time and background current time, it is possible to have a control on metal transfer, allowing only single drops to transfer per pulse, while retaining the background current. In situations like fillet welding of thinner plates where a very precision metal deposition is required pulsed power supply with GMAW can be used to weld in all positions. The heat input in this process being minimum the distortions can also be maintained at a minimum level.

One of the major drawbacks of a pulsed power source is the difficulty in setting the variables for a given welding requirement manually. This, however, is overcome by having electronic and microprocessor controls to set the optimum pulse conditions for a given wire feed speed setting. Such machines are referred to as synergic pulsed power source. The word 'synergic' means 'several things acting as one'. In synergic pulsed GMAW machines, the pulse variables are set automatically based on the required wire feed speed for a given joint geometry. Through the use of electronic controls, it is possible to choose a variety of synergic curves to satisfy particular applications. It can make instantaneous adjustments of the pulse frequency and width depending on the voltage sensed across the arc. Pulsed GMAW power sources typically range up to 500 A peak current over a frequency ranging from 60 to 200 Hz.

13.1.4 Inverter Power Supply

With the availability of affordable and reliable high power insulated-gate bipolar transistors (IGBT), it became possible to build a switching power supply capable of coping with the high loads as is needed in arc welding. These are known as inverter power supplies. These welding power supplies convert the line power to high voltage and store this energy in a capacitor bank. A microprocessor based controller then switches this energy into a second transformer as needed to produce the desired welding current. The switching frequency is very high typically of the order of 10 kHz or even higher. The high frequency inverter-based welding machines are highly efficient and have better control than non-inverter welding machines.

The IGBTs in an inverter based machine are controlled by a microcontroller, so the electrical characteristics of the welding power can be changed by software in real time. Typically the controller software implements features such as pulsing of welding current and current densities through a welding cycle with variable frequencies. The IGBT is a fairly recent invention. With high switching frequencies, smaller high frequency transformers and magnetic components are used in these welding power supply machines. Thus these machines have become lighter, portable and also less expensive [2].

13.2 Metal Transfer Mechanism

The principal metal transfer modes in gas metal arc welding (GMAW) process are as follows:

- Short circuiting transfer
- Globular transfer
- Spray transfer
- Pulsed transfer.

The metal transfer in a GMAW process is influenced by various factors. The most influential of them are the following:

- Welding current, AC or DC and its magnitude
- Electrode wire diameter
- Electrode composition
- Shielding gas composition.

13.2.1 Short Circuiting Transfer

Short circuiting transfer mode is used for low current operation with lower electrode diameters. In short circuiting transfer, also called 'dip' transfer, the molten metal forming on the tip of the electrode wire is transferred by the wire dipping into the molten weld pool thus causing a momentary short circuit. Metal is therefore transferred only during a period when the electrode tip is in contact with the weld pool and no metal is transferred across the arc gap. This type of metal transfer mode produces a small, fast-freezing weld pool that is generally suited for joining thin sections, for out-of-position welding and for bridging large root openings.

The frequency of short circuiting, i.e. electrode dipping in molten pool varies from 20 to over 200 times per second. As the electrode wire touches the molten pool, short circuit takes place causing a sharp drop in arc voltage and rise in the welding current. The sequence of events during a short circuiting metal transfer and the corresponding current and voltages are shown in Fig. 13.5.

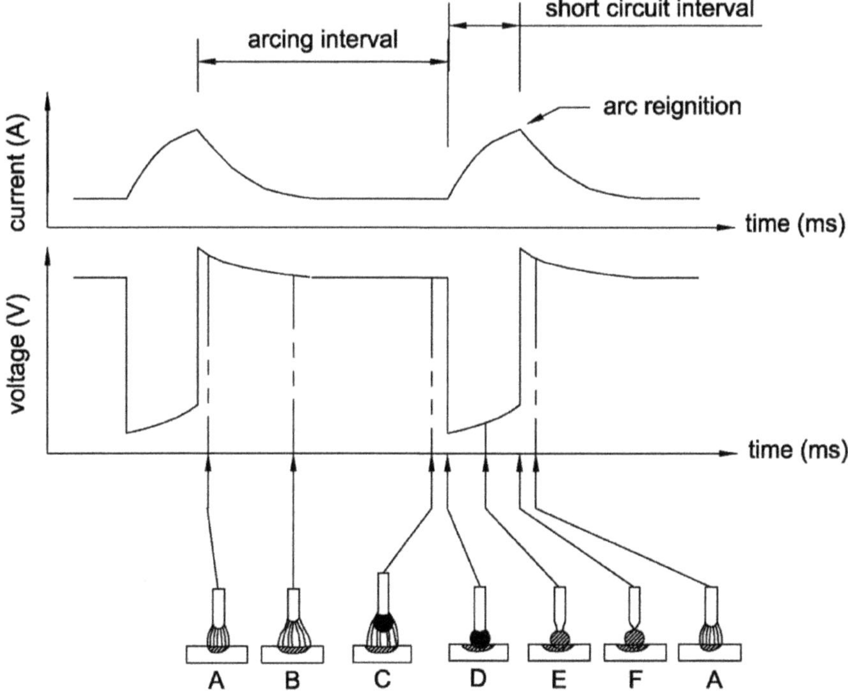

Fig. 13.5 Schematic representation of short circuiting metal transfer

The arc starts little before A. With the arc on, the electrode tip gets heated up and starts melting through B and at C the molten droplet gets fully built up but still adhering to the electrode tip by surface tension forces. At D the molten droplet touches the molten pool below and thus short circuiting takes place causing a sharp drop in arc voltage and surge in the current as shown in the Fig. 13.5. However the surge in current is regulated, i.e. it grows slowly and reaches a peak at F, where the detachment of the droplet occurs. The surge in the current is made gradual to avoid violent detachment of molten droplet to minimize spatter. After the droplet detachment at F as the electrode wire feed continues, again the cycle starts with A as can be seen in Fig. 13.5. The open circuit voltage is kept low enough so as to ensure that the drop of molten metal at the electrode tip does not detach until it touches the molten pool in the base metal.

13.2.2 Globular Transfer

Globular transfer is characterized by drop size having diameter greater than that of the electrode. At the start of the welding as the arc is struck between the electrode and the work piece, the electrode tip gets heated up and starts melting. As the

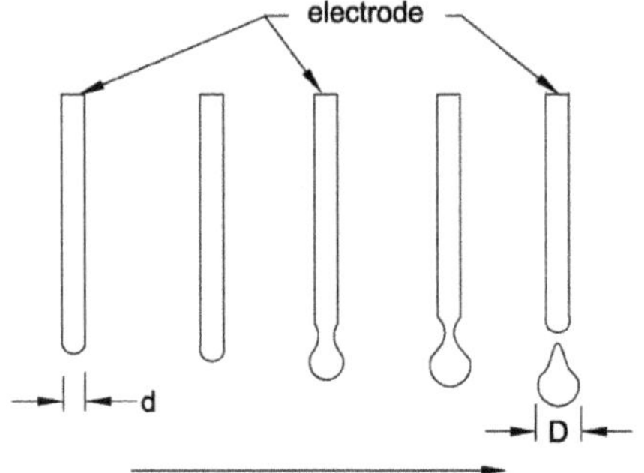

Fig. 13.6 Schematic representation of globular transfer of weld metal

melting progresses, a molten droplet is formed and it gradually grows in size and becomes even larger than the diameter of the electrode. The molten metal holds on to the electrode by surface tension forces. With further increase in size as the weight of the droplet exceeds the surface tension force, the droplet gets detached from the electrode under gravitational force and plunges into the molten pool below on the workpiece as shown schematically in Fig. 13.6. Thus it may lead to spillage of molten metal from the weld pool causing what is known as spatter. Globular metal transfer is characterized by significant spatter all along the weld line.

Since metal transfer takes place under gravity, therefore out of position welding (i.e. vertical, horizontal and overhead) with globular transfer invariably results in irregular and uneven metal deposition leading to a poor weld quality. It is only effective in down hand position welding. In this mode the metal transfer takes place in the form of globules at a rate less than about 10 drops/s. Globular transfer takes place with a positive electrode (DCRP/DCEP) when the current is relatively low regardless of the type of shielding gas. At current levels only slightly higher than those used in short-circuiting transfer, globular axially-directed transfer can be achieved in a substantially inert gas shield.

13.2.3 Spray Transfer

In this mode when the current exceeds a certain critical level known as *Transition Point*, metal transfer takes place in the form of fine droplets at a rate of about few hundred drops per second. These metal droplets are accelerated axially through the arc column. Spray transfer is associated primarily with the use of inert gases. Either pure Argon or argon-rich with 0.5–5 % oxygen shielding gas is used. With such gas mixtures a true spatter-free axial spray transfer becomes possible with DCRP (DCEP) power supply and with current above transition point.

The spray transfer yields a highly directed stream of metal droplets with substantial energy so as to have velocities which overcome the effects of gravity. Because of this, spray transfer mode can be used for welding in any position. The metal droplets being very small, short-circuit does not occur and hence spatter is virtually eliminated. The spray transfer welding is widely used and almost any metal or alloy can be welded because of the inert gas shielding. However, in this process there is a limiting thickness below which spray transfer can not be used unless the arc is pulsed. Spray transfer welding being a high current, high heat input process, it produces a deep penetration and therefore welding thin plates with this mode may become difficult The resultant arc forces can cut through thin plates instead of welding them. It also produces high weld metal fluidity with a large weld pool too large to be supported by surface tension in vertical or overhead position, because of very high temperature of the molten weld metal. This high fluidity again makes it difficult to weld in overhead position. This disadvantage is overcome by using pulsed spray welding process thus reducing the heat input.

Transition Point

For a given electrode diameter and inert gas shielding, the welding current directly controls the size of the droplets and the number of them that are detached from the electrode per unit time.

As already discussed globular transfer takes place at a lower current level, as the current is increased, at a certain point the molten droplet size abruptly decreases substantially as shown in Fig. 13.7. This limiting current level is known as Transition Point. At the same time the rate of detachment of the droplets also increases to several hundred droplets per second. This transition point is associated with a specific current level, for various sizes of electrode diameter and shielding gas composition. With further increase of current, rate of detachment of droplets increases but the rate of increase is less as compared to that of near transition point.

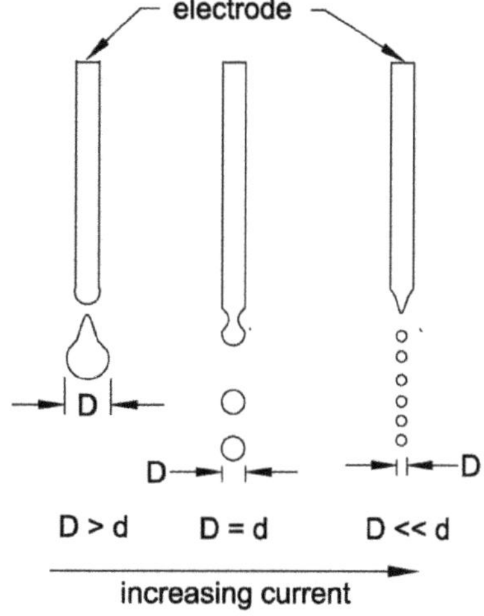

Fig. 13.7 From globular to spray mode of metal transfer

Table 13.1 Typical transition currents for some common metals

Electrode type	Electrode diameter (mm)	Shielding gas	Transition current (A)
Mild steel	0.8	98 % argon—2 % oxygen	150
Mild steel	0.9	98 % argon—2 % oxygen	165
Mild steel	1.1	98 % argon—2 % oxygen	220
Mild steel	1.6	98 % argon—2 % oxygen	275
Aluminum	0.8	Argon	95
Aluminum	1.1	Argon	135
Aluminum	1.6	Argon	180

The transition current is dependent on the following factors:

- Liquid metal surface tension.
- Inversely proportional to the electrode diameter.
- Length of stickout or electrode extension.
- Electrode wire (filler metal) melting temperature.
- Shielding gas composition.

The metal transfer below this transition point is called globular transfer and the transfer mode with a current above this transition point results in spray transfer. In globular transfer, the droplets detach when their weight exceeds the surface tension of the molten metal that holds the drop on the electrode tip. The electromagnetic force that acts to detach the drop is small compared to the gravitational force in the current range of globular transfer mode. Typical transition currents for some common metals are shown in Table 13.1.

13.2.4 Pulsed Transfer

The pulsed transfer mode was developed as a means of stabilizing the arc at low current levels within spray transfer range to avoid short-circuiting and spatter. The difficulties of welding thin plates, vertical and overhead position welding are greatly overcome by using pulsed mode of metal transfer. In pulsed transfer welding, welding current switches automatically from a low level to a higher level in a periodic manner as shown in Fig. 13.4. The power source can be adjusted, for a given shielding gas composition, so that the lower level current also known as background current is set below the transition point and the higher level is set well above the transition point. Thus it derives the advantages of both the modes. In this mode spray type metal transfer is achieved by applying pulses of higher level current, each pulse having sufficient force to detach one droplet. The power supplies are specially designed to produce carefully controlled wave forms and frequencies that "pulse" the welding current. The current wave form is schematically shown in Fig. 13.4. Here a constant background current is maintained which sustains the arc without providing enough energy to melt the electrode wire tip and a superimposed pulsing current higher than the transition current necessary for spray transfer.

Metal transfer takes place only at the pulse current interval. The frequency and amplitude of the pulses control the energy level of the arc and therefore the rate at which the wire melts. Hence with the reduced average arc energy and reduced wire melting rate, thin sections can be effectively welded with the advantages of spray transfer. At the same time with reduced heat input, weld induced distortions are also minimized. All position welding can also be done satisfactorily with this metal transfer mode. Larger diameter electrode wires can be used that cost less than the smaller diameter wires and they also have lesser feeding problems. The optimum combination of all the pulsing parameters is controlled automatically by synergic pulsed power sources.

References

1. Mandal, N. R. (2009). Welding techniques, distortion control and line heating. New Delhi: Narosa Publishing House. ISBN:978-81-7319-971-4.
2. Liberti, A., & Gulino, R. (2015). Welding machines: V and HB series IGBTs on two-switch forward converters. ST Microelectronics.

Chapter 14
Welding Parameters

Abstract Weld quality and weld metal deposition are both influenced by the various welding parameters. These parameters are, welding current, arc voltage, welding speed, electrode feed speed, electrode extension, electrode diameter, electrode orientation, electrode polarity, shielding gas composition. Each of these parameters has influence to a varying degree on the deposition rate, weld bead shape, depth of penetration, cooling rate and weld induced distortion. To achieve a sound welded joint proper selection of welding parameters is needed. Hence a proper understanding of the effects of these parameters or process variables is required. Weld penetration in a welding process is driven by electromagnetic force, surface tension gradient, buoyancy force and impinging force of arc plasma. High temperature gradients within the molten weld pool results in variation of the surface tension over the weld pool surface. This surface tension gradient changes the convection mode of the molten metal in the weld pool. This is known as Marangoni convection. The cooling rate of the weld zone determines the metallurgical structure of the weld metal and the HAZ. As adverse microstructural transformations may take place in the HAZ, it is preferable to choose such welding parameters which also keep the HAZ to a minimum.

Weld quality and weld metal deposition rate both are influenced very much by the various welding parameters as well as the joint geometry. Essentially a welded joint can be produced by various combinations of welding parameters and joint geometries. These parameters are the process variables which control the weld metal deposition rate and weld quality. The process variables are,

- Welding current.
- Arc voltage.
- Welding speed.
- Electrode feed speed.
- Electrode extension (length of stickout).
- Electrode diameter.
- Electrode orientation.
- Electrode polarity.

N.R. Mandal, *Ship Construction and Welding*, Springer Series
on Naval Architecture, Marine Engineering, Shipbuilding and Shipping 2,
DOI 10.1007/978-981-10-2955-4_14

- Shielding gas composition
- Marangoni convection.

Each of these parameters has influence to a varying degree on the deposition rate, weld bead shape, depth of penetration, cooling rate and weld induced distortion. Hence to achieve a sound welded joint with adequate metal deposition rate and minimum weld induced distortion, a proper understanding of the effects of these parameters or process variables is needed. The general effect of these parameters will be dealt with individually in this chapter. In the subsequent chapters their effects for various welding methods will be studied in detail.

14.1 Welding Current

It is the most influential variable in a welding process. It controls the electrode melting rate and hence the deposition rate, the depth of penetration, and the amount of base metal melted. If the current is too high at a given welding speed, the depth of fusion or penetration will be too great. For thinner plates, it tends to melt through the metal being joined. It also leads to excessive melting of electrode resulting in excessive reinforcement. On the other hand, if the welding current is too low, it may result in lack of fusion or inadequate penetration. Too much current means more heat going inside parent metal, leading to more distortion and grain growth.

For any given electrode, the melting rate is directly related to the electrical energy supplied to the arc as well as the polarity of the electrode in case of DC power supply. Part of this energy is used to melt a portion of the base metal, part goes to melt the electrode and burn the flux specially in case of submerged arc welding and welding with coated electrode (q_f) and the rest gets dissipated by way of conduction (q_{cp} and q_{ce}), convection (q_v) and radiation (q_r) losses as schematically depicted in Fig. 14.1. This energy is directly proportional to the welding current as can be seen in the following

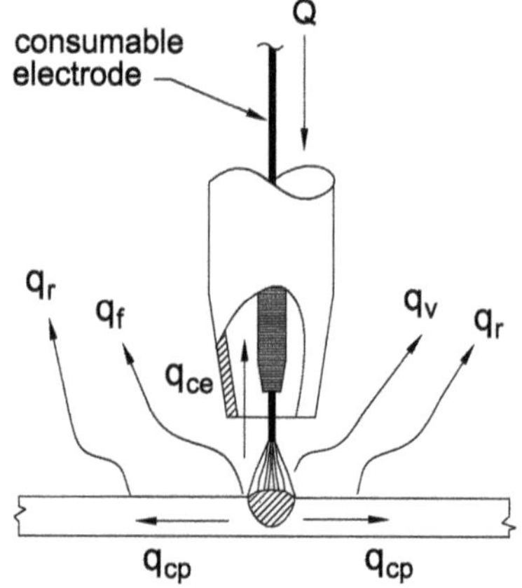

Fig. 14.1 Heat balance in arc welding

$$Q = IV \quad \text{J/s}, \quad \text{or} \quad Q = I^2 R_a \ \text{J/s},$$

where,

Q electrical energy consumed,
I welding current,
V arc voltage,
R_a arc resistance.

Welding current can be either DC or AC, both having their respective advantages and disadvantages. Direct current always provides a steadier arc and smoother metal transfer as compared to AC arc. The DC arc produces good wetting action by the molten weld metal and uniform weld bead size even at low current. For this reason DC is particularly suited to thin section welding. Direct current produces a better weld quality also in out of position welding, i.e. in vertical and overhead welding. The disadvantage of DC welding is the possibility of arc blow when magnetic metals are welded.

The advantages of alternating current over direct current are twofold. One is the absence of arc blow and the other is the cost of power source. In gas tungsten arc welding (GTAW), however AC power is particularly preferred. Because of the polarity reversal in AC power it helps in controlling the heating up of the tungsten electrode as well as helps in breaking the oxide layer in aluminum alloy welding.

The effect of current on welding, keeping other parameters constant, can be summarised as follows:

- Controls heat generation and hence the melting rate of electrode and weld metal deposition rate.
- Depth of fusion.
- Extent of parent metal melting.
- With increasing current, it may result in melt through.
- With increasing current, excessive melting of electrode may take place.
- Too high a current means more heat going in the parent metal, may cause more distortion of the structure.

14.2 Welding Arc Voltage

Arc voltage is the voltage between the electrode tip and the job during welding. It is the voltage drop across the arc column. It is determined by arc length for any given electrode. Open circuit voltage, on the other hand, is the voltage generated by the power source when no welding is done (Fig. 14.2).

Open circuit voltage generally varies between 50 and 100 V, whereas arc voltages are between 17 and 40 V. When the arc is struck, the open circuit voltage

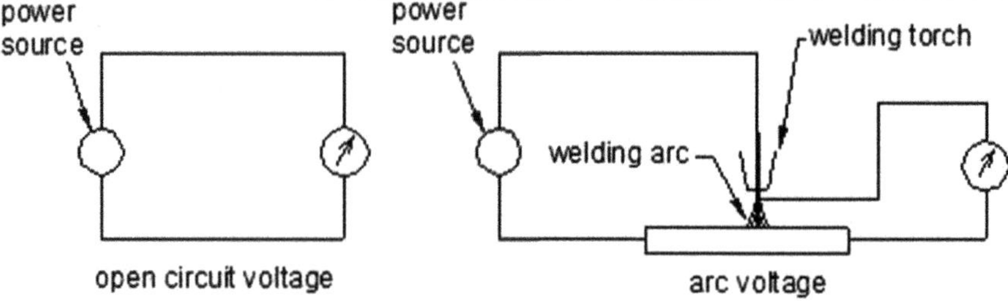

Fig. 14.2 Concept of open circuit voltage and arc voltage

drops to the arc voltage and the welding load comes on the power supply. This arc voltage depends on the arc length and the type of electrode. If the arc length is increased there will be an increase in the arc resistance resulting in higher drop in voltage, or in other words the arc voltage will increase and the current will decrease. This drop in current will depend on the V-A characteristic of the welding power source as discussed in Chap. 13.

Arc length

The arc length is the distance from the electrode tip to the surface of the molten weld pool. Proper arc length is important in obtaining a sound welded joint. Metal transfer from the electrode tip is not a smooth, uniform action. There is variation in instantaneous arc voltage as a metal droplet is transferred through the arc. However such variations will be minimal when welding is done with proper current and maintaining proper arc length. It also depends on the dynamic characteristic of the welding power source. Constant and consistent electrode feed is required to maintain a constant arc length. With increase in electrode diameter and welding current the arc length should also increase. As a general rule the arc length should not be more than the electrode diameter.

A short arc tends to become erratic and may lead to short circuit during metal transfer. On the other hand a long arc may lack direction and intensity. This results in scattering of molten metal as metal transfer takes place from electrode to the weld pool. Thus heavy spatter may occur with a loss in deposition rate. With all variables held constant, arc voltage is directly related to arc length. Truly arc length as a welding parameter needs to be controlled to obtain the desired weld quality. However monitoring and measuring arc length directly is quite difficult and hence arc voltage as a welding parameter is more commonly used.

The weld bead appearance depends on the arc voltage. Keeping all other parameters constant, an increase in arc voltage tends to flatten the weld bead, and increase the width of the fusion zone as schematically shown in Fig. 14.3. As the arc length increases, the arc cone becomes wider resulting in wider and flatter weld bead. At lower arc voltage, the arc length being shorter, the arc is narrow and more focused causing a narrower weld bead. Whereas excessive high voltage may cause porosity, spatter and may lead to undercut. However, arc voltage has a negligible

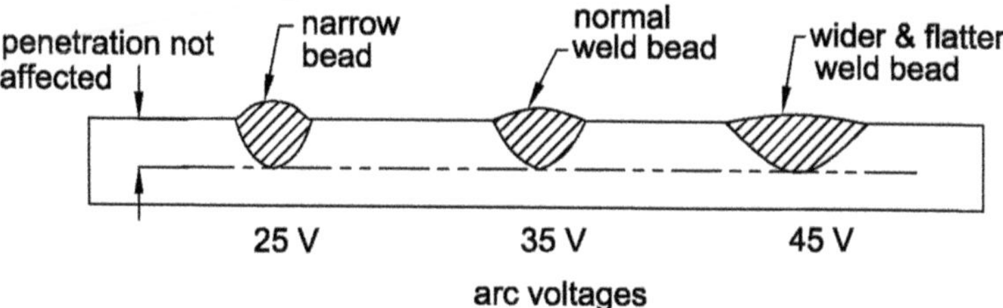

Fig. 14.3 Effect of arc voltage variations on weld bead and fusion zone shape

effect on penetration depth. It is preferable to have trials to obtain an optimum arc voltage. The arc voltage is varied within narrower limits than welding current.

The effect of arc voltage or arc length on weld bead shape and weld quality are summarised below:

- Higher voltages will cause the bead to be wider and flatter.
- Long arc will lack direction and intensity.
- Long arc may cause spattering.
- Long arc may cause porosity, undercutting or uneven deposition.
- Short arc results in narrow bead.
- Short arc tends to become erratic and may lead to short-circuit.

14.3 Welding Speed

Welding speed is the linear rate at which the arc moves along a weld joint. Welding speed is particularly important because it controls the actual welding time and hence it has a direct effect on the production cost. Therefore the speed generally is fixed in mechanized welding while the other parameters like current and/or arc voltage arc varied to control the weld deposit and weld quality. The effects of variation of welding speed for any given combination of welding current and arc voltage are mentioned below:

With increasing welding speed:

- Heat input per unit length of welded joint decreases.
- Less filler metal is deposited, resulting in lack of weld deposition and less weld reinforcement.
- Reduction in distortion and residual stress.
- Possibility of undercut and porosity increases, since the weld solidifies faster.
- May cause uneven deposition resulting in a bad bead shape due to arc instability.

With decreasing welding speed:

- Filler metal deposition rate increases.
- Rate of heat input increases.
- Weld bead gets wider and more convex.
- Large molten pool resulting in a rough bead and possible slag inclusion.
- Increase in heat-affected zone (HAZ).
- Increase in residual stress and distortion.
- Decrease in undercut and porosity.

With all other welding parameters held constant, weld penetration attains a maximum at an intermediate speed of welding. For excessive slow welding speed when the arc strikes a rather larger molten pool, the penetrating force of the arc gets cushioned by the molten pool of metal. Liquid metal has less thermal conductivity. With excessive welding speed, there is a substantial drop in thermal energy per unit length of welded joint, resulting in undercutting along the edges of the weld bead because of insufficient deposition of filler metal to fill the zone melted by the arc. Within limits the welding speed can be adjusted to control weld size, cooling rate and depth of penetration.

14.4 Electrode Feed Speed

The electrode feed speed determines the amount of metal deposited per unit length of weld or metal deposited per unit time. Generally in all welding machines having auto electrode feeding mechanism, the wire feed control is coupled with current control. Increasing electrode feed speed automatically increases the arc current and vice versa.

14.5 Electrode Extension

The electrode extension, also known as length of stickout, is the distance between the end of the contact tube to the electrode tip as shown in Fig. 14.4. An increase in the length of stickout will result in an increase of electrical resistance of the electrode, thus increasing the total resistance in the welding circuit. This leads to resistance heating of the electrode between the contact tube and the arc, at the same time will cause reduction in arc voltage. This gives rise to an additional heat generation and increases the electrode melting rate. Hence, larger length of stickout results in additional heat generation in the electrode leading to additional metal deposition.

Lower arc voltage also will lead to an increased convexity of the weld bead. Hence to maintain the proper bead shape and desired penetration with an increased length of stickout for achieving deposition rate, the voltage setting on the welding power supply should be increased to maintain proper arc length.

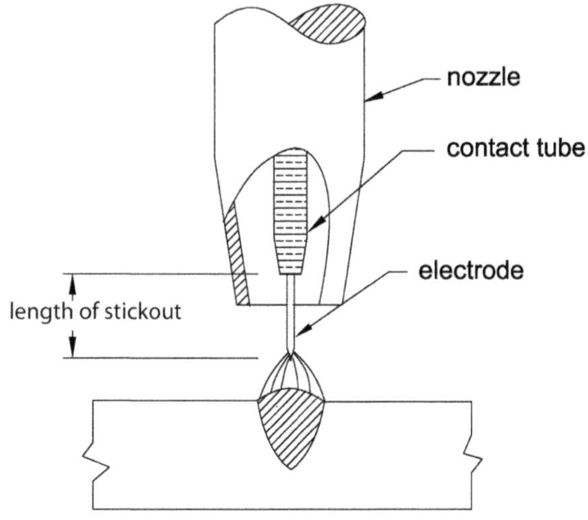

Fig. 14.4 Length of stickout in gas metal arc welding

However, at current densities over 125 A/mm^2, the length-of-stickout becomes a significant variable. An increase of 25–50 % in deposition rate can be achieved by using long electrode extensions without increasing the welding current. This increase in deposition rate is always accompanied by decrease in penetration. Therefore, when deep penetration is desired long electrode extension is not recommended. On the other hand in case of thinner material, when there is a possibility of melt-through, a longer electrode extension will be beneficial. However, as the electrode extension increases, it becomes more and more difficult to maintain the correct position of the electrode tip with respect to the joint to be welded. Typically the length of stickout should not exceed 7 times the diameter of the electrode.

14.6 Electrode Diameter

The electrode diameter influences the weld bead configuration. It has a direct effect on weld penetration as schematically shown in Fig. 14.5.

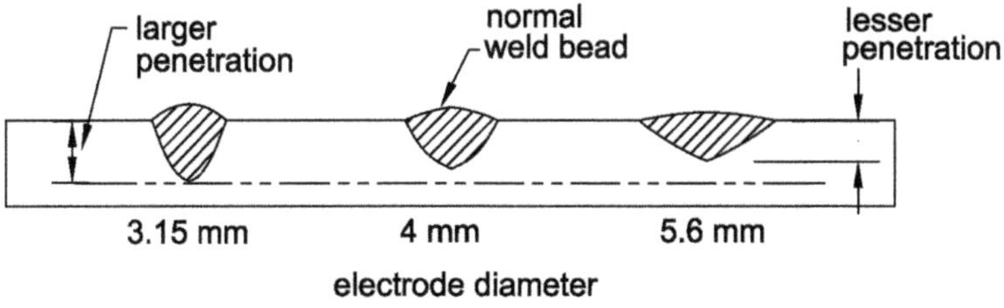

Fig. 14.5 Effect of electrode size on weld bead shape and penetration

At a given current level, a smaller diameter electrode will have a higher current density causing more penetration as schematically shown in Fig. 14.5. As the same amount of current flows through each electrode, the density of current is greater in the smaller diameter electrode compared to that of larger diameter electrode. As a result of this higher current density, the smaller diameter electrode will have greater weld penetration than the larger diameter electrode. However electrodes of different diameter will have a limiting current density, beyond which the welding arc becomes unstable and erratic. Therefore with increasing current to increase deposition rate, the electrode diameter should be increased. In case of poor fit-up, a larger diameter electrode is better than smaller ones for bridging large root openings.

14.7 Electrode Orientation

The electrode orientation with respect to the weld joint affects the weld bead shape as well as weld penetration. The influence of electrode orientation on weld bead shape and its penetration is more than that of arc voltage or welding speed. The electrode orientation is defined in two ways [1],

- The angle of the electrode axis with respect to the direction of travel. This is known as the travel angle.
- The angle between the electrode axis and the adjacent work surface. This is called work angle.

A drag (backhand technique) angle of 10–20° is formed when the electrode axis is inclined in the direction of welding. Whereas pushing (fore hand technique) angle of 10–20° is formed when the electrode axis is inclined against the direction of welding as shown in Fig. 14.6.

Both backhand and forehand techniques have their own merits and demerits as indicated below:

Backhand technique (drag angle):

- Weld penetration is increases.
- Weld bead is narrow and more convex.
- Less spatter and stable arc.
- Produces shorter arc length.
- Welder's view of the welding line is somewhat obstructed by the welding torch.

Forehand technique (push angle):

- Weld penetration decreases.
- Weld bead is wider and flatter.
- Spattering increases
- Produces larger arc length.
- Welder's view of the weld line is unobstructed.

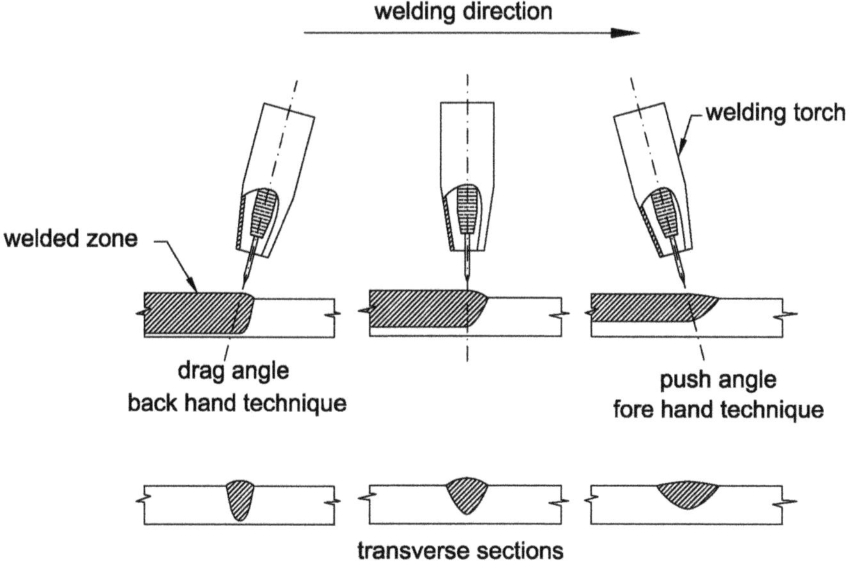

Fig. 14.6 Effect of electrode orientation on bead shape and weld penetration

For all positions, the electrode orientation is generally backhand with a drag angle in the range of 5–15° for achieving good control and shielding of the molten weld pool. However, for aluminum, a forehand technique is preferred. It provides a cleaning action ahead of the molten weld metal, which promotes better wetting and reduces base metal oxidation [2].

14.8 Electrode Polarity

The question of electrode polarity does not arise when welding with AC power supply. However with DC power supply, the polarity of the electrode plays an important role in the fusion process of the plates as well as electrode. When the electrode is connected to the positive pole, it is referred to as DCEP (direct current electrode positive) or DCRP (direct current reverse polarity). Whereas when electrode is connected to negative terminal of a DC power supply, it is referred to as DCEN or DCSP (direct current straight polarity).

With DCEP polarity, the work piece forms the cathode. Here in the work piece, the thermionic cathode develops at the negative end of the arc, which is a very small area of extremely high current density that supplies the electrons. The electrons emitted from the cathode are accelerated through the shielding gas by the positive potential of the anode, i.e. the electrode, producing the plasma by electron collision with the shielding gas atoms [3]. The electrons on collision with the electrode impart its energy thus additionally heating up the electrode tip. The molten droplets at the electrode tip gaining extra energy from the additional heating due to electron bombardment accelerate through the arc column and impinge in the molten weld pool at the work piece. Thus it leads to increased penetration with DCEP (reverse polarity). It also provides better surface appearance of weld bead and resistance to porosity [4].

Whereas with DCEN (straight polarity), it leads to less penetration and may cause porosity and spatter because of globular mode of metal transfer. Hence in GMAW of steel, it is recommended to use only DCEP polarity.

14.9 Shielding Gas Composition

The primary function of shielding gas is to protect the arc and the molten weld pool from atmospheric oxygen and nitrogen. If not properly protected it forms oxides and nitrides and also results in weld defect such as porosity and may cause weld embrittlement. In addition to this, the shielding gas performs a number of important functions,

- Forms the arc plasma,
- Stabilizes the arc on the material surface,
- Ensures smooth transfer of molten metal droplets from the wire to the weld pool.

Thus the shielding gas and its flow rate have a substantial effect on the following,

- Arc characteristics
- Mode of metal transfer
- Penetration and weld bead profile
- Speed of welding
- Undercutting tendency
- Cleaning action
- Weld metal mechanical properties.

General purpose shielding gases for GMAW are mixtures of argon, oxygen and CO_2 and special gas mixtures may contain helium. Carbon dioxide is extensively used in shipyards for welding of shipbuilding quality mild steels, whereas for aluminum welding argon is used.

14.10 Marangoni Convection

Weld penetration in a welding process is driven by electromagnetic force, surface tension gradient, buoyancy force and impinging force of arc plasma. High temperature gradients occur within the molten weld pool surface. This results in variation of the surface tension over the weld pool surface. This surface tension gradient acts as the principle parameter that changes the convection mode of the molten metal in the weld pool. This is known as Marangoni convection. The nature of fluid flow and its direction depend on the magnitude and sign of the surface tension coefficient at a given temperature. Generally with increase in temperature the surface tension decreases i.e. the temperature coefficient of surface tension is negative, $\frac{\partial \sigma}{\partial T} < 0$ for pure metals and some alloys. The temperature in a weld pool is higher at the centre, below the arc and cooler near the pool edge, therefore the liquid metal flows from the centre to the edge resulting in a shallow and wider weld pool as shown in Fig. 14.7a. However it was observed that oxygen, sulfur and selenium, known as surface active elements can change the temperature coefficient of surface tension from negative to positive, i.e. $\frac{\partial \sigma}{\partial T} > 0$. This changes the direction of liquid metal flow in the weld pool as shown in the Fig. 14.7b.

This leads to increase in weld penetration causing a narrow and deep weld pool. The oxygen content in the weld metal changes with the type and quantity of flux

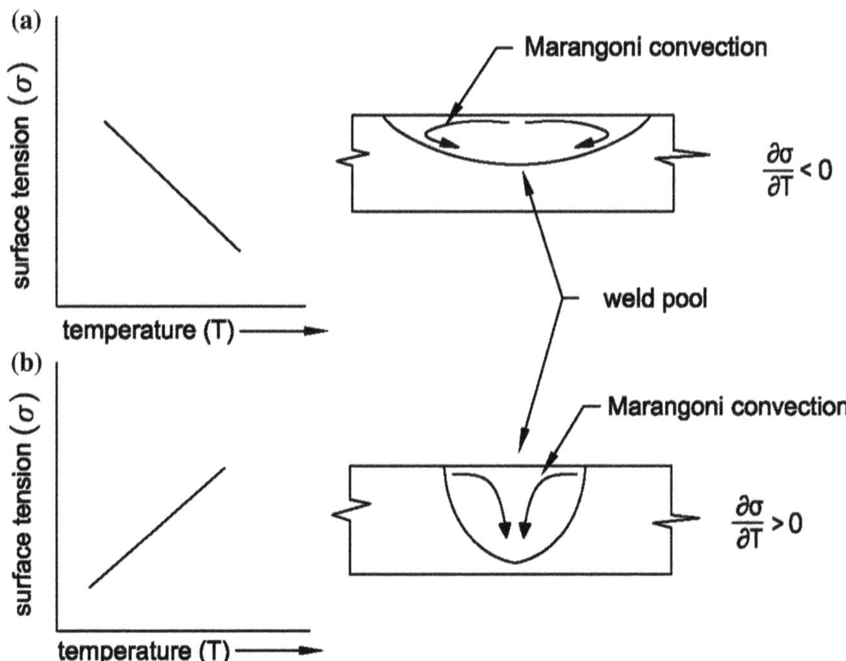

Fig. 14.7 Marangoni convection due to surface tension gradient in weld pool surface

used. The oxygen from the decomposition of the flux in the welding pool changes the surface tension gradient on the weld pool surface. It alters the Marangoni convection direction of the liquid metal resulting in changing the depth/width ratio of the pool. When it is in the range of 70–300 ppm, the depth/width ratio of the weld pool increases by 1.5–2 times [5].

14.11 Cooling Rate and HAZ

In electric arc welding, the heat input, Q in (J/s), is given by the direct conversion of the electrical energy and is given by,

$$Q = VI \quad \mathrm{J/s}$$

where,

V arc voltage (V)
I Welding current (A).

The heat that is available to the weld joint depends on the welding speed v (mm/s) and the way heat is transferred from the electric arc to the base metal. Therefore, the actual heat available to the weld joint is given by,

$$H = \frac{Q}{v}\ \mathrm{J/mm}$$

However, in a real life situation there will always be some heat loss due to various factors as has been indicated in Fig. 14.1. Hence the actual heat (H_{net}) that goes into a welded joint is determined using a heat transfer efficiency η.

$$H_{net} = \eta H = \eta \frac{VI}{v} \quad \mathrm{J/mm} \tag{14.1}$$

Though the net heat as calculated in Eq. (14.1) reaches the weld joint, all of it is not used for melting since part of it would be conducted away from the joint by the base metal as shown schematically in Fig. 14.1. The net heat that is actually utilized for melting can be obtained by assuming another efficiency factor, η_m as melting efficiency as given by the expression (14.2):

$$\eta_m = \frac{\textit{Heat required to melt the joint}}{\textit{net heat supplied}} \tag{14.2}$$

Example 1 Calculate the melting efficiency in arc welding of steel with arc voltage of 20 V and welding current of 200 A. The travel speed is 5 mm/s and the cross-sectional area of the joint is 20 mm^2. Heat required to melt steel may be taken as 10 J/mm^3 and the heat transfer efficiency as 0.85.

Net heat supplied	$\eta VI = 0.85 \times 20 \times 200 = 3400$ J/s
Volume of base metal melted	$20 \times 5 = 100$ mm^3/s
Heat required for melting	$100 \times 10 = 1000$ J/s
Melting efficiency	$\frac{1000}{3400} = 0.2941 = 29.41\,\%$.

Cooling Rate

Welding is an extreme thermal process. Here intense heat is generated and simultaneously it starts to cool down till it attains room temperature as welding progresses. Thus the weld zone is subjected to a typical heating and cooling thermal cycle. The heating cycle or the heating rate is very sharp. The temperature rises beyond the melting temperature very quickly, however the cooling cycle or the cooling rate depends on various factors, namely welding speed, plate thickness, preheat temperature or interpass temperature. The cooling rate of the weld zone is of particular interest as because the metallurgical structure of the weld metal and the adjoining area, the so-called heat affected zone depends on the peak temperature attend as well as the cooling rate for a given material composition.

For relatively thick plates, the cooling rate [6] is given by Eq. (14.3),

$$R_{thick} = \frac{2\pi\lambda(T_c - T_0)^2}{H_{net}} \tag{14.3}$$

where,

T_O initial plate temperature, (°C)
λ thermal conductivity of base metal, (J/mm s °C)
R cooling rate at the weld centre line, (°C/s)
T_c temperature at which the cooling rate is calculated, (°C).

This cooling rate is at the weld centre line, whereas at the weld boundary it may be a little lower, but would be sufficiently accurate in predicting the properties of the heat affected zone. The temperature T_c at which the cooling rate is being calculated should be based on the metallurgical considerations of the base metal. For most of the steels a temperature of 550 °C is satisfactory [7].

In case of relatively thin plates, Eq. (14.4) is used to calculate the cooling rate [8],

$$R_{thin} = 2\pi\lambda\rho C_p\left(\frac{t_h}{H_{net}}\right)^2(T_c - T_0)^3 \tag{14.4}$$

where

t_h thickness of the base metal, (mm)
ρ density of base metal, (g/mm^3)

C_p specific heat of the base metal, (J/g °C).

A plate thickness factor τ is introduced as given by Eq. (14.5) to quantitatively differentiate between thick and thin plates [7]:

$$\tau = t_h \sqrt{\frac{\rho C_p (T_c - T_0)}{H_{net}}} \tag{14.5}$$

for

$\tau \leq 0.75$ thin plate equation is valid,
$\tau > 0.75$ thick plate equation is valid.

The cooling rate in case of thin plate can be expressed in terms of that of thick plate using the thickness factor τ as given by Eq. (14.6):

$$R_{thin} = \tau^2 R_{thick} \tag{14.6}$$

A typical cooling rate of 6 °C/s is found to be the upper limit for most of the steels to provide satisfactory metallurgical quality without martensitic transformations [7]. Hence from metallurgical point of view, for majority of the steel compositions, the welding speed for a given welding current and arc voltage should be so chosen such that, the resulting cooling rate does not exceed the upper limit of 6 °C/s.

Example 2 Find the best welding speed to be used for single side submerged arc welding of 6 mm thick steel plates at an ambient temperature of 30 °C with arc voltage of 28 V and welding current 400 A. The arc efficiency is 0.8 and the possible welding speeds are 10–13 mm/s. The limiting cooling rate for satisfactory performance is 6 °C/s at a temperature of 550 °C. Consider $\lambda = 0.028$ J/mm s °C and $\rho C_p = 0.0044$ J mm^3 °C

Assume the welding speed, v = 13 mm/s
Therefore the net heat input to the weld joint, $H_{net} = \eta \frac{VI}{v}$ = 689 J/mm
Plate thickness factor as given by Eq. (14.5) is, τ = 0.35.

It being less than 0.75, the plate will behave as thin plate, hence thin plate equation is used to calculate the cooling rate. Therefore as per Eq. (14.4), the cooling rate will be,

R_{thin} = 8.2 °C/s, which is higher than the limiting cooling rate.

Considering welding speed, v = 12 mm/s,
The net heat input, H_{net} = 747 J/mm
The plate thickness factor works out to, τ = 0.33.

Therefore R_{thin} = 7.0 °C/s, which is also higher than the limiting value.

Considering welding speed, v = 11 mm/s,
The net heat input, H_{net} = 815 J/mm
The plate thickness factor works out to, τ = 0.32.

Therefore R_{thin} = 5.9 °C/s, which is less than the limiting value of the cooling rate.

Hence the welding can be carried out at a speed of 11 mm/s and the brittle martensite formation is not expected.

Effect of Preheat

The same set of equations can also be utilized to calculate the preheat temperature so as to keep the cooling rate below the limiting value to avoid martensitic transformation in the weld zone.

Example 3 Consider a case of root run of a butt welding of 12 mm thick steel plate with the following parameters:

Arc voltage = 28 V, Welding current = 300 A, Welding speed = 10 mm/s, Arc efficiency = 0.8. The physical properties are taken same as in Example 2.

Without any preheat, i.e. welding at ambient temperature of 30 °C, the cooling rate works out to 34.7 °C/s, which is extremely high. This may cause martensite formation leading to cracking of the root run.

Whereas *with a preheat* of 265 °C, the cooling rate comes down to 5.7 °C/s, thus welding becomes possible without any cracking.

Heat Affected Zone (HAZ)

It is often important to estimate the extent of the HAZ. As adverse microstructural transformations may take place in the HAZ, it is preferable to choose such welding parameters which also keep the HAZ to a minimum. The HAZ is defined as that region where due to the welding heat microstructural transformations may take place. For majority of structural steels 723 °C is the recrystallisation temperature, i.e. microstructural changes will take place only in those regions where the temperature has exceeded the recrystallization temperature of 723 °C. Therefore the HAZ lies between the fusion boundary, i.e. the melting temperature to the region where it reaches 723 °C. Because of recrystallization, grain coarsening may take place in the HAZ. Coarse grain microstructure exhibits inferior mechanical properties. Hence it is preferable to keep the HAZ as small as possible.

The peak temperature attend at different locations away from the fusion zone are to be evaluated for estimating the extent of the HAZ. The peak temperatures can be calculated [6] using the Eq. (14.7):

$$\frac{1}{T_p - T_0} = \frac{1}{T_m - T_0} + \frac{\rho C_p t_h \sqrt{2\pi e}}{H_{net}} y \tag{14.7}$$

where,

T_p peak temperature at a distance of y mm from the fusion boundary (°C)
T_m melting temperature of the base metal, (°C)

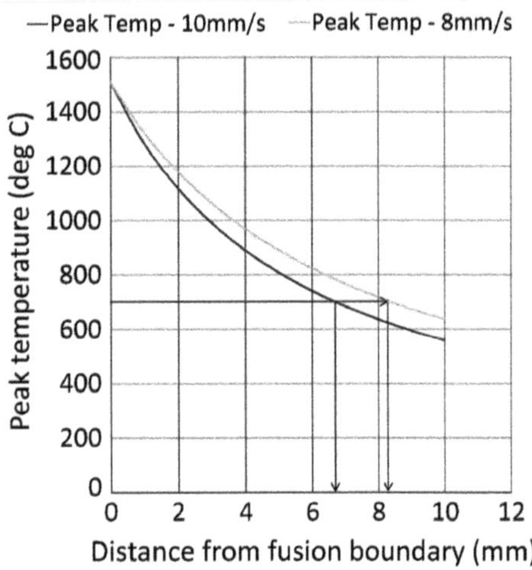

Fig. 14.8 Peak temperature distribution from the fusion boundary and perpendicular to the weld direction

e base of natural logarithm = 2.718218.

At the fusion boundary: y = 0; $T_p = T_m$.

Keeping all other parameters same, increasing welding speed, it leads to decrease in HAZ.

Example 4 Calculate the heat affected zone for Example 2 with two different welding speeds of 10 mm/s and 8 mm/s. Melting temperature of steel be taken as 1510 °C.

Using the Eq. (14.7) the peak temperature attend at intervals of 1 mm each from the fusion boundary were calculated and plotted as shown in Fig. 14.8.

From Fig. 14.8, it can be observed that with decreasing welding speed, the cooling rate decreases and thereby the extent of HAZ increases.

References

1. Essential factors in gas metal arc welding, 2015, Published by Kobe Steel Ltd., Tokyo, Japan.
2. Mandal, N. R. (2005). *Aluminum welding* (2nd edn). New Delhi: Narosa Publishing House. ISBN 81-7319-637-0.
3. Correy, T.B. (1982). What makes an electric welding arc perform its required functions. Prepared for the US Department of Energy under Contract DE-AC06-76RLO 1830.
4. http://www.twi-global.com/technical-knowledge.
5. Lu, S., et.al. (2002). Weld penetration and marangoni convection with oxide fluxes in GTA welding. *Materials Transactions, 43*(11), 2926–2931.
6. Adams, C. M., Jr. (1958). Cooling rates and peak temperature in fusion welding. *Welding Journal Research Supplement, 37*(5), 210–215.
7. Connor, L. P. (1987). Welding technology. In *Welding Hand Book* (Vol. 1, 8th edn.). Miami: American Welding Society.
8. Javeri, P., Moffatt, W. G., & Adams, C. M., Jr. (1962). The effect of plate thickness and radiation on heat flow in welding and cutting. *Welding Journal, 41*(1), 12–16.

Chapter 15
Fusion Welding Methods

Abstract Welding is a joining process by which two separate parts can be joined to make one integral part. Ideally there should be complete continuity between the parts and the joint area should be indistinguishable from the parent metal of the individual parts. The joining of two parts can be achieved if the electrons can be shared by the atoms across the interface. This is achieved by applying external heat leading to fusion of the parts to be joined. All fusion welding processes must satisfy four basic requirements, supply of energy to achieve union by fusion, mechanism for removing superficial contamination from the joint faces, avoidance of atmospheric contamination, control of weld metallurgy to achieve desirable microstructure. This can be achieved in many ways, which give satisfactory service. Not every welding process is equally suitable for all types of metals or all types of joints. It is the knowledge base of a welding engineer which helps him to decide the appropriate welding process which will satisfy all the essential and necessary fabrication and operation requirements.

Welding is a joining process by which two separate parts can be joined to make one integral part. Ideally there should be complete continuity between the parts and the joint area should be indistinguishable from the parent metal of the individual parts. This ideal situation is never achieved. However welds can be made in many ways, which give satisfactory service. Not every welding process is equally suitable for all types of metals or all types of joints. It is the knowledge base of a welding engineer which helps him to decide the appropriate welding process which will satisfy all the essential and necessary fabrication and operation requirements.

The joining of two parts can be achieved if the electrons can be shared by the atoms across the interface. This will result in an ideal welded joint. Hence the simplest welding process would be the one in which the two parts to be joined will be machined with atomic precision. When these two surfaces are brought together in vacuum, bonding between the atoms across the interface takes place. In this case the welding process might be very simple, but the surface preparation with the degree of precision and the required vacuum for the level of cleanliness is not

N.R. Mandal, *Ship Construction and Welding*, Springer Series
on Naval Architecture, Marine Engineering, Shipbuilding and Shipping 2,
DOI 10.1007/978-981-10-2955-4_15

practically feasible in case of industrial structures. However this process might become feasible in space where ultra-high vacuum is already there.

In situations of heavy construction as in shipyards, this problem of atomic contact between the parts to be joined is solved by applying external heat leading to fusion of the parts to be joined. Essentially all fusion welding processes must satisfy four basic requirements:

- A supply of energy to achieve union by fusion.
- A mechanism for removing superficial contamination from the joint faces.
- Avoidance of atmospheric contamination.
- Control of weld metallurgy to achieve desirable microstructure.

These requirements are satisfied by various welding processes which will be dealt with in the following sub sections.

15.1 Manual Metal Arc Welding

In electric arc welding, basic two movements are required, movement of the welding heat source, i.e. the welding torch and feeding of electrode, i.e. feeding of filler metal. In this process of welding both these two operations are done manually. The welder maneuvers the welding torch and the electrode.

This is an arc welding process in which union of metals takes place by fusion. The heat required to produce this fusion is obtained from electric arc that is maintained between the tip of a flux coated electrode and the surface of a base metal along the joint being welded. The welding electrodes are basically composed of a metal core and a flux cover. The metal core can be either of solid metal rod or fabricated by encasing metal powders in a metallic sheath. The metal core acts as the electrode as well as the filler rod. The primary function of the flux cover is to shield the molten metal in the weld pool and during metal transfer from the electrode tip to the weld pool from atmosphere to avoid oxidation or nitride formation. The shielding is provided by the gases generated as the coating burns and decomposes in the heat of the arc (Fig. 15.1). As the electrode melts, the flux coating burns and releases an inert gas around the molten metal that excludes the atmosphere from the weld thus prevents contamination from atmosphere. However it only protects the hot molten metal just at the location of the welding arc. The protective covering of this inert gas moves on with the electrode. However the deposited and just solidified metal still remains at a very high temperature and needs to be protected from atmospheric contamination. This shielding is provided by the molten slag which gets deposited over the molten metal and gets solidified as the welding arc moves on. This slag is generated as a result of burning of the flux coating over the electrode.

The flux coating is composed of such elements, that the slag it forms has a lesser density in molten state than that of the molten metal. It floats to the top of the weld

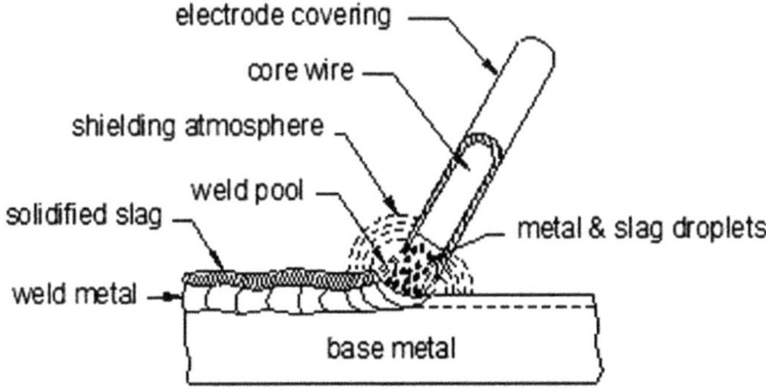

Fig. 15.1 Schematic view of shielded metal arc welding

pool and hardens to protect the weld metal until it sufficiently cools down. The layer of slag thus forms not only prevents the deposited metal from atmospheric contamination but also slows down the cooling rate and produces a more ductile weld deposit.

15.1.1 Types of Electrodes

Arc stability, depth of penetration, metal deposition rate and positional welding capability are greatly influenced by the chemical composition of the flux coating on the electrode. Each of the chemicals in the coating serves a particular function in the welding process. In general, their main purposes are to induce easier arc starting, stabilize the arc, improve weld appearance and penetration, reduce spatter, and protect the molten metal from oxidation or contamination by the surrounding atmosphere [1].

Electrodes can be divided into four main groups based on the composition of flux covering:

- Cellulosic
- Rutile
- Basic
- Metal powder electrodes

Cellulosic electrodes

These electrodes are coated with flux rich in cellulose. During welding it burns and produces hydrogen and carbon monoxide which provide the required shielding. The presence of these gases in the arc having high ionization potentials results in a high arc voltage and therefore a high arc energy. This leads to in a deeply penetrating arc and a rapid burn-off rate calling for high welding speeds. As much of the coating is of carbon origin, there is only little slag left on the weld deposit and this fact, together with strong plasma jet giving a forceful arc, makes the electrode suitable for welding in all positions. The absence of stabilizers in the coating and the high arc voltage necessitates the use of DC power and an electrode-positive polarity while welding. Welding with cellulosic electrodes may produce coarse weld deposit and with fluid slag, slag removal may become difficult.

The basic features of welding with cellulosic electrodes are:

- Deep penetration in all positions.
- Suitability for vertical down welding.
- Reasonably good mechanical properties.
- High level of hydrogen generated-risk of cracking in the heat affected zone.

Rutile Electrodes

These electrodes contain a high proportion of titanium oxide (rutile) in its coating. This with high content of ionizers makes these electrodes easy to use. The titanium oxide promotes easy arc ignition, smooth arc operation and causes low spatter. These electrodes can be classified as general purpose electrodes with good welding properties. Because of the rutile and the ionizers in the coating, these electrodes can be used with either polarity and in all positions. These electrodes are specially suitable for fillet welding in horizontal/vertical position.

The basic features of rutile electrode welding are:

- Moderate weld metal mechanical properties.
- Good bead shape produced because of viscous slag.
- Positional welding possible with a fluid slag (containing fluoride).
- Easy slag removal.

Basic electrodes

The coating of these electrodes contains a high proportion of calcium carbonate and calcium fluoride. This makes the slag more fluid than that with rutile coatings. The slag formed from the basic electrodes is of fast-freezing type. Therefore these electrodes help in welding in vertical and overhead position. The coating composition of these electrodes is produced with very low hygroscopic material. Thus the hydrogen content in the deposited metal is usually less than with other types of electrodes. These are also referred to as low hydrogen electrode. However not all low-hydrogen electrodes are strictly basic type. These electrodes are especially suitable for welding low-alloy steels susceptible to heat-affected-zone (HAZ) cracking.

The weld deposit has high resistance to hot cracking, thereby making it more suitable for welding of thicker steels and steels with higher carbon content compared to other types of electrodes. In addition to these, the weld metal has excellent mechanical properties, particularly impact property. These electrodes are therefore used for welding medium and heavy section fabrications where superior weld quality, good mechanical properties and resistance to cracking are required. As these electrodes are used for high quality applications which call for a low hydrogen content in weld deposit, the moisture content of the electrode coating should be kept to a minimum. To prevent the electrode coating from moisture absorption, they should be carefully stored and dried/backed prior to use, particularly when welding high strength low-alloy (HSLA) steels. If the atmospheric humidity is very high, it is a good practice to store the electrodes in portable driers at the work site, so that the electrodes can be directly used once they are taken out of the drier. This further eliminates the possibility of moisture absorption.

The basic features of welding with basic electrode are:

- Weld deposit with good mechanical properties.
- Low hydrogen content in weld deposit.
- Relatively fluid slag.
- Poor bead profile.
- Slag removal difficult.

Metal powder electrodes

These electrodes contain iron powder in the flux coating. This produces a marked effect on its performance. The amount of iron powder in the flux may vary from 5 to 50 %. The iron powder is added for two reasons,

- to increase deposition rate,
- to improve arc behavior.

In the conventional electrodes, the current is carried wholly by the core wire, whereas with iron powder addition in the flux the coating also becomes conductive near the arc thus providing an additional path to the current. As a result, the arc tends to spread out and metal deposition takes place over a wider area. This however reduces the current density at tip thus reducing the penetrating force of the arc causing less penetration. Welding with metal powder electrode provides a smoother, more stable arc, giving improved sidewall fusion, flatter welds and less undercut and spatter.

Higher deposition rates are achieved by increasing the iron powder content in the flux coating, however iron powder addition beyond 50 % causes deterioration in the behavior of the electrode as the coating fuses away unevenly. In effect the higher deposition rate is actually achieved not because of additional metal in the flux but because of the ability to carry more current for the same core wire diameter. With increase in welding current, generally the extent of spatter increases, however with iron-powder electrodes, the increase in spatter with increasing current is much less as compared to that of conventional electrodes.

The performance of rutile electrodes is significantly improved by addition of iron powder in rutile coatings. All position welding becomes possible with such electrodes which is normally not possible with the fluid rutile slag. It produces a sound welded joint with good mechanical properties. These electrodes can be used both for heavy fabrications as well as for welding thinner plates.

The basic features of welding with metal powder electrodes are:

- Increased deposition rate.
- More stable arc.

- Flatter weld deposit.
- Less penetration.
- Less chances of undercut.
- Less spatter.
- Improved side wall fusion.

15.1.2 Basic Features

This is one of the most widely used versatile welding process, particularly for short welds in production, maintenance and repair work and for field construction. The merits of the process are:

- Simple, portable and inexpensive welding equipment.
- Adequate shielding for the entire welding process is obtained from the electrode coating itself.
- This process is suitable for outdoor work as it is less sensitive to the effect of wind flow.
- Can be used in areas of limited access.
- The process is suitable for most of the commonly used metals and alloys.

The electrodes are produced in lengths of 230–460 mm with varying diameter of the filler core metal. In this process the whole length of the electrode is part of the welding circuit, hence current flows through the entire length of the electrode. The typical welding circuit for SMAW is shown in Fig. 15.2. The maximum current that can be used is, therefore, limited by the electrical resistance of the core wire of the welding electrode. Excessive current may overheat the electrode burning and damaging the flux coating. This will deteriorate the arc behavior and the shielding. The deposition rate of this welding process is generally lower compared to other welding process such as GMAW.

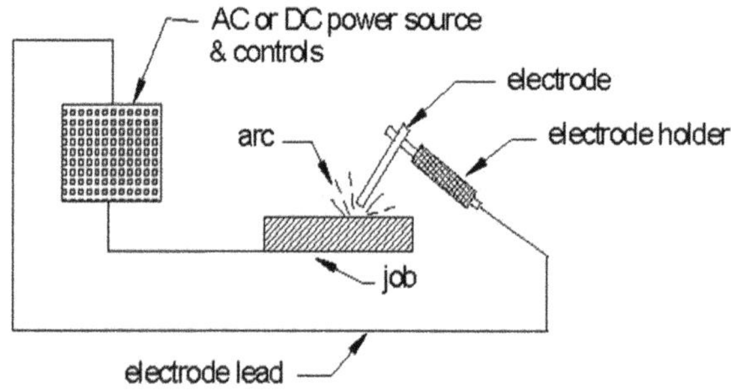

Fig. 15.2 Typical SMAW circuit

15.1.3 Operating Variables

A constant-current power source is preferred for SMAW. It being a manual welding process, there can be variations in arc length causing variation in arc voltage, however with CC power source, it will keep the current fairly constant. Thus uniform metal deposition can be obtained. However in out of position welding and also in cases where root gap is not uniform due to bad fit-up or edge preparation, a CP power source is preferable, where the highly skilled welder can vary the deposition rate depending on the requirement by simply changing the arc length. Both AC and DC power can be used depending on requirement.

The operating variables and other factors that need attention are addressed below:

Welding current
Direct current provides better operating characteristics with a stable arc especially with lesser diameter electrodes and at low currents. Most coated electrodes operate better on DCEP (reverse polarity). Deeper penetration can be achieved with DCEP, whereas DCEN gives higher electrode melting rate, thus a higher deposition rate can be achieved.

Voltage drop
The voltage drop in welding cables is less with AC as compared to DC. Therefore for carrying out welding at a site away from power source, AC is preferred.

Arc initiation
Initiating the arc is generally easier with DC, particularly if small diameter electrode is used. With AC, striking the arc presents a problem because the current drops to zero once in each half cycle.

Magnetic arc blow
Magnetic arc blow is experienced generally when direct current is employed for welding. When current flows through a conductor, it produces a magnetic flux that circles around the conductor in a plane perpendicular to the conductor. The buildup of the flux causes a deflection of the arc column as it pulls away from this heavy concentration of magnetic force. The arc column at the electrode acts as a flexible conductor. This concentration of magnetic flux deflects the arc from its path. This phenomenon is called arc blow. Magnetic fields created by the flux can never be removed but they can be controlled.

This can be avoided/reduced by various methods, e.g. relocating the ground connections to the structure being welded; by welding away from the ground connection; welding toward a heavy tack or portion of the weld already completed; etc.

Welding position
DC performs better than AC for vertical and overhead welds. However with suitable electrodes satisfactory welds can be produced in all positions with AC.

Metal thickness
Both thin and thick sections can be welded with DC. However for thin section AC is not very suitable because arc conditions at low current levels as required for thin sections are less stable with AC than on DC.

15.2 Gas Metal Arc Welding

In this welding process an external gas is used to shield the arc and the molten metal pool from atmospheric oxygen and nitrogen. An arc is maintained between a continuous filler electrode and the weld pool in the base metal. The filler metal is a continuous bare electrode wire fed through a wire feeder and a welding "gun". The gun delivers both the shielding gas and the electrode wire. Gas metal arc welding (GMAW) can be operated either in semiautomatic or automatic modes. In automatic mode the welding torch is mounted on a trolley which moves at the required speed. In semi-automatic mode this process can be used for welding in all position with appropriate shielding gas, electrode and welding parameters. A GMAW nozzle is schematically shown in Fig. 15.3.

15.2.1 Process Characteristics

In GMAW, the operator needs to set the voltage controlling the arc length, wire feed speed which controls the current, direction and positioning of the welding torch. Once set, the arc length and the current are automatically maintained. Generally the GMAW equipment provide for self-regulation of the electrical

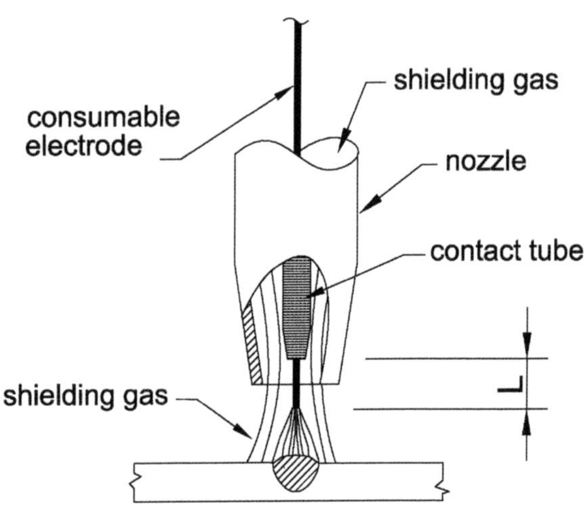

Fig. 15.3 Schematic of a GMAW nozzle

characteristics of the arc. In GMAW of steel, DCRP (DCEP) is preferred as it gives sufficient localized heating of both filler and base metal resulting in good bead characteristics and superior weld penetration.

Arc-length control
The self-regulation of the arc length is achieved through constant potential power source. In GMAW generally a CP power source along with a constant-speed electrode feed unit is used to achieve the desired self-regulation. There is another option of using a constant-current power source with electrode feed unit which is arc-voltage controlled. In first case the current adjusts to the changed situation and in the second case, the wire feed speed is changed to maintain constant arc length, thus achieving self-regulation of arc length.

However, in case of aluminum welding, a combination of a constant-current power source and a constant-speed electrode feed unit is used. This combination provides a small degree of self-regulation and requires a high degree of operator's skill. However, because of the high thermal conductivity of aluminum a highly skilled welder can have a wide range of control over the current which in turn controls the arc energy.

15.2.2 Metal Transfer Characteristics

The mechanism of weld metal transfer from the electrode through the arc column to the weld pool has been dealt with in detail in Chap. 13, Sect. 13.2.

In ship construction primarily normal and some higher strength steels are used extensively. Along with this to save in structural weight in weight restricted construction, marine grade aluminum alloy is also used. GMAW as a method of joining both steel and aluminum is widely used both in semiautomatic and automatic mode. In case of steel welding CO_2 is used as the shielding gas, mainly because it is inexpensive and at the same time it gives desired weld quality. However using CO_2 as the shielding medium, the metal transfer remains in the globular mode only, as because the transition current with CO_2 is very high. Hence for a given welding condition, the transition current becomes much higher than the current required for achieving necessary fusion for welding. The limitations of welding with globular transfer mode are excessive spattering and inability to do all position welding, mainly vertical and overhead. Hence the only option remains is to use short circuit or dip transfer mode of metal transfer in GMAW of steel using CO_2 as shielding gas.

For aluminum alloy welding argon is used as shielding medium. Here spray transfer is obtained leading to a spatter free welding.

Optimum short circuiting transfer conditions
Various combinations of open-circuit voltage, and wire feed rate can produce short-circuiting transfer of weld metal. However for best results, an optimum combination of voltage and wire feed rate need to be chosen. Plotting the various combinations of arc voltages and wire feed speed for good dip transfer situation,

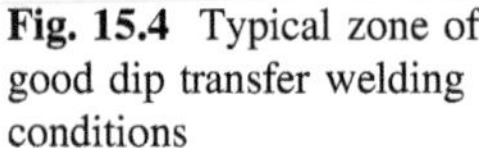

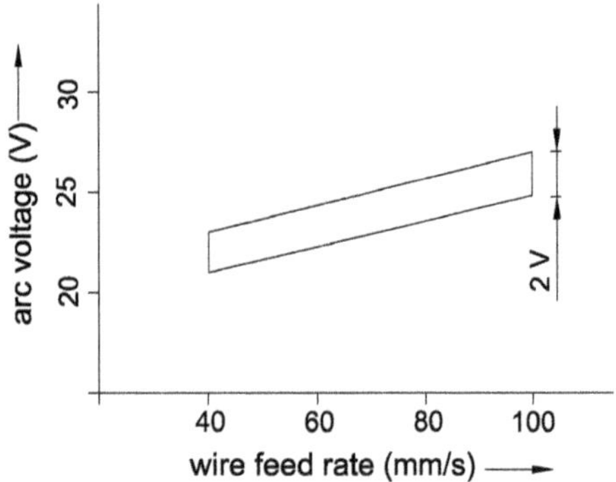

Fig. 15.4 Typical zone of good dip transfer welding conditions

one can see that the zone of good dip-transfer arc welding conditions is about 2 V wide only as shown in Fig. 15.4.

These data are needed to set the welding parameters to get the optimum condition for dip transfer of weld metal in GMAW. However the other practical ways of setting these optimum parameters are as follows:

At a given voltage, the wire feed speed is slowly increased until the electrode wire begins to "stubout" on the workpiece. From that point, the feed rate is slowly decreased till the erratic action of the arc disappears. The ratio of voltage to wire feed speed will then be just about optimum. The other option would be, for a given wire feed rate, to slowly reduce the voltage output of the power source till the arc begins to behave erratically and stubout. At this point the voltage is increased by 2–3 V such that the arc becomes stable. This method also will give the optimum parameters for dip-transfer welding.

Typical welding currents for short-circuiting transfer in steel are 50–150 A for 0.76 mm wire, 75–175 A for 0.89 mm wire, and 100–225 A for 1.1 mm steel wire using DCRP (DCEP) power source.

Spray transfer

Spray transfer occurs at welding currents above the transition point as explained before. In this mode the metal transfer occurs in the form of very small droplets that are accelerated axially across the arc gap from the electrode tip at the rate of several hundreds of droplets per second. Spray transfer is achieved with argon or argon with 0.5–5 % oxygen as shielding gas. A true spatter-free metal transfer becomes possible with DCRP (DCEP) power supply.

The spray transfer yields a highly directed stream of metal droplets with substantial energy so as to have velocities which overcome the effects of gravity. Because of this, spray transfer mode can be used for welding in any position. The metal droplets being very small, short-circuit does not occur and hence spatter is virtually eliminated. The spray transfer welding is widely used and almost any metal or alloy can be welded because of the inert gas shielding.

15.2.3 Operating Variables

The weld bead geometry, depth of penetration and overall weld quality depends on the following operating variables:

- Welding current (wire feed speed)
- Polarity
- Arc voltage
- Welding speed
- Length of stickout
- Electrode orientation
- Electrode diameter

Welding current
The control of welding current is normally coupled to wire (electrode) feed speed. Current is changed with change of wire feed speed. In gas metal arc welding the electrode is consumable, i.e. the arc is struck between the electrode wire and the job. In the heat of the arc plasma, the electrode melts and gets deposited, so the electrode is the filler material here. At low current levels, for different electrode size, the variation of current with wire feed speed is fairly linear particularly for lower diameter electrodes. However with increase in current level this gradually becomes non-linear. This is due to the resistance (joule) heating of the electrode extension beyond the contact tube.

With all other variables held constant, increasing welding current (electrode feed speed) will increase:

- Depth and width of the weld penetration,
- Deposition rate,
- Size of weld bead.

Arc force and deposition rate being exponentially dependent on current, operation above the transient current may make the arc unstable in the vertical and overhead positions. By reducing the average current with pulsing, the arc forces and deposition rates can both be reduced, allowing welds to be made in all positions and also in thin sections. With metal cored wire additional advantage is that it produces an arc which is less sensitive to changes in electrode stickout and voltage compared to solid wires. Thus the process becomes more tolerant to operator's torch fluctuations.

Polarity
In GMAW applications mainly direct current electrode positive (DCEP) polarity is used. This gives a stable arc, smooth metal transfer, good weld bead, less spatter and higher depth of penetration for a wide range of welding currents. Whereas in case of direct current electrode negative (DCEN) polarity, the metal transfer mode becomes globular causing excessive spatter and porosity in the deposited weld bead. That is why DCEP is almost always used in GMAW.

Arc length (Arc voltage)

The arc length is a critical variable in GMAW. The arc length determines the current and arc pressure distribution on the weld pool, which in turn determines the size and the shape of the fusion zone. When the length is too short, the electrode can contact or short circuit to the weld pool, which will result in a reduced base metal melting, a highly convex and narrow weld deposit. On the other-hand excessive arc length causes a wider and shallow deposit having less penetration. It makes the arc to wander increasing spatter and may cause porosity due to inadequate gas shielding because of too long an arc. Increasing arc length makes the weld bead wider due to the widened arc area at the weld surface and consequently reduces the reinforcement height because the same volume of the filler metal is involved. Conversely reducing the arc length makes the bead narrower and increases the height of reinforcement. Arc length is an independent variable. Arc voltage depends on the arc length as well as many other variables, such as the electrode composition and dimensions, the shielding gas, welding technique and even the length of the welding cable. The voltage across the contact tube and the work piece is measurable and is generally referred to as arc voltage.

The power in the arc column increases with increase in welding current for a given arc length. At the same time, power increases with increase in arc length at a constant current. The power in the arc column directly affects the melting of the electrode as well as the formation and detachment of the droplets from the electrode. Therefore, the arc power affects the size and frequency of the droplet detachment, and the solidification of the weld pool, along with the resulting microstructure and mechanical properties of the weld. Since the arc power varies with both arc length and current, both current and arc length (arc voltage) are to be controlled, specially in case of constant current applications in order to maintain a constant arc power and thereby achieving a consistent weld quality.

Welding speed

The speed of movement of the welding torch is the welding speed. As this speed is increased initially the thermal energy input per unit length of weld transmitted to the base metal increases. This happens because the arc at every instant acts directly on the base metal and not on the molten pool. However with further increase in welding speed, less thermal energy gets transmitted to the base metal per unit length of deposited weld metal. Therefore, melting of base metal first increases and then decreases with increase in weld speed. If the speed is further increased, this will cause undercutting along the edges because of insufficient deposition of filler metal being deposited along the welding path.

On the other hand when the welding speed is decreased, the filler metal deposition per unit length increases with increase in rate of heat input. On further slowing down, the arc impinges on the molten weld pool, rather than on the base metal. In the process, the weld penetration does not increase, as the molten metal has lower thermal conductivity and results in a wider weld bead and higher rate of heat input. This increases the heat affected zone (HAZ), weld induced deformation and residual stresses. Hence, with all other welding parameters held constant, maximum weld penetration is achieved at an intermediate welding speed.

Length of stickout

Length of stickout is the length of electrode extension, 'L' from the end of the contact tube as shown in Fig. 15.3. Higher length of stickout will offer higher electrical resistance and will cause a higher voltage drop. This higher resistance will also cause a resistance heating of the electrode resulting in a small increase in electrode melting rate. However, the increase in the length of stickout increases the overall resistance in the welding power circuit, thereby reducing the welding current. The power source generally used in GMAW systems has a flat or CP characteristics and the electrode feed rate is constant. This constant potential system compensates for the variation in the length of stickout by automatically supplying the increased or decreased welding current to attain equilibrium. The power source provides the proper current so that the melting rate is equal to the electrode feed rate.

Increased length of stickout increases electrode melting rate, resulting in an wider weld bead, whereas depth of penetration increases with decrease in length of stickout. This occurs due to the constriction of the distributed heat and current flux by the decreased arc length, which results in a larger electromagnetic force, which promotes the convection effect in the weld pool. Thus the length of stickout is an important operating variable which can control the weld deposit and fusion zone geometry.

Electrode orientation

The electrode orientation with respect to the weld joint affects the weld bead shape as well as weld penetration. The influence of electrode orientation on weld bead shape and its penetration is more than that of arc voltage or welding speed. The electrode orientation is defined in two ways,

(i) By the relationship of the electrode axis with respect to the direction of travel. This is known as the travel angle.
(ii) The angle between the electrode axis and the adjacent work surface. This is called work angle.

When the electrode points against the direction of travel, the technique is called backhand welding with drag angle. When the electrode points in the direction of travel, it is called forehand welding with a lead angle. The effect of electrode orientations and their effect on weld profile and penetration are shown in Fig. 14.6. Backhand welding with a drag angle of about 25° in the flat position of the job, gives the maximum penetration. When it is changed to forehand welding with a lead angle, with all other parameters held constant, the penetration decreases and the weld bead becomes wider and flatter. The drag technique apart from deeper penetration also produces a more convex and narrow bead. It offers a more stable arc with fewer spatters. For all positions, the electrode orientation is generally backhand with a drag angle in the range of 5°–15° for achieving good control and shielding of the molten weld pool. However, for aluminum, a lead technique is preferred. The lead technique provides a cleaning action ahead of the molten weld

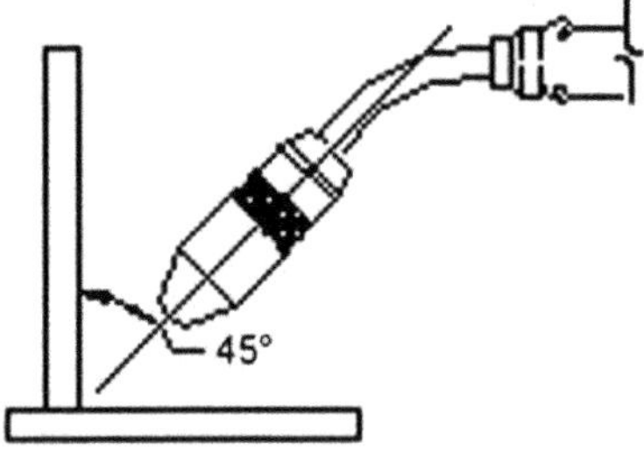

Fig. 15.5 Normal work angle for filet welds

metal, which promotes better wetting and reduces base metal oxidation. For fillet welds in horizontal position, the electrode should be positioned about 45° to the vertical member, thus the work angle is 45° as shown in Fig. 15.5.

Electrode diameter
The weld bead shape also depends on the electrode diameter. Higher current is required for higher diameter electrode to achieve the same metal transfer characteristics. It results in higher heat generation and therefore higher electrode melting causing increased weld metal deposition. However overhead and vertical welding are usually done with smaller diameter electrode with lower currents. The transition current for spray transfer depends on electrode diameter. As the welding current should be more than the transition current to have a spray transfer, welding of thin materials requiring low current should be done with wires of small diameter.

Operating variable selection
Proper selection of operating variables becomes more difficult because of the interdependence of several variables. The practical way of selection of these parameters is to carry-out some trial runs to determine an acceptable set of conditions.

15.2.4 Shielding Gases

The primary function of shielding gas is to protect the arc and the molten weld pool from atmospheric oxygen and nitrogen. If it is not properly protected, it may cause oxide and nitrite formation along with causing porosity and weld embrittlement. In addition to this, the shielding gas and its flow rate also contributes to a number of important functions,

- Forms the arc plasma,
- Stabilizes the arc on the metal surface,
- Ensures smooth transfer of molten metal droplets from the wire to the weld pool.
- Arc characteristics
- Mode of metal transfer
- Penetration and weld bead profile

Table 15.1 Typical shielding gas composition for GMAW of steels and non-ferrous metals

Steels	Non-ferrous metals
100 % CO_2	
Argon + 2 to 5 % O_2	100 % Argon
Argon + 5 to 25 % CO_2	100 % Helium

- Speed of welding
- Undercutting tendency
- Cleaning action
- Weld metal mechanical properties.

The shielding gases mainly used for welding of steels and non-ferrous metals are shown in Table 15.1.

Carbon dioxide

Carbon dioxide is the only active gas used in gas metal arc welding. It is extensively used in its pure form for shielding of metal arc welding of carbon and low alloy steels used in shipbuilding. With CO_2, higher welding speeds with greater joint penetration can be achieved. However, the metal transfer mode with CO_2 shielding is either short circuiting or globular. With globular transfer, the arc produces a high level of spatter. Hence to reduce the spatter in CO_2 welding, the welding parameters are to be such that it produces a very short arc with the tip of the electrode actually below the surface of the work. The CO_2 shielded arc produces a weld bead with high penetration and rough surface profile. It gives a sound weld deposit, however the mechanical properties of the weld deposit may get adversely affected due to the oxidizing nature of the arc.

One of the major advantage of CO_2 as a shielding medium is, it is much less expensive compared to other inert gases.

Argon-oxygen

It is difficult to strike an arc in pure argon atmosphere [2] and an unstable metal transfer is often observed in these arcs [3]. The addition of oxygen to argon generally improves the initiation of arc and enables a more stable metal transfer. The addition of oxygen mainly alters the transport phenomena at the anode (electrode) and the cathode (work piece weld pool). Jönsson et al. [4] have observed that the addition of oxygen leads to improved arc stability by:

- Forming nonmetallic inclusions of oxides, that help in arc initiation.
- Forming films that improve the wetting conditions of the weld pool,
- Affecting the surface tension gradient such that deeper and narrow weld pools are formed.

The addition of oxygen does not affect the electrode melting rate significantly [5]. The addition of up to 5 % oxygen to an argon shielding gas affects the characteristics of the arc column to only a small extent. However the influence of O_2 addition on transport phenomena at the anode needs further investigation.

Argon-Carbon dioxide

The droplet diameter in a shielding medium of 95 % argon with 5 % carbon dioxide is smaller and the frequency of droplet transfer is higher, compared to that of with a gas mixture of 98 % argon with 2 % oxygen at all current levels [6].

5 % CO_2 in 95 % argon provides for a higher oxygen content than that in the gas mixture of 98 % argon with 2 % oxygen. With increase in oxygen content, the surface tension of the liquid metal reduces. The liquid metal droplet at the electrode tip detaches when the static detaching forces (gravitational, electromagnetic and plasma drag forces) exceed the static retaining force due to surface tension. Therefore with reduction in surface tension forces in argon-CO_2 gas mixture, it results in smaller droplet sizes with higher droplet transfer frequencies compared to the argon-O_2 gas mixture.

Merits of GMAW

The advantages of GMAW can be summarized as follows:

- It provides a continuous feeding of electrode thus reducing the number of starts and stops over a certain weld length as encountered in SMAW.
- It can be used to weld almost all commercial metals and alloys.
- All position welding can be done with GMAW.
- With GMAW, significantly higher deposition rates can be achieved as compared to SMAW.
- A higher welding speed can be achieved because of continuous feeding of electrode and higher metal deposition rates.
- Deeper penetration compared to SMAW is possible in spray transfer mode.
- Smaller size fillet welds of adequate strength can be achieved.
- No necessity of slag removal, hence minimum post weld cleaning is required.

These advantages make GMAW an efficient welding process for high production welding and also make it a suitable candidate for automated welding applications.

Limitations of GMAW

Like any other process, GMAW also has certain limitations as mentioned below:

- The welding equipment is less portable and more complex and expensive than that of SMAW.
- It is difficult to reach congested or hard-to-reach places with a GMAW torch than a SMAW electrode holder.
- For proper shielding of the molten metal, the GMAW torch should be close to the joint (1–20 mm).
- Air draft may disperse the shielding gas. This limits its outdoor use.

15.3 Submerged Arc Welding

This welding method derives its name from the fact, that the welding arc is submerged in a pool of flux. Thus it is a welder friendly process, since the extremely intense arc remains altogether hidden under flux. It is an automatic down hand welding method. In this, both feeding of electrode and movement of welding nozzle are motor driven. Once the speed of the carriage holding welding wire spool and the wire feed rate are set, on switching on of the machine the welding continues in the pre-determined path. Along with this a small hopper is also mounted on the carriage containing the flux material, which continuously fed to the welding zone shielding the welding arc as well as the deposited metal. A typical submerged arc welding (SAW) set up is shown in Fig. 15.6.

As the arc is completely covered by the flux, heat loss is minimum. This provides a thermal efficiency as high as 80–90 %. It produces no visible arc light, welding is spatter free and there is no need for fume extraction. The flux, apart from shielding the arc and the molten pool from atmospheric contamination, plays an important role on:

- providing a stable arc,
- controlling the composition of the deposited metal by providing alloying elements to the weld pool. Thus enhancing the mechanical properties of the deposited weld metal.

Fig. 15.6 A typical SAW setup

15.3.1 Process Characteristics

SAW is usually operated as a fully mechanized or automatic process. After the initiation of the arc, the electrode as a filler wire is continuously fed by an wire-feeder to the weld pool at a set feed rate through a feeding nozzle and the same is moved automatically along the weld seam. Simultaneously flux is continually fed from a hopper in front of and around the electrode sticking out of the nozzle. The heat generated by the electric arc progressively melts and burns some of the flux, the end of the electrode wire and the adjacent edges of the base metal, creating a pool of molten metal below a layer of liquid burnt flux (slag). The molten slag floats over the molten metal and thus completely shields the molten zone from the atmosphere.

As the welding arc advances along the seam, the deposited metal and the liquid slag covering it cools down and solidifies, forming a weld bead with a protective slag shield over it. This slag shield results in a slower cooling rate for the deposited weld metal and thus provides an annealing effect to the weld deposit. Figure 15.7 shows a schematic representation of a submerged arc welding process.

It is a high deposition, down hand welding process capable of making welds with current up to 2000 A, AC or DC, using single or multiple wires or strips of filler metal. Normally, the submerged arc welding process is known as a process

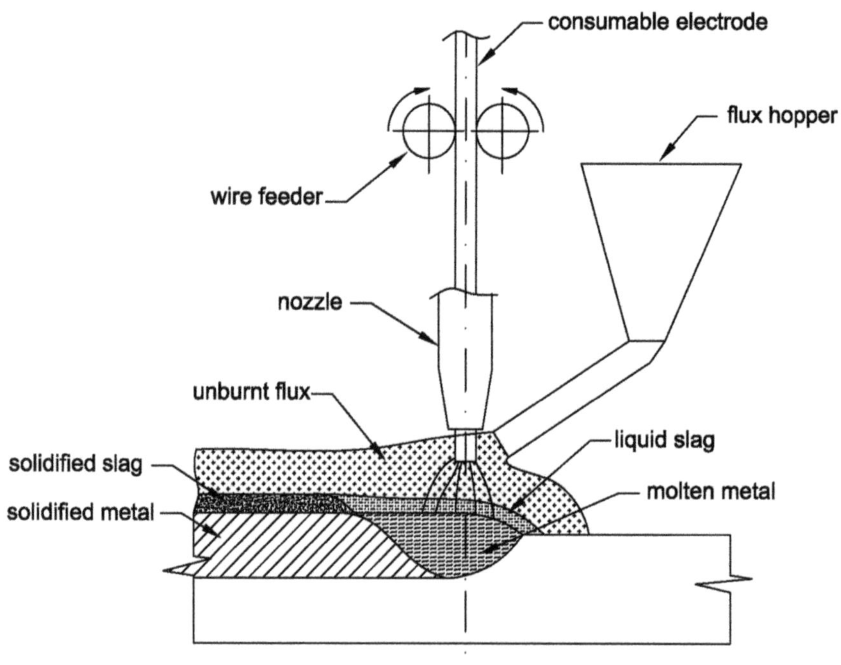

Fig. 15.7 Schematic representation of submerged arc welding process

that achieves deep penetration. The submerged arc process produces high-quality welds having excellent bead profile. The elimination of fumes, smoke, and any visible arc column adds to its ease of operation and efficiency, which, in turn, encourages its application in the shipbuilding industry.

Power sources

Several types of power supply are suitable for submerged arc welding. SAW being generally a high current process with high duty-cycle, a power supply capable of providing high amperage at 100 % duty cycle is recommended. A constant-voltage (CP) power supply, being self-regulating, can be used with a constant speed wire feeder. The current in the welding circuit is coupled with the speed control of the wire feeder. As the wire feed speed is increased, automatically suitable increase in welding current takes place. With increase in current, the arc voltage also needs to be increased suitably. This is achieved by increasing the open circuit voltage of the power supply. Mostly DCEP polarity is used for it gives deeper penetration and uniform weld deposit with a good bead shape.

15.3.2 Operating Variables

Proper control of the operating variables is very much essential to achieve high production rates and welds of good quality. The operating variables are:

- Welding current
- Arc voltage
- Welding speed
- Electrode diameter
- Electrode extension (length of stick out)
- Electrode polarity

Welding current

Welding current is the most important parameter of SAW. Since the amount of heat generated is directly proportional to the square of the current. It controls the melting rate of the electrode and thereby controls the deposition rate. It also directly influences the depth of penetration and the extent of base metal fusion.

If the welding current is too high at a given welding speed it will lead to:

- excessive depth of penetration resulting into melt through the metal being joined,
- excessive reinforcement (bead height),
- increased heat affected zone,
- increased weld induced distortion.

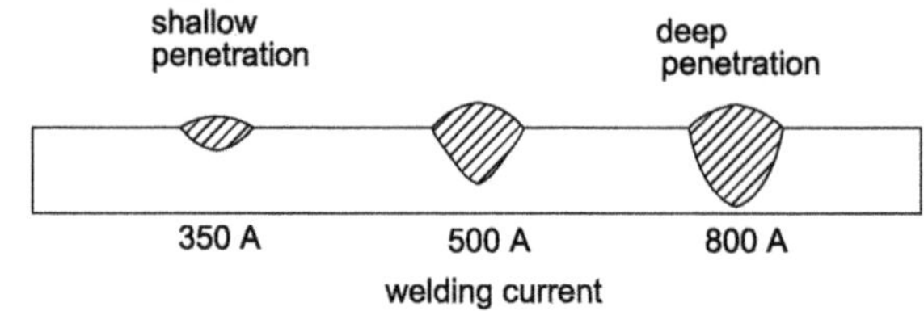

Fig. 15.8 Effect of current variation on weld bead shape and penetration

If the current is too low for the same welding speed, it may cause:

- inadequate penetration or incomplete fusion,
- unstable arc.

The effect of increasing welding current on depth of penetration is shown in Fig. 15.8.

The metal transfer process from the electrode in SAW is much less known as compared to that of GMAW. Different forms of transfer are known to occur with different welding current and other process variables. Experimental investigation [7] revealed that at 500 A, a very chaotic, nonaxial globular metal transfer involving frequent explosions and bursts takes place in both AC and DC polarities. In none of the experimental investigations, spray transfer of metal was observed. At 1000 A current, the electrode tapered off in a similar way to that of spray transfer in GMAW, but no steady stream of small droplets was observed.

Arc voltage

The voltage drop across the arc length between electrode and base metal is the arc voltage. With increase in the open circuit voltage, the arc length increases causing increase in the arc voltage. The electrode melting rate and hence the deposition rate is not much affected by arc voltage. It has a direct influence on the shape of fusion zone and external bead appearance.

If the arc voltage is increased keeping current and welding speed constant, it will result in,

- a flatter and wider bead,
- increased flux consumption,
- reduction of porosity caused due to rust or scale on the base metal,
- increased tolerance of root opening arising out of poor fit-up. It will help to bridge an excessive root opening,
- increase in pickup of alloying elements from flux.

If the arc voltage is increased excessively, it will:

- produce a very wide weld bead which is more prone to cracking,
- make slag removal difficult in groove welds,
- increase undercut along the edges of fillet welds.

If the arc voltage is low, it will:

- improve penetration in a deep weld groove,
- resist arc blow.

However, excessive reduction in arc voltage will produce a high and narrow bead and may affect arc stability. The effect of welding voltage on the weld bead shape is shown in Fig. 14.3.

Welding speed

The welding speed is another important variable, which has a pronounced effect on weld size and penetration for a given combination of welding current and arc voltage. If the welding speed is increased:

- heat input per unit length decreases,
- less filler metal is applied per unit length of weld leading to less weld reinforcement,
- weld bead becomes smaller.

Excessively high welding speeds may lead to:

- undercut, porosity and uneven bead shape,
- arc blow.

If the welding speed is decreased:

- Heat input per unit length of weld is increased.
- More filler metal is consumed.
- Leads to more weld reinforcement.
- Increased penetration.
- Provides more time for gases to escape thus reducing porosity.

If the welding speed is further reduced, i.e. excessively slow speed will produce:

- A convex bead shape, more prone to cracking.
- A large molten weld pool resulting in a rough bead and slag inclusions.
- Weld penetration is not further increased because much of the molten weld pool will be beneath the arc. The penetrating force of the arc gets cushioned by the molten pool thus restricting further penetration.

Effects of welding speed on the bead shape and penetration is schematically shown in Fig. 15.9.

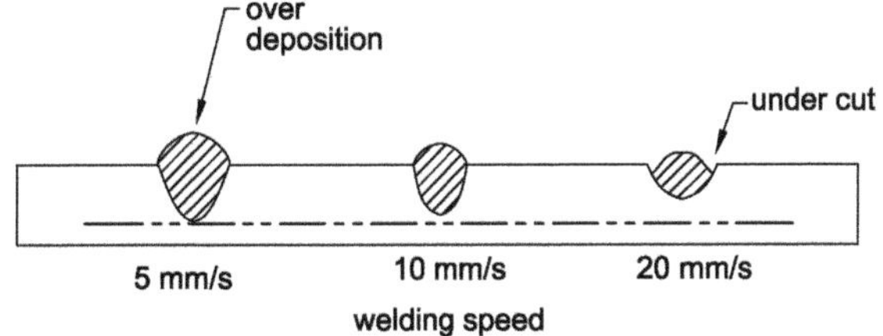

Fig. 15.9 Effect of welding speed on weld bead shape and penetration

Electrode diameter
At any given current, a smaller diameter electrode will have a higher current density. It gives higher penetration as compared to that of a larger diameter electrode. However, a higher diameter electrode can carry much higher current than lesser diameter electrode, thus it can produce higher deposition rate at higher welding current. In case of poor fit-up, a large diameter electrode is better suited compared to thinner one for bridging large root openings.

Electrode extension
Electrode extension or length of stickout becomes an important operating variable for current densities above 125 A/mm^2. The melting rate increases because of resistance heating of the electrode between the contact tube and the arc at higher-current densities. Thus deposition rates can be increased from 25 to 50 % by using long electrode extensions with no change in welding current. For procedure development an electrode extension of about seven times the electrode diameter is a good starting point. The length is to be suitably adjusted to achieve the desired melting rate. The melting rate of the electrode comprises that due to the heat generated by the arc at zero electrode extension and that due to the resistance heating effect of the electrode extension. For design purpose the electrode melting rate m_r (g/s) can be estimated from the Eq. (15.1) as given in [8]:

$$m_r = \frac{I}{60}\left[0.35 + \frac{d^2}{645} + 2.08 \times 10^{-7}\left(\frac{I \times L_e \times \pi}{d^2}\right)^{1.22}\right] \mathrm{g/s} \quad (15.1)$$

where,

d electrode diameter (mm),
L_e length of stickout (mm),
I welding current (A).

Since increase in electrode extension adds to the resistance component of the electrode in the welding circuit, it causes an increased voltage drop across the electrode thus reducing the arc voltage. Hence the bead shape changes with change in electrode extension. It is increased to take advantage of higher melting rate, however the voltage setting of the power supply should be accordingly increased to maintain proper arc voltage to achieve proper bead shape.

However, an increase in deposition rate is accompanied by a decrease in penetration. Therefore, where deep penetration is required it is not recommended to have a long extension of electrode. On the other hand, where melt through may occur as in case of thin plates, a longer electrode extension is preferable. A long stickout is normally only used in cladding and surfacing applications where there is greater emphasis on deposition rate and control of penetration, rather than accurate positioning of the wire.

15.3.3 Flux

Fluxes used in SAW are granular fusible minerals containing oxides of manganese, silicon, titanium, aluminum, calcium, zirconium, magnesium and also other compounds such as calcium fluoride. The flux is specially formulated to be compatible with a given electrode wire type so that a given wire-flux combination will yield a metal deposition having the required chemical composition providing desired mechanical properties. The flux under the arc heat melts, burns and reacts with the molten metal in the weld pool, resulting in diffusion of alloying elements from the flux to the weld metal, thus enhancing the chemical composition and mechanical properties of the weld deposit. It is a common practice to refer to fluxes as 'active' if they add manganese and silicon to the weld deposit. The amount of manganese and silicon that gets added is influenced by the arc voltage and the welding current level. Fluxes, used in SAW, are primarily of two types:

Bonded flux—This type of flux is produced by drying the ingredients, then bonding them together with a low melting point compound such as sodium silicate. These bonded fluxes generally contain metallic deoxidisers, which help in preventing weld porosity. That is why these fluxes are effective in case of welding of plates having rust and mill scale.

Fused flux—This is produced by mixing the ingredients, then melting them in an electric furnace to form a chemically homogeneous product, subsequently cooled and crushed and ground to the required particle size. This type of flux provides for smooth and stable arc with consistent weld metal properties even at welding currents as high as 2000 A.

Width and depth of flux

The bead appearance and soundness of the finished weld also depend on the width and depth of the granular flux layer. If the layers are too deep, the gases generated during welding can not readily escape and may result in a rough weld bead. If the layer is too shallow, the arc may not get fully submerged causing flashing and spattering of molten metal as well as exposing the molten metal to atmospheric contamination. It will result in a poor bead appearance and may cause porosity. The optimum flux depth can be achieved by slowly increasing the flow of the flux until the arc is fully submerged and flashing no longer occurs. The unburnt flux is recycled back to the flus hopper. The flux prior to use should be backed at appropriate temperature for a recommended period of time to remove moisture. Moisture, if present in the flux, will cause porosity in the weld deposit and also may cause hydrogen embrittlement.

15.3.4 Estimation of Welding Parameters

Square butt

For carrying out welding of square butts as shown in Fig. 15.10, the following conditions are to be satisfied,

$$b_r/h_r \geq 6 \text{ and } b_r/h_f \geq 2 \tag{15.2}$$

The joint geometry satisfying the above conditions provides better means of gas escape and also reduces the possibility of development of cracks.

The first approximations of the SAW parameters for developing a welding procedure can be made from the following as proposed by E. Richter [9].

Welding Current (electrode positive)

$$I = 112\sqrt{(h_f d)} \quad \text{A} \tag{15.3}$$

where,

h_f expected fusion depth (mm)
d electrode diameter (mm).

Arc voltage

$$V = I/4(d+1) \quad \text{V}. \tag{15.4}$$

Deposition rate

$$D_r = I^{1.36}/(400\sqrt{d}) \quad \text{kg/h} \tag{15.5}$$

Welding speed

$$S_w = (35D_r)/G \quad \text{mm/s} \tag{15.6}$$

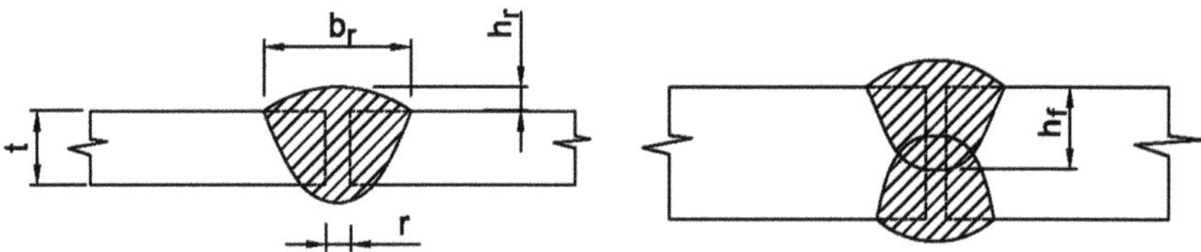

Fig. 15.10 Fusion zone geometry of square butt

where,

$$G = \text{sectional area of fusion zone}\,(\text{mm}^2)$$
$$= (rh_f) + \frac{2}{3}(b_r h_f)$$

The depth of penetration achieved with the welding parameters calculated above is given by,

$$(h_f)_{cal} = 3.931 \times 10^{-3} \frac{I^2}{\sqrt{V} \times (S_w)^{0.4} \times d} + 0.8r \qquad (15.7)$$

If $(h_f)_{cal} > h_f$, then the welding current is to be reduced or the root gap should be decreased and again $(h_f)_{cal}$ should be checked and accordingly the parameters are to be set.

V-groove butt and fillet weld
For carrying out butt joint with V-groove or a fillet joint at a predetermined welding speed with a given electrode diameter it is recommended that the sectional area of deposited metal per run should not exceed 50 mm^2.

The ratio of the sectional area of required weld deposit to the sectional area of the deposited metal in a single run gives the number of welding runs N_w.

i.e. $N_w = G/G_S$.
where, G_S = sectional area of weld deposit in a single run ≤ 50 mm^2.
Deposition Rate

$$D_r = G_S \times S_w$$

Welding Current *(electrode positive)*

$$I = (110 D_r \sqrt{d})^{\frac{1}{1.36}}$$

(electrode negative)

$$I = (130 D_r \sqrt{d})^{\frac{1}{1.2}}$$

Welding Voltage

$$V = \frac{I}{4(d+1)}$$

The above mentioned formulae as proposed by Richter [9] are valid under the following conditions:

Electrode diameter, d = 2–5 mm
Welding current, I = 200–1000 A
Arc voltage, V = 20–50 V

Welding speed, S_w = 3–16 mm/s
Length of stick out L_e = 35 mm
Electrode orientation: perpendicular to the plate.

15.3.5 Cracks in Submerged Arc Welds

Submerged arc welding is a high deposition and an efficient method of welding. I t is widely used in panel lines in shipyard. However adequate care needs to be taken such that cracks do not occur in these welds. A case study is given below to illustrate the causes of cracking.

It involves a panel line in a shipyard where single side submerged arc welding was carried out on a 4 m long, 10 mm thick C-Mn steel plates using glass fibre tape backing. The tape lay within a flat groove in a copper backing bar which was maintained in close contact with the plates by air pressure. A square edge preparation was used with a root gap of 4 mm. The weld was sound apart from the final stretch of 400 mm, where a centre line crack occurred.

The welding conditions were as follows:

Flux: Agglomerated alumina, semi-basic type; Wire: 3 mm diameter, 0.5 %Mn;
Welding current: 520 A; Arc voltage: 36 V; Electrode polarity: DCEN;
Electrode extension: 30 mm; Travel speed: 6.7 mm/s.

Two transverse samples were cut; one each from sound part and cracked part of the weld. These samples were polished and etched to observe the fusion zone cross section. Measurements of the cross sections showed that the size and shape of the weld in both the section were same. Therefore, it could be concluded that the dilution and chemical composition of the weld were identical. Spectrographic analysis of the cracked specimen showed that the composition of the weld was: 0.11 % C, 0.020 % S, 0.022 % P, 0.25 % Si, 0.56 % Mn and 0.002 % Nb.

The factors that affect solidification cracking are: weld metal composition, weld solidification pattern (which depends on the shape of the weld), and the strain on the solidifying weld. A parameter UCS (units of cracking susceptibility) [10] based on the weld metal composition is used to check the cracking susceptibility of a weld deposit:

$$UCS = 230C + 190S + 75P + 45Nb - 12.3Si - 5.4Mn - 1$$

UCS levels less than 10 indicate a high resistance to cracking and levels above 30, a low resistance. Within this range of UCS, the risk of cracking increases with increasing depth to width ratio, made at higher welding speeds and also having poor fit-up.

In case of fillet welds, having depth to width ratio of about one, UCS values of 20 and above will indicate a risk of cracking. For butt welds, UCS of 25 and above are critical. If the depth to width ratio is decreased from 1 to 0.8, the allowable UCS can be increased by about nine. However, very low depth to width ratios, such as obtained when penetration into the root is not achieved, it also promotes cracking [11].

In butt welds, with acceptable bead shape, UCS values less than 25, cracks are not expected. The shape of the weld influences the solidification pattern and to minimize cracking the columnar grains of the solidifying metal should appear in an upward pattern rather than inwards [11].

In the present case the UCS was 23.7, as such cracks were not much expected, however cracks did appear towards the end of the welded panel. In single side welding with glass fibre backing, the solidification pattern of the weld leads to an unfavourable grain structure. The columnar grains tend to grow inwards causing a more pronounced centreline segregation of impurity elements and also concentrate the contraction strain in the same region. This appears to be the reason behind the cracks that developed in the weldment.

To avoid cracking in the present case, special care is to be given to choose the right kind of wire flux combination having low carbon and sulphur, and high manganese and silicon contents. For this case study it was recommended that the wire be replaced by one having higher manganese content so that the weld metal would contain approximately 1.0 %Mn. The cause of center line cracking may have been the increasing stress on the weld due to contraction of the weld metal further back in the joint leading to development of excessive tensile stress nearing the end of the weld line. The exact cause of the problem, however, was not identified, but a change in wire composition solved the solution.

15.4 Electrogas Welding

Electrogas welding (EGW) is a further development of electroslag welding as well as gas metal arc welding process. Unlike resistance heating of molten slag as in case of electroslag welding, the electrode is melted by the arc heat in an inert gas environment, very similar to that of GMAW process. This method proved to be very useful for vertical welding of thick plates up to a thickness of about 100 mm in a single pass. It led to large cost savings compared to conventional manual GMAW process for welding of very thick steel plates.

Similar to that of gas metal arc welding, solid or cored wire electrodes are used. This method produces a smaller heat-affected zone (HAZ) and somewhat better notch toughness as compared to electroslag welding. In this method the length of electrode stick-out is comparatively much greater. It gives an advantage of extra melting of electrode wire leading to high deposition rate and produces less molten base material thereby resulting in lesser HAZ. In ship building use of thicker plates and their requirement of vertical welding is well known. With the advent of larger and larger container ships, bulk carriers and alike, the welding requirement of higher and higher thickness of steel plates is increasing. This calls for welding processes with higher production efficiency as well as higher weld quality. In such situations EGW proves to be a very viable option because it is capable of a single pass, vertical position welding that is highly efficient.

15.4.1 Salient Features

The salient features of electro gas welding are as follows:

- It is a fully automatic welding process.
- It welds only in vertical direction.
- The weld pool is bound either by two sliding water cooled copper shoes or a sliding water cooled copper shoe in the front and ceramic backing strip at the back.
- The filler wire electrode is fed through a copper nozzle.
- The electrode is melted by a shielded arc.
- Welding starts from a run-up plate and ends on a run-off plate.
- Constant current DC power supply is generally used.
- DCEP polarity is used.

15.4.2 Process Characteristics

With initiation of the arc, the electrode and the plate edges along the vertical groove melts forming a weld pool covered sideways with copper shoes. With further feeding of electrode in the weld pool the molten volume increases. The melting rate of electrode and thereby the metal deposition rate is synchronised with the upward movement of the welding nozzle and the copper shoes holding the molten weld pool. In the process the molten material at the lower end of the groove solidifies and the process continues till it reaches the upper end.

In order to avoid lack of fusion at the beginning of the welding, the process is started on a run-up plate which also closes the bottom end of the groove. Once the welding is started, it continues till the upper end of the plates onto the run-off plate. Synchronisation of the feeding of electrode and moving up of the shoes is extremely important. Any mismatch will lead to arc instability or stubbing of electrode in the molten pool. In either of the cases it might lead to stoppage of the arc. On restarting the arc it may give rise to inadequate fusion causing welding defects locally. Hence any interruptions of the welding process should be avoided. Suitable power sources are rectifiers with a slightly dropping static characteristic. The electrode has a positive polarity, i.e. DCEP polarity.

The conventional EGW operates with rather higher diameter of filler electrode about 2.4–3.2 mm as compared to that of GMAW process. For welding higher thickness material this often requires high heat input resulting in wider sectional area of the groove. This leads to coarse microstructure in the heat affected zone

(HAZ). The application of conventional EGW for low alloyed steels is not recommended, as the heat affected zone with coarse grain structure may not satisfy the required toughness standards. This happens due to long exposure to higher temperatures of more than 1500 °C along with low crystallisation rates. This can be overcome with high speed EGW. In this process, the gap cross-section is reduced and metal powder is added to increase the deposition rate. By increasing welding speed, the dwell times of weld adjacent regions above critical temperatures is reduced. Thus the brittleness effects are significantly decreased.

15.4.2.1 Single Electrode EGW

Further development took place in the 1970s in the form of narrow groove welding. To reduce the heat input, a fine diameter wire is used with a lower welding current and a narrow groove. Also, to attain a sound weld bead formation near the front and back surfaces of the joint, the welding wire (nozzle) is oscillated in the plate thickness direction (Fig. 15.11). The development of this welding process enabled

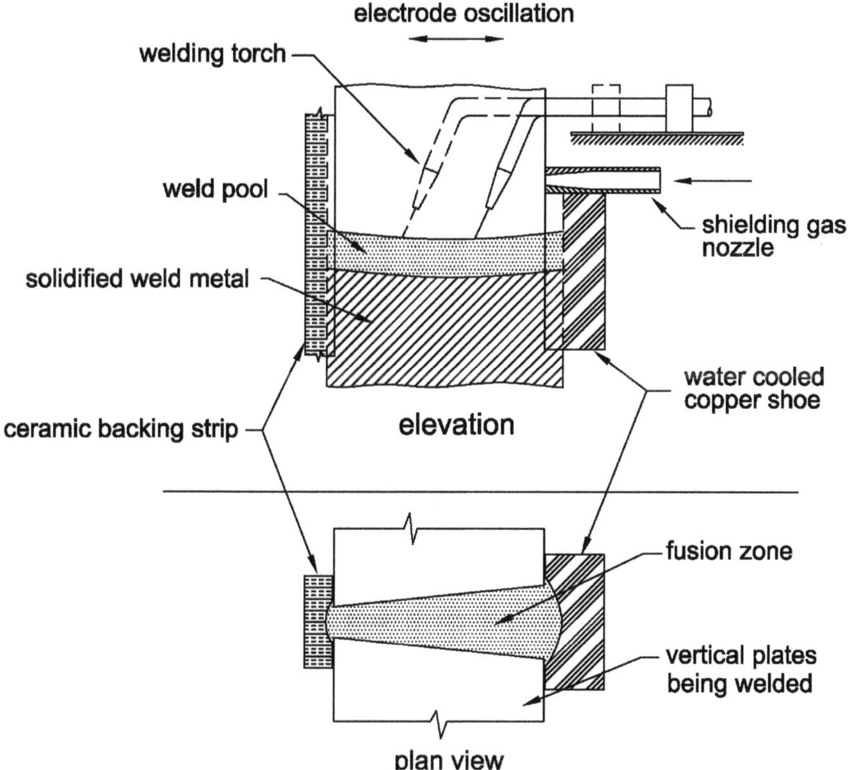

Fig. 15.11 Schematic description of single electrode EGW

proper welding with adequate reinforcement on both sides of the plate. At the same time it reduced the welding heat input and the amount of weld metal, thereby HAZ toughness and welding efficiency substantially increased. It gives a narrow weld bead and HAZ. This method can be gainfully used to weld steel plates up to 50 mm thick.

However, with the use of thicker steel plates beyond 50 mm, as frequently used in large container ships and oil tankers, it becomes extremely difficult to attain the required joint quality satisfying the desired mechanical properties with oscillating single electrode EGW. It has been observed as the thickness of the steel plates exceeds about 50 mm, welding defects (such as a lack of fusion) are generated when using the oscillating single electrode welding process. This happens as the dwell time of arc becomes insufficient to fully fuse the groove faces using oscillating single-electrode EGW process [12].

15.4.2.2 Double Electrode EGW

To address the welding requirement of such thick plates a 2 electrode oscillating EGW process is used (Fig. 15.12). The 2-electrode EGW process applies one more electrode to the single-electrode welding process. Depending on the plate thickness, the additional electrode can be added to change the configuration into two-electrode EGW device.

Direct current power source is used for 2-electrode EGW process. One of the electrode is made positive and the other is kept negative. This provides for a stable arc. If both electrodes have the same polarity, then the arc becomes highly unstable and leads to high amount of spatter. Also with electrode positive wider bead is formed with less penetration whereas negative polarity gives narrow bead with deeper penetration [13]. Hence to weld a wide single V groove, the electrode on the wider grooved surface is made positive and the electrode on the rear side of the groove is made negative. Typical welding parameters and the joint geometry are shown in Table 15.2 [12] and Fig. 15.13 respectively.

The shape of backing strip and the consumable welding electrodes play an important role in the subsequent quality of the weld deposit. The type of welding wires (consumable electrodes) used also determines the amount of slag generated during EGW. The geometry of the ceramic backing strip, as shown in Fig. 15.14, is generally used.

The groove made in the backing strip accommodates the slag that is generated from the use of flux cored welding wire. However to keep the slag generation to a minimum and at the same time taking advantage of flux cored wire, the first electrode near the face is of flux cored type and the second electrode at the back is of solid wire type. The shielding gas used is CO_2. With the use of double electrode, the amount of shielding gas consumed per welding length is approximately halved compared to that of single electrode welding. Thus this method of welding works out to be more economical.

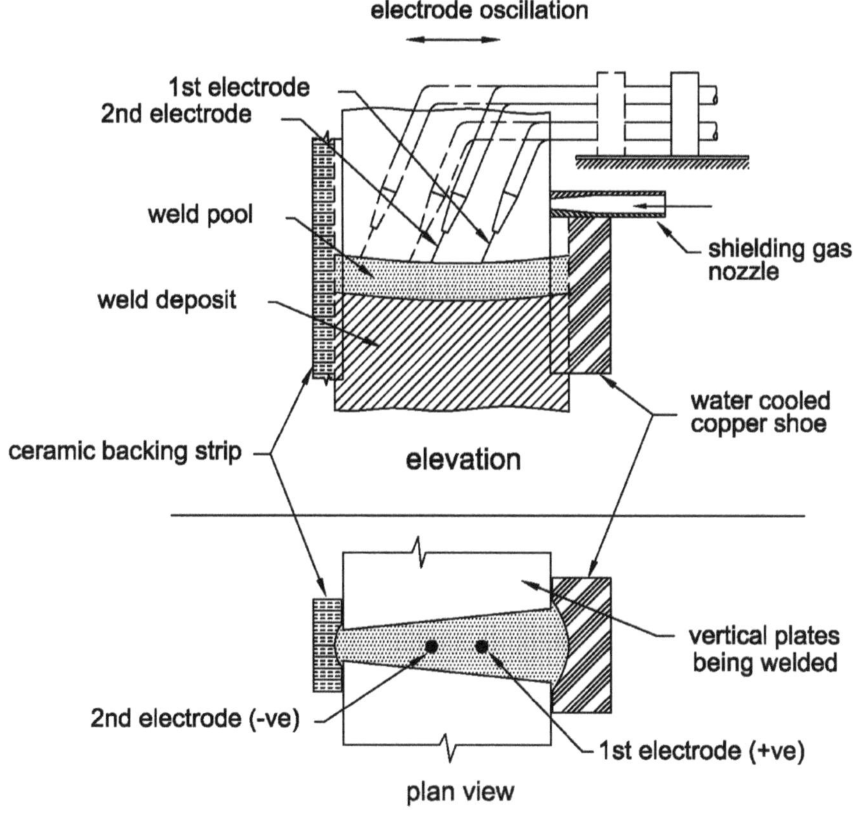

Fig. 15.12 Schematic description of single electrode EGW

Table 15.2 Typical double electrode EGW parameters

t (mm)	Electrodes	Welding wire	Welding parameters			Electrode spacing (mm)	Osc. width (mm)	Shielding gas	
			I (A)	V (V)	S (mm/s)			CO_2 (%)	Flow rate (l/min)
50	1st 2nd	Flux cored wire Solid wire	410 400	41 40	1.2	15	5	100	30
60	1st 2nd	Flux cored wire Solid wire	410 400	41 40	1.03	15	17	100	30
70	1st 2nd	Flux cored wire Solid wire	420 460	42 42	0.97	15	35	100	30

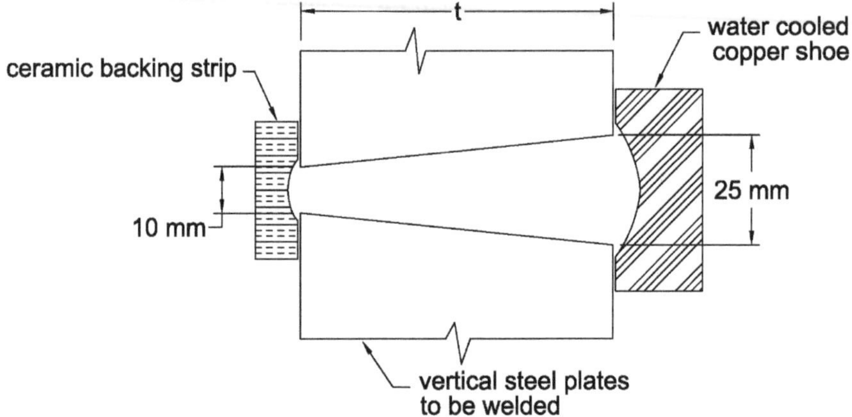

Fig. 15.13 A typical V-groove edge geometry for EGW

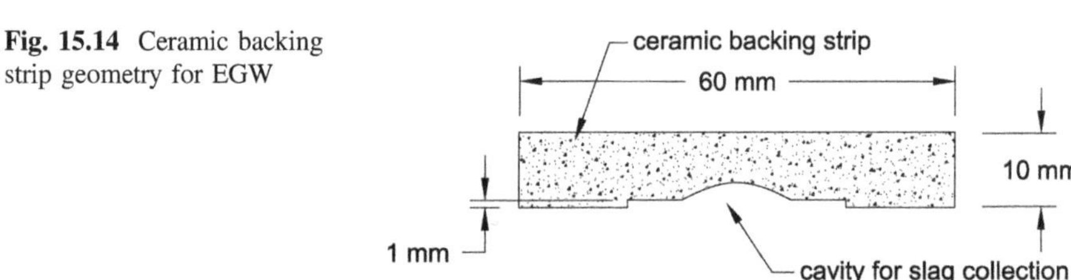

Fig. 15.14 Ceramic backing strip geometry for EGW

15.4.3 Application and Welding Performance

Double wire EGW with appropriate welding consumables and backing strip is a highly efficient process for single pass welding of 50–70 mm thick steel plates. This method of welding produces weld beads with adequate top and bottom reinforcements along with proper penetration to the groove surface and without the defects of poor penetration or lack of fusion. The deposition rate that can be achieved compared to that of the single electrode EGW is almost twice and at the same time the set up time required for both the processes remains more or less the same. Thereby the process efficiency of double electrode process is nearly double to that of the single electrode process. The welding length that can be carried out at one continuous stretch also is doubled in case of double electrode welding. In double electrode EGW, the welding speed is nearly double to that of the single electrode process thereby the amount of shielding gas consumed per unit weld length becomes nearly half. Thus the double electrode process works out to be more economical.

The double electrode EGW is a highly efficient automatic welding method for vertical position welding of steel plates having thicknesses of 50–70 mm. This process results in stable weld metal penetration in single-pass. With these qualities this method of welding is gainfully applied in shipbuilding applications especially for ultra large vessels where the plate thicknesses are 50 mm and beyond.

15.5 Electroslag Welding

Electroslag welding (ESW) unlike other fusion welding processes is not an electric arc welding process. Here fusion of base metal as well as filler metal is achieved from the heat generated by the electrical resistance of the molten slag. The weld pool is shielded by this floating molten slag, which moves along the joint as welding progresses. The process however is initiated by an electric arc. Under the heat of this arc, the granulated flux burns and melts to form the slag. The arc then gets extinguished because the molten slag provides the conducting path to the current. The slag remains in the molten state by the heat generated by its resistance to electric current passing through it between electrode and the base metals.

The consumable electrode is continuously kept dipped in the molten slag. It thus melts in the heat of the molten slag and gets deposited as the welding progresses. Electroslag welding is a vertical position automatic welding process, once started, it should continue uninterrupted to completion. If for some reason, the welding stops in between, it becomes very difficult to restart from that point and also likely to induce weld defect in the form of lack of fusion at that zone.

Since during welding no arc is involved, the welding action is quiet and spatter free. It is essentially a single pass welding process irrespective of plate thickness. Extremely high deposition rates can be achieved with ESW and thus very thick sections can be welded in a single pass. A high quality weld deposit can be achieved through this process. Since the heat is somewhat evenly distributed over the entire plate thickness, it does not lead to any form of angular distortion.

The molten slag bath over the weld pool acts both as the heat source as well as a shielding medium. The molten slag with a temperature up to 2000 °C washes the joint edges, melting into them to produce a weld with a dilution as high as 50 %. The melted electrode and base metals form a molten pool beneath the molten slag and slowly solidify to from the weld. A progressive solidification takes place from bottom upwards, and always there is a molten metal pool over the solidifying weld metal.

In this method of welding, the electrode wire is generally fed from top in between the two edges of the plates being welded. Thus a suitable guide system is needed to prevent the wire from short circuiting with the plate edges. This is achieved with either a non-consumable guide or a consumable guide.

One uses a wire electrode with a non-consumable guide cum contact tube and the other uses an electrode with a consumable guide extending down to the molten slag.

15.5.1 Process Characteristics

Electrode with non-consumable guide

In this method one or more electrodes are fed to the molten slag pool depending on the thickness of the plates being welded. Water-cooled copper shoes are normally used on both sides of the plates being joined to contain the molten metal and the molten slag. These shoes move vertically upwards as the welding progresses. The speed of this vertical movement determines the welding speed and it is consistent with the electrode melting and deposition rate. The contact surfaces of the shoes are shaped concave, which provides the required amount of reinforcement on the weld deposit. As the shoes move up, the solidified weld deposit gets exposed. The exposed surfaces remain covered with a fine layer of slag. Hence small amounts of flux must be added at intervals to compensate for this consumption of slag. The ESW is shown schematically in Fig. 15.15. A constant potential DC power source is used with current rating of 750–1000 A. Load voltages generally range from 30 to 55 V.

In this method the metal deposition takes place from electrode as well as from the consumable guide if it is used. The consumable guide provides approximately 5–15 % of the filler metal. As the welding progresses and the molten slag pool rises, the consumable guide melts and becomes part of the weld deposit. The electrode wire is directed to the molten slag pool by the guide tube extending the entire length of the joint.

In this method, the welding equipment remains stationary and non-sliding shoes are used. For welding over short lengths, shoes are used having length equal to that of the joint. For longer joints, several sets of such shoes are used. As the welding progresses and the weld metal solidify, the shoes from below are removed and

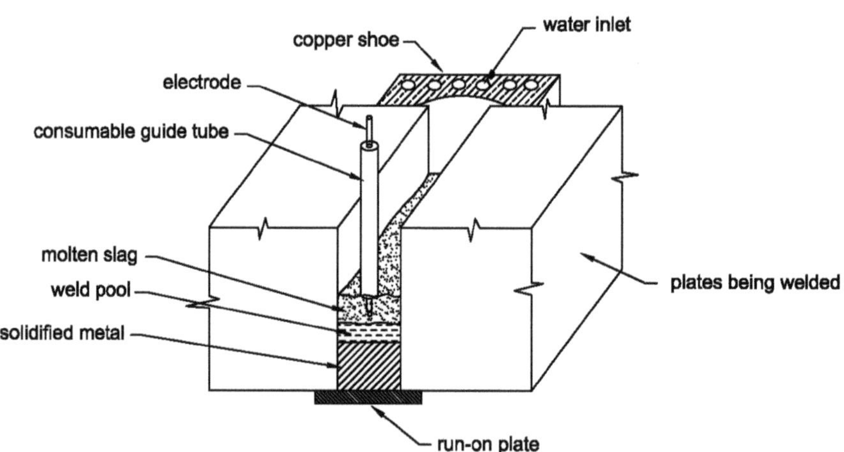

Fig. 15.15 Schematic presentation of ESW process

placed above the top shoes. It is repeated till the completion of welding of the entire joint.

This method can be very effectively used to weld plates ranging in thickness from approximately 20–460 mm [14]. As the welding current is carried by the guide tube, it needs to be insulated. The required insulation may be provided by one of the following means:

- Flux coating of the outer surface of the guide tube.
- Doughnut shaped insulators.
- Fibreglass sleeves.
- Insulating tape.

15.5.2 Operating Variables

In electroslag welding it is essential to understand the effect of each operating variable since they differ from those with conventional arc welding processes.

Form factor

In electroslag welding generally a higher volume of molten metal is present in the weld pool. At the same time the welding speed being slow, the solidification of the molten metal takes place at a slower rate. The welding starts from the bottom and moves upward with progressive solidification of the molten weld pool. The resulting grain structure, determined by the angle at which the grains meet at the centre, of the solidified metal depends on the shape of the molten weld pool. Weld pool shape is expressed by a "form factor". It is the ratio of the weld pool width (w) to its maximum depth (d). Width is calculated as the root opening plus the side wall fusion, i.e. penetration into the base metal as shown in Fig. 15.16. The depth is

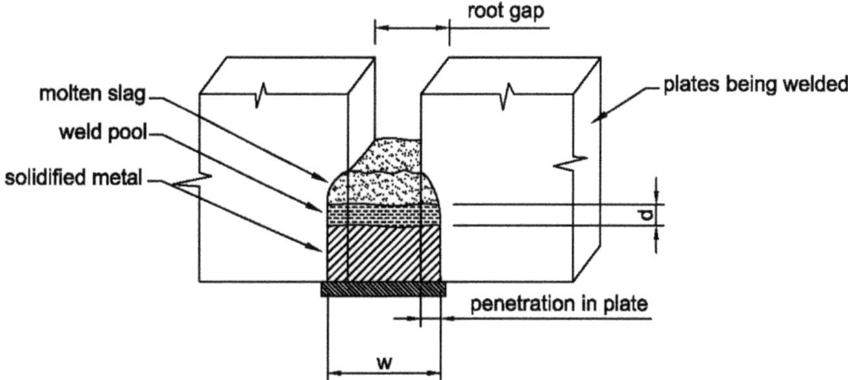

Fig. 15.16 Molten pool width and depth

calculated as the distance from the lowest level of the molten metal to the top of the molten weld metal. $Formfactor = \frac{w}{d}$

A weld deposit having high form factor will solidify with the grains meeting at an acute angle. Whereas grains will form an obtuse angle in case of low form factor. Solidified metal with grains at acute angle has high resistance to cracking. The same is low in case of grains meeting at obtuse angles. Therefore, high form factor gives maximum resistance to cracking. However, form factor alone does not control cracking. The base metal composition (especially the carbon content), the filler metal composition and joint restraint have a significant effect on crack formation. The shape of the molten pool and hence the form factor also depends on the other welding variables.

Welding current

Increasing welding current implies increased heat input and results in increased depth of molten weld pool. Thus with increasing current the form factor decreases lowering its resistance to cracking. Generally currents in the range of 500–700 A are used with 3.2 mm diameter electrode. For metals more susceptible to cracking, current below 500 A is used.

Welding voltage

In electroslag welding process, welding voltage is an extremely important variable. It has a significant effect on base metal fusion (depth of penetration) as well as on the process stability. As voltage increases, the base metal fusion increases (sideways depth of penetration), thus increasing the weld width. Therefore it increases the form factor and thereby increases resistance to cracking. However, for process stability, voltages within 32–55 V per electrode are generally used. Too high a voltage may lead to slag spatter and arcing may take place on top of the molten slag pool. With inadequately lower voltage, short circuiting or arcing to the weld pool may occur. Welding of thicker sections, higher voltages are used.

Electrode extension

Electrode extension is the length of the electrode extending out of the electrode guide tube. It is measured as the end of the guide tube to the surface of the molten slag pool. With constant-voltage power source having constant-electrode feed speed, resistance will increase with increase in the electrode extension. This results in some drop in welding current, which slightly increases the form factor.

Electrode extensions of 50–75 mm are generally used. Overheating of guide tube may take place with electrode extension less than 50 mm. Whereas electrode extensions higher than 75 mm will cause increase in electrical resistance of the electrode leading to overheating of the electrode. Hence with long electrode extensions, the electrode may melt at the surface of the molten slag pool instead of inside the slag pool. This will lead to improper slag pool heating and may result in process instability.

Slag pool depth

A minimum slag pool depth is necessary to ensure adequate electrode immersion and melting of electrode inside the slag pool. A shallow slag pool will cause slag spitting and arcing at the surface, whereas too deep a slag pool offers a larger area for heat transfer into the retaining shoes and base metal. This reduces the overall temperature of the molten slag pool which reduces the weld width and hence the form factor. Higher slag pool depth also may lead to poor circulation of the cooler slag at the bottom which may tend to solidify at the bottom of the slag pool causing slag inclusion. Optimum slag pool depth is about 40 mm, but it can also be as low as 25 mm or as high as 50 mm without much significant effect.

Root gap

Root gap should be enough to provide adequate clearance for the guide tube and its insulation. However for sufficient slag pool size and good slag circulation, a minimum root gap is required.

Increased root gap does not affect the weld pool depth. However, it increases the weld width and hence the form factor. Excessive root gap will require extra amount of filler metal which may become uneconomical. Also it may give rise to lack of adequate edge fusion. Depending on base metal thickness, number of electrodes and the use of electrode oscillation, generally 20–40 mm of root gap is used.

As welding progresses up the joint, the shrinkage forces developed may pull the plates reducing the root gap. Therefore for long welds, the root gap at the top of the joint should be approximately 3–6 mm more than that at the bottom to allow for this shrinkage effect. The effect of shrinkage depends on material type, joint thickness and joint length. Thus proper shrinkage allowance needs to be provided. The use of proper root gap and shrinkage allowance is important for maintaining the dimension of the welded structure. However, if an incorrect root gap is applied, the same may be rectified within limits by changing the welding parameters. For example, if the initial root gap is too small, the wire feed speed may be reduced, to reduce deposition rate and to increase penetration. If the root gap is too large, the wire feed speed may be increased. Regardless of the type of base metal, the voltage should be increased for wider root gaps.

For excessive root gaps, for achieving adequate metal deposition, additional electrode may be added if space is available. Electrode oscillation also helps in compensating for wider root gaps. On the other hand, when the root gaps are too small, the joint may fill too fast, causing weld cracks or lack of edge fusion. In case of small root gaps, it may further become smaller due to shrinkage effect and causing stoppage of oscillating guide tube from traversing.

Electrode oscillation

Oscillation of electrode is generally carried out horizontally across the thickness. The oscillation of electrode distributes the heat and thus helps in achieving more uniform heat distribution to obtain better edge fusion. Oscillation speeds vary between 8 and 40 mm/s. Higher oscillation speed is needed with higher plate thickness. Generally the oscillation speeds are based on a traverse time of 3–5 s.

Increase in oscillation speed reduces the weld width and hence the form factor. Therefore the oscillation speed needs to be balanced with other parameters. A dwell period of 2–7 s is used at each end of the oscillation travel. It is required to achieve complete fusion at each end and to also overcome the chilling effect of the retaining shoes.

15.5.3 Merits and Limitations

Merits

- Ability to weld very thick plates in single pass.
- Very high deposition rate. It can achieve a deposition rate of 15–25 kg per hour per electrode.
- Preheating, even for materials with high hardenability, is normally not required.
- Produces high quality weld deposit.
- Minimum joint preparation and fit-up requirements.
- High-duty cycle.
- Minimum materials handling, once started it continues to completion.
- Elimination of weld spatter, which results in 100 % filler metal deposition efficiency.
- Low flux consumption 1 kg of flux consumed for each 20 kg of weld metal.
- Minimum distortion and no angular distortion.
- ESW is the fastest welding process for large and thick material.

Limitations

- Only carbon and low alloy steels and some stainless steels can be welded by ESW.
- Only vertical or near vertical joints can be made.
- Once the welding has started, it must continue to completion otherwise defects are likely to occur.
- ESW cannot generally be used on materials thinner than about 12 mm.

15.5.4 Application

Electroslag welding is used in ship building mainly for joining the vertical side shells of two adjacent blocks. Plate thicknesses of 12–40 mm are commonly found in side shells and the weld length varies from 12 to 20 m depending on the size of the ship. The plate thickness will also vary with ship size.

15.6 Single Side Welding

One of the primary activities in shipyards, is butt welding of flat or curved panels. The conventional method of welding of such panels consists of several operations like welding from top, turning over of the panel, gouging of root followed by final welding. Usually these panels have very large overall dimensions making it difficult for turning over operation. In such cases overhead back gouging followed by overhead welding is carried out. This results in an uneven weld deposit and slower rate of production. The uneven bead shape calls for additional operation of grinding resulting in further manhour requirement. This difficulty can be overcome by implementing single side submerged arc welding using suitable backing strip.

Implementation of single side welding will result in improved productivity and this productivity will be further enhanced if the required welding can be achieved by single run. However this has to be achieved with desired level of quality of bead profile and mechanical properties of the welded joint.

15.6.1 Process Variables

To achieve a sound welded joint, adequate metal deposition and required dilution of the parent metals have to be done. This is achieved by suitably choosing the operating variables. From the point of view of production and distortion control the preferred joint geometry is a square butt. Welding of a given joint geometry can be carried out by various combinations of the operating variables. However, to obtain adequate metal deposition with proper top and bottom reinforcement, an optimum combination of the variables along with suitable wire flux combination is to be selected.

As an example [15], the process variables and joint geometries for single side single pass submerged arc welding of low carbon micro alloyed steel of thickness range from 3 to 12 mm are detailed in Table 15.3.

Table 15.3 Process variables used in the single side single pass submerged arc welding

Process variable	Unit	Plate thickness		
		3 mm	5 mm	12 mm
Welding current	A	250–350	500–550	675–700
Arc voltage	V	22–24	22–24	30–34
Welding speed	mm/s	11–12	9–10	5–6
Electrode diameter	mm	4	4	4
Length of stick out	mm	20	25	30
Root gap	mm	0–0.5	1–2	3.5–4.25
Electrode polarity		DCEP	DCEP	DCEP

Root gap
Root gap plays an important role in achieving adequate root fusion and bottom reinforcement. The root gap is varied from 0 to 4.25 mm. Lesser than required root gap may lead to incomplete penetration. Whereas a wider root gap will ensure full penetration but it will be accompanied with excessive bottom reinforcement. The optimum root gap for single aide single pass submerged arc welding of 3, 5 and 12 mm thick low carbon micro-alloyed steel plates are given in Table 15.3.

Heat input
Studying the combined effect of welding current, voltage and speed, it is observed that, for achieving full penetration with adequate top and bottom reinforcement in single side single pass submerged arc welding, heat input can be kept within 0.65–0.70 kJ/mm for 3 mm, 1.25–1.30 kJ/mm for 5 mm and 3.70–3.75 kJ/mm for 12 mm thick low carbon micro alloyed steel plates.

Effect of welding speed
For the plate thickness range of 3–12 mm, the welding speed variation can be from 5 to 12 mm/s. At relatively higher speeds though full penetration could be achieved, but it may result in inadequate top reinforcement. Whereas, at comparatively lower speeds, fairly good top and bottom reinforcement can be achieved along with full penetration. Thus a welding speed of 6 mm/s is recommended for the welding of 12 mm thick plates.

Also it was found that at relatively lower speeds for 5 and 3 mm thick plates, it resulted in excessive bottom reinforcement. Whereas at higher speeds fairly good top and bottom reinforcement were achieved along with full penetration for 3 and 5 mm plates. Hence higher values of welding speeds 9 and 11 mm/s are recommended for 5 and 3 mm thick plates respectively.

15.6.2 Backing Strip

In single side welding, the root of the joint needs to be melted to achieve proper root fusion as well as bottom reinforcement. Therefore at the root of the joint a large volume of molten metal is formed which has to be supported till it gets solidified. The support is provided by means of backing strips. Various types of backing strips have been successfully used to carry out single side welding in various configurations. Ceramic backing strip is the most widely used. However these ceramic backing strips are of single use type.

Aluminum backing strip filled with flux to provide support to the molten metal proved to be very effective compared to the conventional ceramic backing strip. Under the action of the arc, the flux at the root burns and forms the slag, which in molten state provides a fluid support to the molten metal at the root of the joint. For sufficient slag formation below the root of the weld, somewhat higher

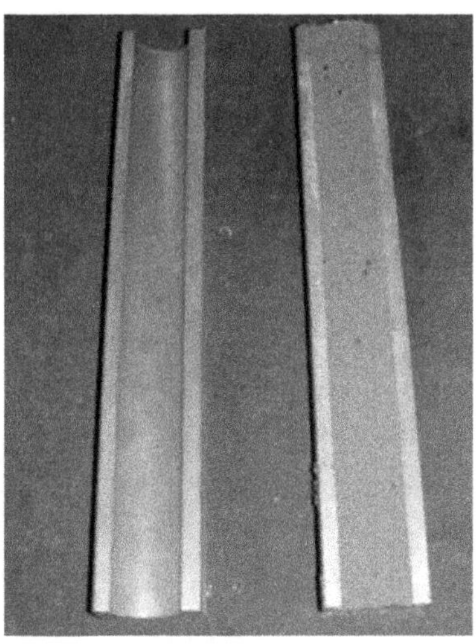

Fig. 15.17 Aluminum backing bars with deep groove, one filled with flux

current, than otherwise required, becomes necessary. Thus it helps in proper root fusion as well as formation of adequate bottom reinforcement. The conventional ceramic backing strips are for one time use, whereas the aluminum backing strips can be used repeatedly, virtually infinite number of times. Because here the molten metal is actually supported by the flux and thus the aluminum bar does not get damaged by the heat of the molten metal. 25 mm thick aluminum bars are used, wherein the layer of flux gives adequate protection to the backing strip, thereby these backing strips could be reused several times, unless accidental damage occurs.

Aluminum backing bar having deep groove is used as shown in Fig. 15.17. The groove is filled up with the flux prior to welding and it is held beneath the plates as shown in Fig. 15.18.

This flux in the trough of the backing strip supports the molten metal. The flux backing being of sufficient depth, the backing bars could be reused without any damage occurring to them during welding.

Examples The bottom and top reinforcements achieved in welding of 12 mm thick low carbon micro alloyed steel by a single run using SAW are shown in Fig. 15.19a, b respectively. Thus one can observe that this form of backing strip holds a great promise towards achieving single side welding without the requirement of any additional fixture or additives and at the same time the backing strips can be reused virtually infinitely.

Fig. 15.18 Single side submerged arc welding with flux filled aluminum backing bar

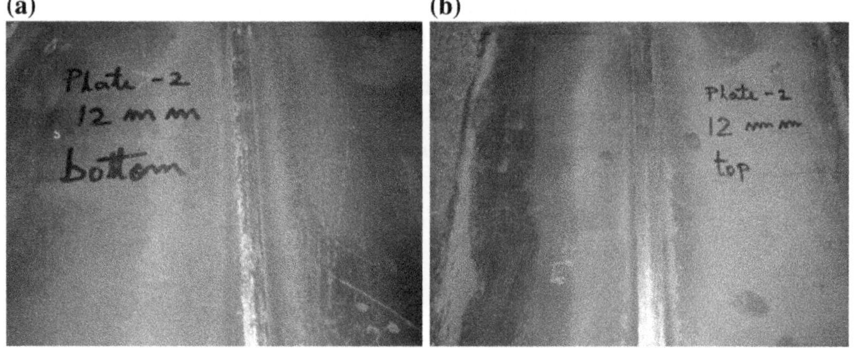

Fig. 15.19 **a** Bottom reinforcement and **b** Top reinforcement of 12 mm thick welded plate

15.7 Multi Electrode Welding

Increased demands for higher productivity in the production of welded structures have led to developments of new higher-performance welding procedures. Submerged arc welding is already one of the efficient, high deposition processes, but with certain improved variants, its performance can be further increased. The variants used for this are:

- Multiple head welding,
- Double electrode welding,
- SAW with metal powder addition.

To achieve still higher levels of deposition welding with two or three electrodes in tandem is done. Submerged arc welding with metal powder addition has also been used to enhance deposition rates. A further improvement over these techniques is the application of multiple electrodes in a joint contact tube.

15.7.1 Characteristic Features

A multiple electrode welding unit essentially is very similar to commonly used SAW units. A setup for triple wire welding system is shown schematically in Fig. 15.20. All the wires are fed at the same rate through a joint contact tube. Power is supplied from a single power source. A single weld pool is formed where all the wires melt together and is covered by flux. The wires can be of the same diameters or different diameters and also can have different chemical composition.

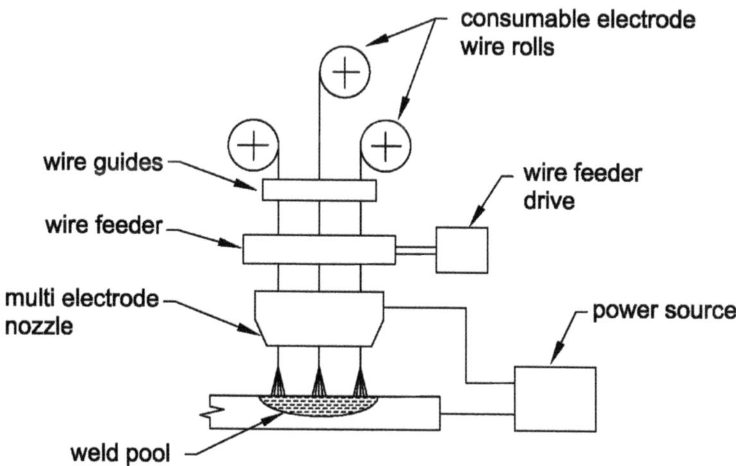

Fig. 15.20 Schematic of triple wire SAW

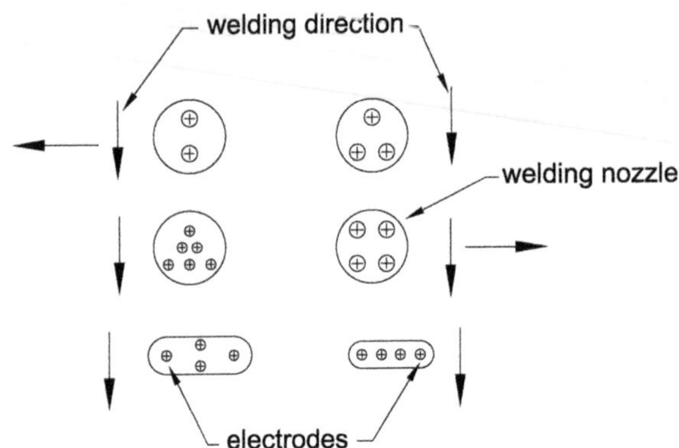

Fig. 15.21 Various alternatives of multi-electrode arrangement

The arrangement of the electrode in the contact tube can have different configurations. Some of the typical wire arrangements in the contact tube are shown in Fig. 15.21.

The electrodes can be arranged in a line, side by side or at an angle with respect to the welding direction. They can also be arranged in triangle or rectangle or other forms depending on the number of electrodes. The electrode arrangement and distance between them can influence the weld deposit, and the final property of the welded joint. The distance between electrodes and their arrangement depend upon the diameter of the wires and the welding parameters. All the welding parameters should be selected such that a single weld pool is formed.

Deposition rate

Deposition rate is defined as the mass of metal deposited in a unit time. Deposition rate primarily depends on current density, electrode extension and polarity. In case of multi electrode welding, the metal deposition rate also depends on the number of electrodes as well as distance between them, whereas the other parameters like arc voltage and welding speed do not have any significant effect on deposition rate. The deposition rate does not increase linearly with increase in the number of electrodes, the increase is rather exponential. Welding with triple electrodes, each of 3.2 mm in diameter, with welding current of 700 A and electrode negative, a maximum deposition rate of 35 kg/h can be achieved with an optimum distance of 8 mm between the electrodes and an optimum arrangement of electrode in the contact tube. The deposition rate thus achieved is almost 30 % higher than what could be achieved through three welding runs using a single electrode.

This happens due to the several metallurgical and dynamic thermal processes taking place in a multi electrode welding and it results in a rise of heat generation leading to a higher deposition rate.

Deposition Pattern

The arrangement of the electrodes in the contact tube can alter or control the deposition pattern. Choosing a suitable number of electrodes and a proper arrangement in the contact tube will enable welding of various types of joints with large root opening and weld grooves. For example, in a three electrode welding, the electrodes can be arranged in a triangular form such that, the arcs from the outer wires will melt the edges of the two side pieces, while the third trailing electrode will fill up the root opening at the middle of the joint. Thus the metal deposition with the required bead shape can be achieved with a single run.

In case of bridging an unusually wide root opening, welding with a single electrode will not be possible without weaving the electrode transversely. Whereas by selecting a tween electrode welding process with appropriate distance between the electrodes, wire diameters and welding parameters, a good quality weld deposition with proper bead shape can be achieved.

In all these cases it can be observed that joint geometries with wider root openings can be efficiently and quickly filled simply by using welding with multiple-wire electrodes. All that is needed is to select the optimum welding parameters, the right number and arrangement of the wires and a suitable wire diameter.

15.7.2 Salient Features

- Any submerged arc welding machine can very easily be adapted to welding with multiple electrodes.
- Using suitable contact tubes, required electrode configuration can be achieved.
- Deposition rate in multiple-electrode welding increases exponentially if the number of wires is increased.
- By adjusting the distance between the wires, it is possible to influence the weld shape, deposition rate and heat input into the workpiece.
- Utilisation of electrical energy and shielding flux is much higher in welding with multiple-wire electrodes than in welding with a single-wire electrode.

References

1. Mandal, N. R. (2009). Welding techniques, distortion control and line heating. New Delhi: Narosa Publishing House. ISBN: 978-81-7319-971-4.
2. Doan, G. E., & Myer, J. L. (1932). Arc discharge not obtained in pure argon gas. *Physical Review, 40*(4), 36–39.
3. Ries D. E. (1983). Gas metal arc welding of titanium. *MS Dissertation*. Cambridge, Mass: Massachusetts Institutes of Technology.
4. Jonsson, P. G., Murphy, A. B., & Szekely, J. (1995). The influence of oxygen additions on argon-shielded gas metal arc welding processes. *Welding Journal Research supplement* 48-s to 58-s.
5. Kim, Y. S. (1989). Metal transfer in gas metal arc welding. Ph.D. Dissertation, Massachusetts Institute of Technology, Cambridge, Mass.

6. Jonshon, P. G. (1993). Arc properties and metal transfer in gas metal arc welding. *D.Sc. Dissertation*, Massachusetts Institute of Technology, Cambridge, Mass.
7. Mendez, P. F., Goett, G., & Guest, S. D. (2015). High speed video of metal transfer in submerged arc welding. *Welding Journal, 94*, 326s-333s.
8. Houldcraft, P. T. (1979). *Welding process technology*. Cambridge University Press.
9. Poradnik Inzyniera, Spawalnictwo, Tom - 2, Wydawnictwa Waukowo - Techniczne, 1983.
10. Bailey, N., & Jones, S. B. (1978). The solidification cracking of ferritic steel during submerged arc welding. *Welding Research Supplement, Welding Journal* 217s–231s.
11. Defects—solidification cracking, Job Knowledge 44, The Welding Institute, http://www.twi-global.com/technical-knowledge/job-knowledge/defects-solidification-cracking-044/.
12. Sasaki, K., et al. (2004). Development of two-electrode electrogas arc welding process. Nippon Steel Technical Report No. 90, July 2004.
13. Japan Welding Society. (1993). *Welding and joining technology* (1st ed., p. 388). Tokyo: SANPO Publications Incorporated.
14. *Welding Handbook* (8th edn., Vol. 2). Miami: American Welding Society.
15. Mandal, N. R., & Rajiv, M. (2005). Study of single side single pass submerged arc welding using reusable backing strip. *Journal of Science and Technology of Welding and Joining, 10* (3), 319–324.

Chapter 16
Solid State Welding

Abstract The welding processes that produce sound joints at temperatures much below the parent metal melting temperature are referred to as solid state welding. Since there is no melting taking place, solid-state welded joints are usually free from all the defects that may occur in case of fusion welding, like, porosity, slag inclusion, hot cracking, undercut, etc. At the same time it does not require any shielding medium as well as any kind of filler metal. Using solid state welding, dissimilar materials can also be effectively joined which may not be possible through fusion welding. Friction stir welding is one such method of solid state welding. In the heart of this method lies a cylindrical shouldered tool with a profiled pin. The pin while rotating is inserted into the joint line between the two pieces, clamped rigidly to a base plate. The welds are created by the combined action of frictional heating and mechanical deformation due to the rotating tool. The majority of the heat is generated from the friction. It softens the material but does not attain the melting temperature. The pin traverses the entire joint resulting in the required bonding.

The welding processes that produce sound joints at temperatures much below the parent metal melting temperature are referred to as solid state welding. Unlike fusion welding here no melting takes place. Here the bonding is achieved through extreme deformation and diffusion under the action of mechanical, electrical or thermal energy. Since there is no melting taking place, solid-state welded joints are usually free from all the defects that may occur in case of fusion welding, like, porosity, slag inclusion, hot cracking, undercut, etc. At the same time it does not require any shielding medium as well as any kind of filler metal. Using solid state welding, dissimilar materials can also be effectively joined which may not be possible through fusion welding.

There are several types of solid state welding processes, e.g. forge welding, ultrasonic welding, diffusion bonding, explosion welding, friction stir welding, etc. However each of them is different from each other in their principal characteristics

N.R. Mandal, *Ship Construction and Welding*, Springer Series
on Naval Architecture, Marine Engineering, Shipbuilding and Shipping 2,
DOI 10.1007/978-981-10-2955-4_16

and application. Here friction stir welding process will be taken up in detail as it is a relatively new method and at the same time it is found to be an effective alternative to fusion welding in aerospace and automotive industries. It is also slowly gaining foothold in shipbuilding industry.

16.1 Friction Stir Welding

Friction stir welding, unlike traditional fusion welding methods, is a solid state welding process. The process of friction stir welding (FSW) was invented in 1991. Since it is essentially solid-state process, i.e. welding is carried out without melting, high quality weld can generally be achieved with absence of solidification cracking, porosity, oxidation, and other defects resulting from traditional fusion welding. In the heart of this method lies a cylindrical shouldered tool with a profiled pin. The pin while rotating is inserted into the joint line between the two pieces, clamped rigidly to a base plate. The welds are created by the combined action of frictional heating and mechanical deformation due to the rotating tool. The majority of the heat generated from the friction, i.e., about 95 %, is transferred into the workpiece and only 5 % flows into the tool [1]. The heat thus generated softens the material but it does not attain the melting temperature and the pin traverses the entire joint resulting in the required bonding.

The two plates are held firmly in position without any gap in between them and the FSW tool rotates and moves along the butt line with its shoulder rubbing against the plate top surface. The frictional heat generated plasticizers the material and under the action of the rotating pin, the metal gets sheared from the front face of the pin to the back continuously as it travels along the joint line. Through this shearing action the metal from two edges gets consolidated together and the bonding is achieved in solid state.

This welding process does not require any filler metal, thereby the weld is nominally identical in composition to the base material, unless, of course, dissimilar metals are joined. However the extreme stirring action causes formation of the nugget zone having significantly different microstructure compared to that of the parent metal. It often results in a fine grain structure, generally assumed to be a consequence of recrystallisation processes. Because of the extreme shearing action, primary particle fragmentation may also occur. Friction stir welding offers several advantages over arc welding. The lower apparent energy inputs of FSW are expected to limit distortion and residual stress in the welded structures. Additionally, problems of porosity, slag inclusion, hydrogen cracking, etc. are eliminated since FSW being a solid-state process. A schematic of friction stir welding operation is shown in Fig. 16.1. The plates are placed on a base plate and the plate edges are clamped rigidly to prevent lateral movement during the welding operation.

A significant advantage of friction stir welding is that it has fewer variables that need to be controlled. The variables are: tool rotation speed, travel speed, and

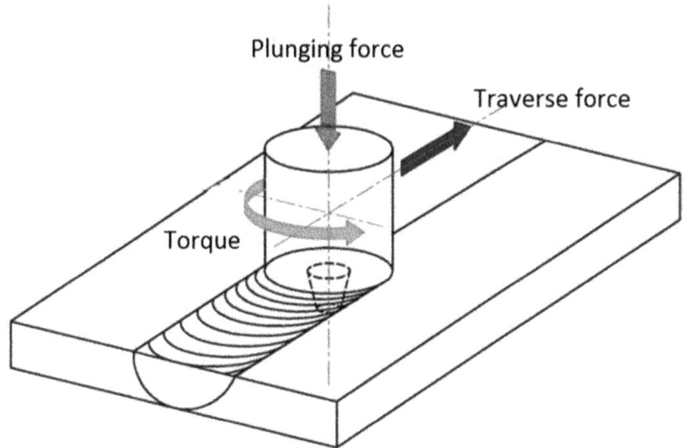

Fig. 16.1 Schematic presentation of a FSW process

plunging force. Important process parameters include the tool rpm, travel speed, tool dimensions and the downward force on the tool. Heating is caused by rubbing of the tool shoulder and the tool pin against the work piece and by visco-plastic dissipation of mechanical energy.

16.1.1 Features of FSW Joint

The peak temperature in the process is generally of the order of about 80 % that of the liquidus temperature of the material being welded. The joint produced in this process is asymmetric about the weld line as the material in a highly plastic state flows differently at the two sides of the welded joint. On one side of the tool the rotational direction is same as that of the tool travel direction, it is referred to as 'advancing side' and the other side is referred to as the 'retreating side'. Four distinct microstructural zones can be observed in FSW. Moving towards the weld line from the parent metal, the four zones are, (i) parent metal zone, (ii) heat affected zone, HAZ, (iii) thermo-mechanically affected zone, TMAZ, and (iv) nugget zone.

A complex interaction of tool rpm, welding speed, heat generation and dissipation along with shearing action of the FSW tool influences the microstructure of the nugget zone. This in turn determines the strength of the welded joint. The material flow takes place due to shearing action at the same time in case of threaded tool, it tends to push the material in downward direction closer to the tool [2–4].

In friction stir welding the aspects of heat generation and metal stirring are very important as they strongly influence the entire process of welding and thereby the resulting mechanical properties of the welded joint. The heat generation primarily depends on tool shoulder diameter and tool plunging force. In FSW the plasticized material in the weld nugget remain in a highly viscous state and therefore the effect

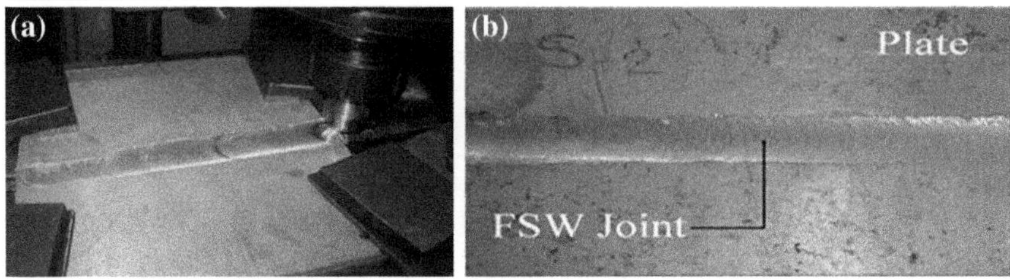

Fig. 16.2 **a** and **b** A typical friction stir welded samples

of convective heat transfer is not expected to be significant. Hence in the heat flow analysis conduction mode of heat flow is considered. Temperature distribution in the transverse direction is found to be non-symmetric with higher temperatures located on the advancing side of the weld. Tool rotational speed contributes to the stirring effect. The heat flow and the cooling rate of the welded nugget depend on the welding speed. All these finally determine the grain structure of the weld nugget which controls the strength of the welded joint. Consolidation of the thermally plasticized material takes place behind the tool that forms the welded joint. This metal transfer takes place through the retreating side. Worm hole defect often is observed at the advancing side near the nugget/TMAZ interface [5–9].

Though plunging force is responsible for heat generation, but much variation of this is not possible and therefore extent of heat generation cannot be controlled through this parameter. It is also observed that variation of tool rpm does not significantly affect heat generation, however it does have an effect on metal stirring and thereby straining rate of the weld nugget.

In case of FSW of marine grade 5083 aluminum alloy, it was observed that increasing tool rotational speed adversely affected weld quality. Near consistent results were achieved at tool rotational speeds in the range of 500–710 rpm and weld speed range of 40–56 mm/min. About 78–90 % of parent metal strength was achieved in FSW of AA5083. Typical friction stir welded joint in 5083 aluminum alloy is shown in Fig. 16.2a, b.

16.1.2 Tool Geometry

The FSW tool is at the heart of the friction stir welding process. As already observed, that in FSW there are only three process variables, which need to be controlled for getting a quality weld. However one can see that the weld quality is highly influenced by the tool geometry [10, 11]. It has a direct bearing on heat generation and metal movement within the weld nugget. Therefore the FSW tool should be such that it will produce the desired joint quality as well as enable higher welding speed and longer tool life. Applications in high melting point alloys, such as mild steel, stainless steel, titanium remain limited due to problems in terms of required force and tool life.

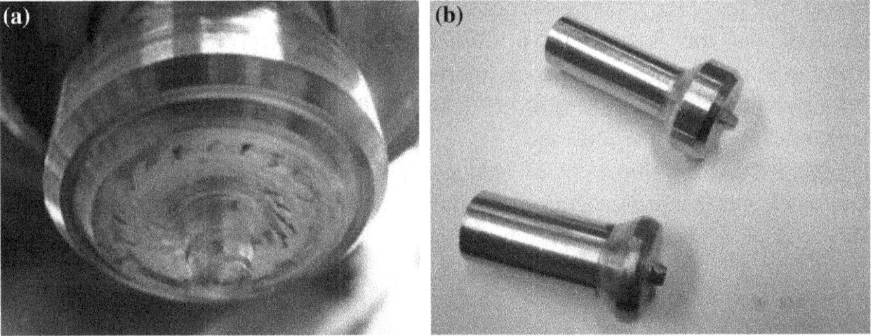

Fig. 16.3 **a** and **b** Typical FSW tool having threaded pin and a concave grooved shoulder

The shape of the bottom of the tool shoulder affects material flow around the tool pin and contributes to preventing the escape of plasticized material. They can be flat or concave, smooth or grooved, with concentric or spiral grooves. A concave shoulder has the advantage, when compared with a flat shoulder, of directing material flow to the center close the tool pin. The tool pin may have a diameter one-third of the cylindrical shoulder and typically has a length slightly less than the thickness of the work piece. Typical FSW tools are shown in Fig. 16.3a, b.

16.1.3 Tool Material

The tool material, the shoulder and pin material in particular plays an important role in the heat generation and heat dissipation process of friction stir welding. The welding efficiency is affected by the heat loss by way of conduction through the tool as well as the base plate. Using a tool material and base plate having low thermal conductivity, this heat loss can be substantially reduced. To further reduce heat loss through base plate, suitable fibre glass cloth can be used as insulation between the work piece and the base plate.

This will enable increasing welding speed. Thus a combination of low thermal conductivity tool material, e.g. SS310 or even better SS660 and base plate insulation can significantly improve process efficiency through increase in welding speed. If full advantage is to be taken regarding welding efficiency, an increase in welding speed necessitates a similar increase in rotational speed to get sound welds without line porosity and weld defects ensuring proper material flow [12].

16.1.4 Heat Generation

In friction stir welding, heat is generated due to mechanical friction and the material movement in the shear layer at the tool-work piece interface. The heat generation due to friction is the product of frictional force and the velocity difference between the tool and the interface layer. Whereas the heat generated due to material movement at the tool-work piece interface is the product of shear stress and the velocity of the interface shear layer sticking to the tool. This velocity is actually the tangential speed of the tool.

The frictional component of heat generation on an elemental area *dA* at the tool-work piece interface considering high rotational speed compared to traverse speed of the FSW tool, is given by,

$$dQ_f = \left(v_{friction}\right) \mu p \, dA \tag{16.1}$$

where, $v_{friction}$ is the velocity difference i.e. the slippage between the tool surface and the shear layer in the plate being welded. This is given as,

$$v_{friction} = (1 - \delta) \, \omega \, r$$

where,

μ coefficient of friction,
p tool plunging pressure applied on the elemental area *dA*,
δ contact parameter defining extent of slip,
ω tool angular rotational speed,
r length along tool radius.

The component of heat generation due to work piece material sticking to the tool is given by,

$$dQ_s = \left(v_{sticking}\right) \tau_y \, dA \tag{16.2}$$

where

τ_y Shear yield stress,
$v_{sticking}$ velocity of the shear layer at the tool-work piece interface.

This is given as,

$$v_{sticking} = \delta \, \omega \, r \tag{16.3}$$

where δ is the velocity field at the tool-work piece interface. It is the ratio of the velocity of the shear layer formed due to shearing of the plate material at the tool-work piece interface to the velocity of the FSW tool.

Therefore,

$$\delta = \frac{v_{shear\,layer}}{v_{tool}} \tag{16.4}$$

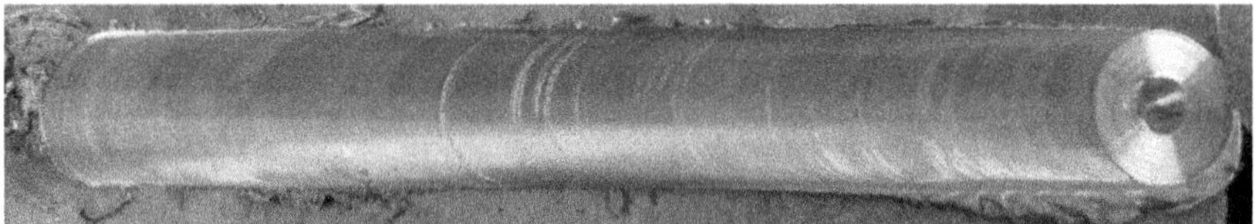

Fig. 16.4 A friction stir welded sample of AA5083

Now, $v_{tool} = \omega\, r$, therefore from Eqs. (16.3) and (16.4), $v_{shear\, layer} = \delta\, \omega\, r = v_{sticking}$.

A case of pure friction implies non-existence of shear layer. Therefore $v_{shear\, layer} = 0$. Hence from the above relation in case of pure friction, $\delta = 0$.

At the other extreme when there is no friction and the metal is fully sticking to the tool surface, i.e. a case of full sticking, one obtains a shear layer attaining the speed of the tool, i.e. $v_{shear\, layer} = \omega\, r$. Therefore in case of pure sticking condition, $\delta = 1$. Thus the contact condition at the tool-work piece interface is defined by the contact parameter δ.

There is no straight forward mechanism to estimate this contact parameter. From experimental observation it was seen that very smooth friction stir welded surfaces were achieved as shown in Fig. 16.4 provided suitable tool material, tool geometry and process parameters are used.

The welded surface in Fig. 16.4 appears as if the FSW tool slided over the plate surface without any material sticking to the tool shoulder. This condition resembles that of pure friction. Hence considering $\delta = 0$ using Eq. (16.1) the heat generated due to the shoulder friction works out to,

$$dQ_f = \omega\, r\, \mu p\, dA \tag{16.5}$$

Two distinct tool-work piece interface surfaces are tool shoulder and tool pin side. The contribution of the tool pin tip surface towards total heat generation is neglected [13]. Q_1 and Q_2 are the components of the respective heat generated from these interfaces as shown in Fig. 16.5.

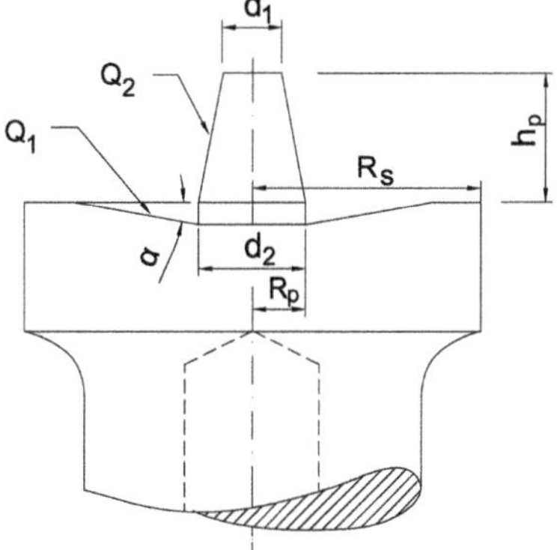

Fig. 16.5 FSW tool dimensions and heat generation components on different parts of the tool

Tool shoulder—Work piece interface

The expressions were derived considering a concave shoulder surface as shown in Fig. 16.5. The purpose of this geometric feature is to act as an escape volume for the expunged metal as the tool pin is plunged into the plate during the welding operation.

The concave shoulder surface is represented by its vertical and horizontal projected surfaces as A_v and A_h respectively. Therefore for an elemental segment,

$$dA_v = r\,d\theta\,dz$$

Now

$$dz = \tan\alpha\,dr$$

Therefore

$$dA_v = r\,d\theta\,\tan\alpha dr \quad \text{and} \quad dA_h = r\,d\theta\,dr$$

The forces acting on the tool shoulder surface can be written as,

$$dF_s = \mu\,p(dA_h + dA_v) = \mu\,p\,r\,d\theta\,dr(1 + tan\alpha) \tag{16.6}$$

From Eq. (16.6) one can observe that the concave shoulder surface actually contributes to increased frictional area by a factor of *tan α*.

Therefore combining Eqs. (16.5) and (16.6) the heat generation from the elemental shoulder surface is given by,

$$dQ_f = \omega r\,dF_s = \omega r^2\,\mu p\,d\theta\,dr\,(1 + \tan\alpha) \tag{16.7}$$

The heat generated through friction of tool shoulder with the plate surface is obtained by integrating Eq. (16.7) from pin root radius to the outer radius of shoulder surface. Therefore considering pure friction the heat generated at the tool shoulder and the plate interface is given by,

$$Q_1 = \int_0^{2\pi}\int_{R_p}^{R_s} \omega\mu p\,r^2(1 + \tan\alpha)dr\,d\theta = \frac{2}{3}\pi\mu\,p\,\omega\left(R_S^3 - R_p^3\right)(1 + \tan\alpha) \tag{16.8}$$

where

R_s shoulder radius,
R_p pin radius at the root.

For welding of 6 mm thick 5083 aluminum alloy plates, the shoulder concavity $\alpha = 6°$ gives satisfactory results.

Tool pin—Work piece interface
From the phenomenological view point to achieve friction stir welded joints material flow has to happen between the two mating plate edges. Hence it points to the existence of flow in the shear layer around the tool pin side surface. Therefore a state of sticking condition is considered to estimate the extent of heat generation at the tool pin side surface to plate interface.

As per Eq. (16.2), the heat generation at this interface due to material movement in the shear layer is given by Eq. (16.9).

$$Q_{2s} = \left(\delta\omega\frac{(d_1 + d_2)}{4}\tau_y\right)\left(\pi\frac{(d_1 + d_2)}{2}h_p\right) \tag{16.9}$$

At the same time as the FSW tool traverses along the joint, the forward half of the tool pin experiences a reaction force F [14] as shown in Fig. 16.6. It is given by the product of the projected area of the tool pin and the yield stress of the aluminum alloy at the prevailing temperature of pin-plate interface as given by Eq. (16.10). The temperature is approximated about 80 % of the liquidus temperature of (570 °C) of the AA5083 plate material i.e. 456 °C in the present case.

The reaction force on the tool pin is given by,

$$F = \frac{[h_p \times (d_1 + d_2)]}{2} \times (\sigma_y)_{456\,^{\circ}\mathrm{C}} \tag{16.10}$$

Therefore the frictional force experienced by the tool pin surface is given by,

$$F_f = (\mu)_{456\,^{\circ}\mathrm{C}} \times F \tag{16.11}$$

Hence the heat generated due to friction of the tool pin front surface,

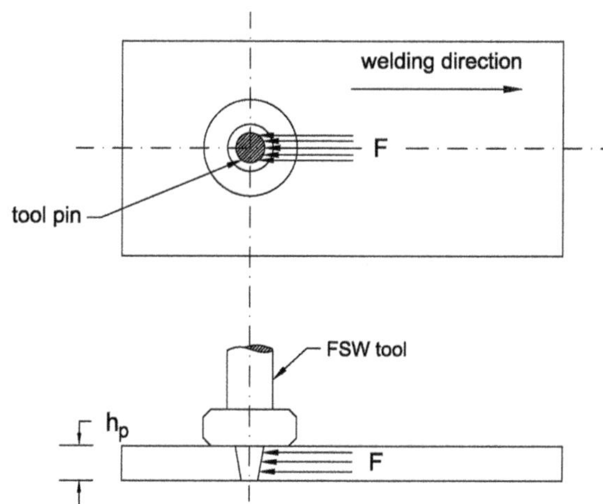

Fig. 16.6 Lateral force acting on FSW tool pin during welding

$$Q_{2f} = (1 - \delta)\omega \frac{(d_1 + d_2)}{4} F_f \quad (16.12)$$

Therefore the total heat generation considering both friction and sticking condition at the pin side surface interface is given by,

$$\begin{aligned} Q_2 &= Q_{2f} + Q_{2s} \\ &= (1 - \delta)\omega \frac{(d_1 + d_2)}{4} F_f + \pi\left(\delta\,\omega \frac{(d_1 + d_2)}{4} \tau_y\right)\left(h_p \frac{(d_1 + d_2)}{2}\right) \end{aligned} \quad (16.13)$$

Since the tool traverses forward as it rotates, it is logical to consider the effects of shearing of metal as well as friction remaining limited to only front half of the pin surface. Hence the contact parameter δ was taken 0.5. Therefore Eq. (16.13) becomes,

$$Q_2 = 0.5\,\omega \frac{(d_1 + d_2)}{4} F_f + \pi\left(0.5\omega \frac{(d_1 + d_2)}{4} \tau_y\right)\left(h_p \frac{(d_1 + d_2)}{2}\right) \quad (16.14)$$

Substituting from (16.10) and (16.11) in Eq. (16.14) one obtains,

$$Q_2 = \omega h_p \left(\frac{d_1 + d_2}{4}\right)^2 \left[(\mu\sigma_y)_{456\,^\circ\mathrm{C}} + \pi(\tau_y)_{456\,^\circ\mathrm{C}}\right] \quad (16.15)$$

Heat generated in a FSW process increases with increasing tool rotation speed and tool downward force. Travel speed influences the heat input per unit weld length (rate of heat input), affecting metal flow around the tool pin. Other relevant parameters are the dwell time of the tool at the start of the weld after plunging the pin in the plates, shoulder diameter and shoulder concavity angle of the FSW tool.

The dwell time is the period between the instant the tool pin contacts the work-piece and the instant the tool begins to move along the joint. During this period, the generated heat spreads in the vicinity of the tool pin, softening the adjacent material and stabilizing material flow around the tool pin. If this period is too short defects can appear in the initial part of the weld. Time can range usually from 5 to 30 s. The tool shoulder angle allows a gradual increase of the pressure on the top surface of the plates being welded and helps to direct the material flow. Tool angles up to 6° are common.

16.1.5 Basic FSW Metallurgy of Aluminum Alloy

In friction stir welding of aluminum alloys three effective zones with differing material flow and associated temperature changes can be identified. These zones

Table 16.1 Typical microstructural zones in friction stir welds

Zone	Material flow	Temperature
Weld nugget	High	High
Thermo-mechanically affected zone (TMAZ)	Low	Medium
Heat affected zone	None	Medium

consist of differing levels of material flow and associated temperature changes as described in Table 16.1.

Weld nugget
The region swept by the tool pin as it rotates and traverses along the welding seam is referred to as the weld nugget. Here during welding the metal is heavily stirred and thus grain refinement takes place leading to formation of fine grained microstructure. The majority of the grain boundaries within the nugget zone are believed to form through dynamic recrystallisation during the stirring process. The width of the nugget zone mainly depends on tool design, welding parameters. The tensile strength of the joint attains about 80–90 % that of the parent metal.

Thermo-mechanically affected zone (TMAZ)
Outside the nugget zone, aluminum alloy friction stir welds contain two other distinct microstructural zones. In the region immediately surrounding the nugget zone, the parent alloy grain structure becomes both heated and deformed, although to a lesser extent than for the nugget zone. This zone is referred to as the Thermo-Mechanically Affected Zone (TMAZ).

Heat affected zone (HAZ)
Surrounding the TMAZ is the heat affected zone.

16.1.6 Defects and Their Detection

The types of defect resulting from the solidification process associated with conventional fusion welding methods are not expected in a solid-state joining technique. However defects can occur in friction stir welds if any of the process variables which control the quality of the weld deviate from the optimum. The process variables are,

- Tool rotation speed
- Weld travel speed
- Tool plunging force
- Tool geometry.

The possible defects that may occur due to improper process variables are,

- Lack of penetration.
- Wormholes.

- Root toe defects.
- Improper joint strength due to formation of adverse nugget microstructure.

The rate of heat input decreases with increasing travel speed, which reduces material softening in the vicinity of the tool pin, making plastic flow more difficult. High travel speeds therefore may cause defects, such as cavities. For low tool rotation speed, low downward force and high travel speed external defects may form for welds in some aluminum alloys.

The ratio of tool rotation speed *versus* travel speed is sometimes used to distinguish between hot welds having high ratio, and cold welds with low ratio. Hot welds are less sensitive to defect formation but may exhibit more significant changes in microstructure and mechanical properties than cold welds in aluminum alloys.

Too fast a welding speed or insufficient plunging force applied to the weld can lead to formation of voids [15]. Whereas either too short a pin depth or tool plunge can cause joint line defects at the weld root. It is preferable to clean and remove the oxide layer from the plate surfaces coming under the tool shoulder to avoid possible inclusion of the oxides in the weld nugget making it brittle.

For various reasons internal defects primarily in the form of worm hole may occur mainly in the weld nugget. This can be detected by conventional X-ray and ultrasonic NDT methods. However external defects like surface cracks and lack of proper root coalescence can be effectively detected by dye penetrant test.

16.1.7 Merits and Limitations of FSW

In general, FSW is claimed to produce stronger, lighter and more efficient welds than other joining processes. It has the ability to join dissimilar materials, and materials previously thought to be unweldable, or not recommended for fusion welding. An obvious merit of friction stir welding is the absence of fusion, which automatically eliminates all defects associated with fusion welding process. Specific merits and limitations of FSW can be summarised as follows [16, 17]:

Merits

- Absence of melting resulting in less weld contamination.
- No consumables.
- Minimal edge preparation required.
- Very controllable process with high productivity.
- High integrity welds with low shrinkage and little porosity.
- Relatively simple process, with low running costs.
- Fine-grained microstructure in the weld region, leading to improved mechanical properties.
- Low distortion and residual stress levels, compared to arc welding processes.
- Environmentally friendly—relatively quiet process, with absence of fumes and spatter.

- No gas shielding required.
- Able to join dissimilar materials.

Limitations with the present state of technology

- Welding speeds slower compared to fusion welding.
- Plates require rigid clamping.
- Keyhole at the end of each weld.
- Post weld machining required.

References

1. Chao, Y. J., Qi, X., & Tang, W. (2003). Heat transfer in friction stir welding–Experimental and numerical studies. *Transaction of the ASME, 125*, 138–145.
2. Jata, K. V., & Semiatin, S. L. (2000). Continuous dynamic recrystallization during friction stir welding. *Scripta Materialia, 12*, 743–748.
3. Nandan, R., Roy, G. G., & DebRoy, T. (2006). Numerical simulation of three dimensional heat transfer and plastic flow during friction stir welding. *Metallurgical and Materials Transactions A: Physical Metallurgy and Materials Science, 37*, 1247–1259.
4. Nandan, R., Roy, G. G., & DebRoy, T. (2007). Three-dimensional heat and material flow during friction stir welding of mild steel. *Acta Materialia, 12*, 883–895.
5. Nandan, R., Lienert, T. J., & DebRoy, T. (2008). Toward reliable calculations of heat and plastic flow during friction stir welding of Ti-6Al-4 V alloy. *International Journal of Materials Research, 9*, 434–444.
6. Nandan, R., DebRoy, T., & Bhadeshia, H. K. D. H. (2008). Recent advances in friction-stir welding—process, weldment structure and properties. *Progress in Materials Science, 53*, 980–1023.
7. Russell, M. J., & Shercliff, H. (1999). Analytical modelling of friction stir welding. In M. J. Russell & R. Shercliff (Eds.), *INALCO_98: Seventh International Conference on Joints in Aluminium* (pp. 197–207). Cambridge, UK: TWI.
8. Sato, Y., Urata, M., & Kokawa, H. (2002). Parameters controlling microstructure and hardness during friction-stir welding of precipitation hardenable aluminum alloy 6063. *Metallurgical and Materials Transactions A: Physical Metallurgy and Materials Science, 33* (3), 625–635.
9. Peel, M. J., Steuwer, A., Withers, P. J., Dickerson, T., Shi, Q., & Shercliff, H. (2006). Dissimilar friction stir welds in AA5083-AA6082. Part I: process parameter effects on thermal history and weld properties. *Metallurgical and Materials Transactions A: Physical Metallurgy and Materials Science, 37*(7), 2183–2193.
10. Biswas, P., Kumar, D. A., & Mandal, N. R. (2012). Friction stir welding of aluminum alloy with varying tool geometry and process parameters. *Proceedings IMechE, Journal of Engineering Manufacture Part B, 226*(4), 641–648.
11. Venkateswarlu, D., Mandal, N. R., Mahapatra, M. M., & Harsh, S. P. (2013). Tool designs effects for FSW of AA7039. *Welding Journal, 92*, 41s–47s.
12. Midling, O. T., & Rørvik, G. (1999). Effect of tool shoulder material on heat input during friction stir welding. In *The First International Symposium on Friction Stir Welding*. Thousand Oaks, CA, USA, 14–16 June 1999.
13. Lohwasser, D., & Chen, Z. (2010). *Friction stir welding* (pp. 282 & 309). Woodhead Publishing Limited.

14. Biswas, P., & Mandal, N.R. (2011) Effect of tool geometries on thermal history of FSW of AA1100. *Welding Journal, 90,* 129s–135s.
15. Leonard, A. (2001). *Flaws in aluminum alloy friction stir welds.* PM Member Report 726.
16. Kallee, E. D., Nicholas, & Thomas, W. M. (2001). Friction stir welding—invention, innovations and applications. In *INALCO 2001, 8th International Conference on Joints in Aluminum*, Munich, Germany, 28–30 March 2001.
17. Thomas, M. (1998). Friction stir welding and related friction process characteristics. In *Proceedings of INALCO 98*, TWI, Cambridge, U.K.

Chapter 17
Welding Residual Stress and Distortion

Abstract The panels, subassemblies and assemblies being welded are subjected to thermal cycles of heating followed by cooling. It causes shrinkage forces to develop in the welded panels. The shrinkage forces tend to cause different degrees of distortion. Non uniform shrinkage forces across thickness may lead to angular deformation, whereas the inplane compressive forces in plate panels made of thinner plates tend to buckle. In a welded joint, the base metal away from the weld zone remains at room temperature throughout the welding operation and is not subjected to any expansion or contraction. This 'cold' part of the base metal restrains the welded zone and the adjacent heated base metal from free expansion and contraction. This leads to stresses of near yield point magnitude in the weld. Under this stress the weld deposit and the adjacent heated base metal yields resulting in plastic strains. As the weld metal and the base metal cool down to the room temperature residual stresses are formed. If some of the external restraints such as clamps or welded lugs are removed, the residual stresses may find partial relief by causing the base metal to further deform.

Shipbuilding is primarily an assembling industry. The panels, subassemblies and assemblies being welded are subjected to thermal cycles of heating followed by cooling. Because of the phenomenon of thermal expansion and nonuniform heating and cooling cycle, it causes shrinkage forces to develop in the welded panels. The shrinkage forces tend to cause different degrees of distortion. Non uniform shrinkage forces across thickness may lead to angular deformation, whereas the inplane compressive forces in plate panels made of thinner plates tend to buckle.

In a welded joint, the base metal away from the weld zone remains at room temperature throughout the welding operation and is not subjected to any expansion or contraction. This 'cold' part of the base metal restrains the welded zone and the adjacent heated base metal from free expansion and contraction. As a result as the weld metal cools down and solidifies, it attempts to contract to the volume it would normally occupy at the lower temperature. But because of the restraints from the

N.R. Mandal, *Ship Construction and Welding*, Springer Series
on Naval Architecture, Marine Engineering, Shipbuilding and Shipping 2,
DOI 10.1007/978-981-10-2955-4_17

adjacent cold metal, it cannot do so. This leads to stresses of near yield point magnitude at the prevailing temperature in the weld. Under this stress the weld deposit and the adjacent heated base metal yields resulting in plastic strains. The stresses which exceed the yield point only get relieved by this phenomenon. As the weld metal and the base metal cool down to the room temperature residual stresses are formed. If some of the external restraints such as clamps or welded lugs are removed, the residual stresses may find partial relief by causing the base metal to further deform.

Ships designed for high speed operation are generally weight sensitive and calls for thin plates and sections. The increased use of thin plates in panel fabrication results in significantly increased distortion as shown in Figs. 17.1 and 17.2. The net result is increased man-hours for fitting, flame straightening and rework. Therefore, in-process control of welding distortion is more desirable than post welding rectification from the point of manufacturing efficiency. The hull distortion may

Fig. 17.1 Buckling of side shell panels

Fig. 17.2 Panel buckling bounded by stiffeners

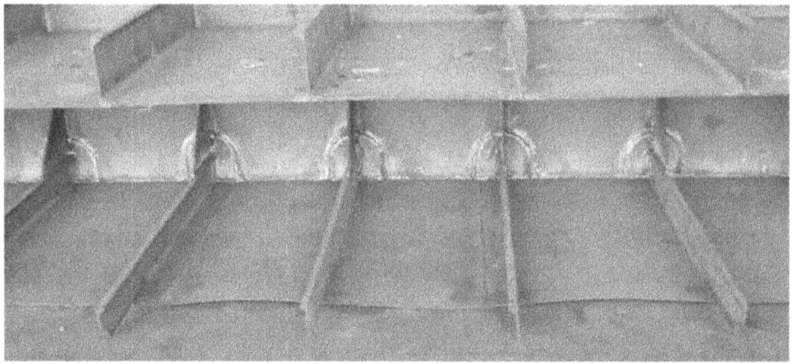

Fig. 17.3 May lead to poor fitup

Fig. 17.4 Unacceptable appearance

adversely affect the hydrodynamic performance of a vessel as well. The deck and bulkhead undulations may give rise to problem of equipment installation.

As a result of welding distortion the finished product may not be able to perform its intended purpose because of poor fitup (Fig. 17.3), reduced buckling strength and unacceptable appearance (Fig. 17.4).

As the stiffened panels are fabricated by welding longitudinal and transverse to plates, residual stress are generated within the plate and when the magnitude of these residual stresses exceeds critical buckling strength of panels it results in buckling of panels. Buckling can have severe effects on the ultimate strength of panels as it may decrease the ultimate strength of panels and hence resulting in decrease in load carrying capacity of ship, mismatch during block assembly and aesthetic loss of structures. In the shipyard almost 3–5 % of cost is incurred in the correction of this buckling distortion.

The extent of distortion depends on welding parameters, plate thickness, thermo-physical properties of plate material, structural restraints and welding sequence. Competitiveness in cost and time can be increased by eliminating or mitigating these distortions during the fabrication process rather than allowing them to accumulate and then removing them. This can be achieved through proper understanding of the distortion mechanism, its type and causes, which will help in developing suitable predictive tools and control mechanisms.

The arc welding process is a very complex phenomenon involving extremely high temperatures, leading to micro-structural changes and formation of thermal stresses. These stresses may cause distortions and produce high levels of residual stresses. These tend to reduce the strength of the structure, which may become vulnerable to fracture, buckling and fatigue and may suffer enhanced corrosion. Distortion is one of the major problems of welding.

Weld induced distortion can be broadly classified under five heads:

- Longitudinal shrinkage
- Transverse shrinkage
- Buckling
- Angular distortion
- Bowing.

The welded plate tends to shrink. The edges of the welded plate tend to come closer towards the welded region, either longitudinally or transversely. Buckling is the phenomenon in which the whole plate bends 3-dimensionally. Angular distortion is the defect, in which the two plates to be joined instead of making 180° with each other, form an angle with each other less than 180°.

Transverse shrinkage of butt joints: Dimensional reduction in a direction perpendicular to the welding line.

Longitudinal shrinkage of butt joints: Dimensional reduction in a direction parallel to the welding line. The amount of the longitudinal shrinkage is smaller than the transverse shrinkage, about 1/1000 of the weld length as reported by several researchers.

Angular distortion of butt joints: Angular change that occurs due to a non-uniform thermal contraction through the thickness of the plate. The non-uniform thermal contraction originates from the uneven heating through the thickness during welding.

Angular distortion of fillet welds: Similar to the angular distortion of welds in butt joints, the non-uniform thermal contraction through the thickness of the flanges creates a moment about the flange neutral axes.

Longitudinal bending distortion: Produced by bending stresses induced by the longitudinal shrinkage forces of the welds not coinciding with the neutral axis of the weldment.

Buckling: Produced by compressive stresses developed due to the welding of stiffeners, especially in case of thin stiffened panels.

Welding being a fully transient phenomenon, its numerical analysis is highly time consuming and hence analysis of large size structural panels as in ships or offshore structures is simply not feasible using conventional FE analysis. Hence different modeling approaches are needed for analysis of such structures.

At the same time control of welding distortion assumes great importance in ship's stiffened panel fabrication. This can be achieved by suitably designing and also implementing one or more of the distortion control measures as is suitable for a particular manufacturing situation.

The weld distortion problems in shipbuilding got compounded due to the requirement of more and more lighter-weight structures [1, 2]. It naturally calls for usage of thinner steel plate. This is being achieved by using higher-strength steel material. A further effect of this has been the increased difficulty to control unit/block dimensions when thin plate is used due to variations in shrinkage [3].

Researchers all over, for more than last almost 4 decades, are working on thin plate distortion. But how much of that research has actually percolated in the shipyards leading to reduction in the distortion is the key question in this whole scenario.

17.1 Key Issues

The key issues in tackling the problem of distortions can be identified as follows:

- Role of designers.
- Degree of initial deformation.
- Plate handling.
- Steel cutting.
- Stiffener welding.

17.1.1 Role of Designers

It is highly unlikely that ships are designed with potential distortion effects as a major consideration. Structural distortion is traditionally regarded as a problem generated by welding. However this is not always the case. It is very pertinent to look into the influence that 'design' has on distortion. This is to be noted that distortion has to be addressed at the concept design stage followed by detail design. The following are the key factors related to design which influence extent of distortion.

- Distortion increases with decrease in plate thickness.
- With decrease in panel width (frame spacing), buckling strength increases for same plate thickness.
- The heat input at which buckling is likely, decreases as plate thickness decreases and the free span increases.

- The critical buckling strength decreases with decrease in Young' modulus and increase in Poisson's ratio.
- The critical buckling strength reduces as the size of panel increases for the same thickness of panel.
- Increase in stiffener scantlings increases critical buckling strength for same plate thickness.

The role of designers in adopting a "design to reduce distortion" philosophy is also significant in the overall scheme of this problem.

17.1.2 Degree of Initial Deformation

Initial deformations (out of flatness) in plates reduce buckling strength of thin plates [4]. Greater the degree of out of flatness in a plate, the lower will be the critical buckling stress [4]. The dominant distortion mechanism in them is local buckling. Plate straightening through cold rolling significantly improves flatness. However the effect of this straightening operation is not well investigated. It could be that a badly out-of-flat plate, which has been straightened, may be very susceptible to heat induced distortion at a later stage. This is a particularly difficult area to obtain realistic data, but it is quite evident that it needs to be addressed to effectively control weld induced distortion [5].

Stiffening of plates is done using rolled steel sections. The initial deformations of these sections may significantly affect the distortion level of the stiffened panels. A stiffener section with initial deformation may pull a plate into the shape of the section thus introducing a "distortion". The other situation could be that the section is pulled down onto the plate and creates a stress situation that may relieve itself during the welding of the stiffener.

Shot blasting operation done to remove mill scale may cause a redistribution of surface residual stresses leading to deformation.

17.1.3 Plate Handling

This is often very much neglected. Transportation of plates by securing them heavily on lorry trailers may introduce a camber across the plates. Further tightening down may result in localized edge distortion/damage.

17.1.4 Steel Cutting

Plate cutting if done through thermal means may also lead to significant distortions. The best method for cutting plate in terms of minimizing distortion, would be abrasive waterjet cutting, followed by laser cutting, underwater plasma cutting, dry plasma cutting, and oxyfuel flame cutting.

Cutting of openings at the plate cutting stage may cause enhanced distortion especially on thinner plates. The better option is to make the necessary openings after the plate has been adequately stiffened. The scallops those are cut near the plate edges can cause significant deformations to the plate flatness.

The accuracy of cut edge and the bevel angle if any is also very important. If the root gap or the bevel angle is more than the designed joint geometry, it will require more weld metal to be deposited causing more heat going into the joint. This will increase the possibility of distortion. A change in bevel angle from 30° to 35° will cause an increase of about 25 % in deposited weld metal volume. At the same time the tack welds are to be kept consistently smaller to keep heat input to a minimum.

17.1.5 Stiffener Welding

In welding of stiffeners, the fillet leg length relates directly to the heat input. Hence to keep distortion to a minimum it is essential that the fillet size is kept to the minimum as required by classification rules. Depositing extra metal in fillet is a gross case of over welding, e.g. in place of 5 mm leg length if 7 mm is deposited, it amounts to an over welding of about 92 % [6]. This is equivalent to increasing the heat input by almost two times in the same structure. Therefore controlling fillet leg length in fabrication of stiffened panels assumes paramount importance. Hence it is equally important to make the welders aware of the effect of over welding which otherwise may appear to be insignificant.

The fitup between the stiffener web and the plate should be such that the gap between the web and the plate is minimum. This is to ensure that the volume of deposited weld metal is also kept to a minimum. At the same time excessive penetration is to be avoided as it also leads to increased heat input.

17.2 Residual Stresses

To understand how residual stresses are formed physically during welding, the simple case of a bead-on-plate welding may be considered. Essentially residual stresses are formed due to constrained expansion and contraction of the heated metal during welding. Figure 17.5 shows schematically the effect at various stages of heating and cooling cycle produced by welding.

The welding arc in Fig. 17.5 is moving at a given speed from section D-D towards section A-A. Along section A-A, which is ahead of the welding arc, the temperature change due to welding is almost zero. Therefore there is no question of metal expansion or contraction and hence no strains and stresses. Whereas at the location of the welding arc at section B-B, the rise in temperature is extremely high

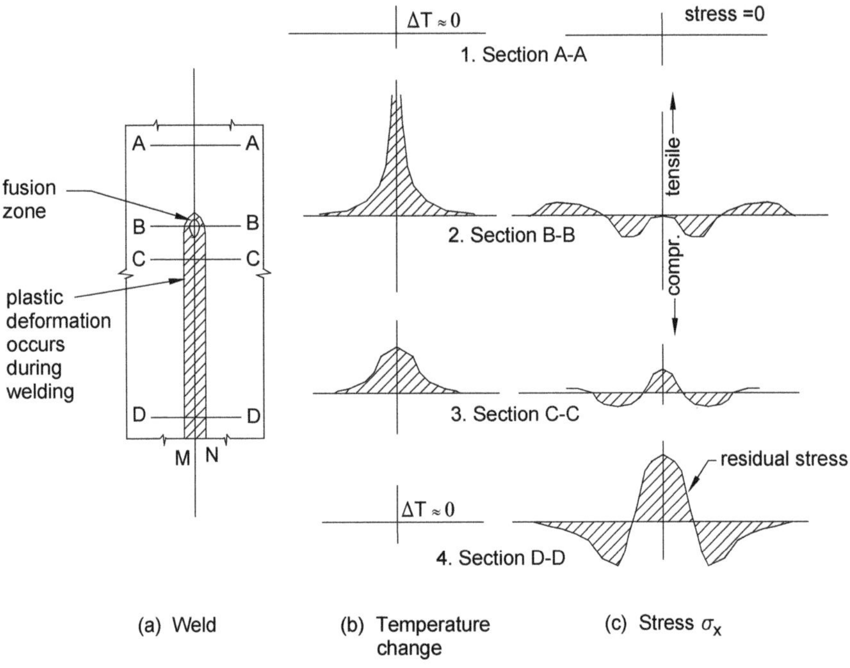

Fig. 17.5 Schematic representation of changes in temperature and stresses during welding

causing a high temperature difference ΔT. As the arc approaches section B-B, the heating phase at B-B starts. This will cause the hot metal surrounding the molten weld pool to expand. However the far away cold metal (unaffected by the welding heat) will restrict (oppose) this expansion causing compressive stresses to form. The stress distribution along section B-B is shown in Fig. 17.5c-2. Since molten metal can not sustain load, stress underneath the welding arc is close to zero. Compressive stresses are formed in regions, a short distance away from the arc, as the expansion of these areas is restrained by the surrounding cold metal. The yield stress being low at high temperatures, plastic strains occur in the near vicinity of the weld pool under the action of this compressive stress.

Along section C-C, which is some distance behind the welding arc, it is under the cooling phase and the temperature is naturally lower than that at section B-B. With solidification of the molten pool, contraction of the weld deposit along with the surrounding metal in the close vicinity as shown by the hatched portion MN in Fig. 17.5 has started. Once again this contraction process is restrained by the surrounding cold metal causing tensile stresses to form in regions close to the weld. As the distance from the weld increases, the stresses first change to compressive and then become tensile as shown in Fig. 17.5c-3.

As the cooling continues, the contraction of metal also continues causing the magnitude of the tensile stress to increase and finally at section D-D, which is far

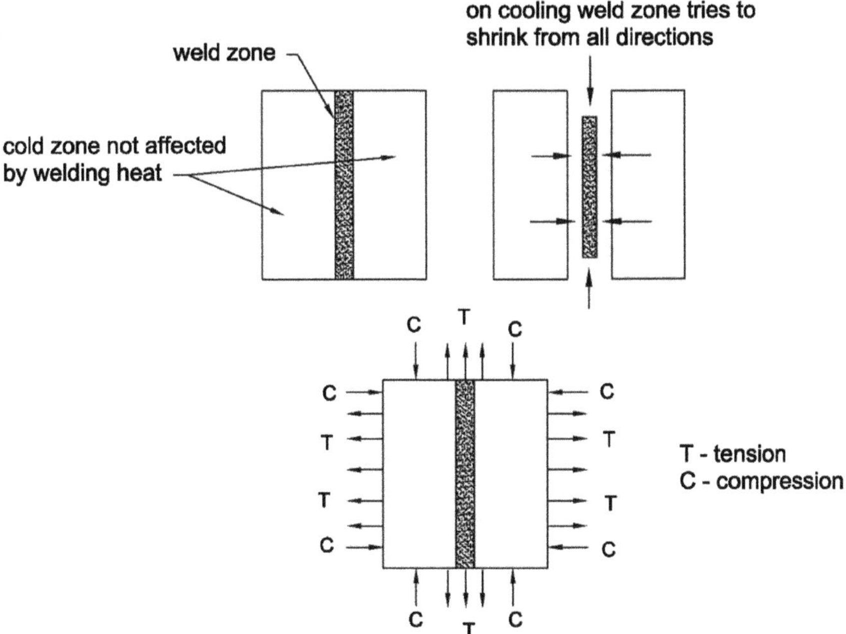

Fig. 17.6 Conceptual representation of development of residual stresses in butt welding

away from the welding arc where the meal has cooled down to ambient temperature, high tensile stresses are produced in regions near the weld. To balance this tensile stress field, compressive stresses are produced away from the weld as shown in Fig. 17.5c-4. This is the usual distribution of longitudinal residual stress that remains after welding is completed.

Conceptual presentation of residual stress development due to butt welding is shown in Fig. 17.6. If the weld zone is considered to be free from the rest of the plate as shown in the Fig. 17.6, it would then tend to shrink from all directions on cooling. But in reality it is an integral part of the remaining portion of the plates which remained unaffected by the welding heat, i.e. remained cold. This will result in development of tensile forces along the longitudinal as well as along the transverse direction of the plate causing plastic strains to develop.

As no external forces are working, for the plate to be in equilibrium condition, these internal forces must be in self-equilibrating condition. Thus the longitudinal tensile force will get balanced by compressive force along both sides of the welded region and the transverse tensile force will get balanced by compressive force developed at both ends of the plate as schematically shown in Fig. 17.6.

The typical welding residual stress distribution in the butt-welded joint is shown in Fig. 17.7. The positive and negative sign indicates tensile and compressive residual stress respectively. As the transverse (along Y axis) thermal gradient is larger than the longitudinal thermal gradient, the transverse plastic strain is about

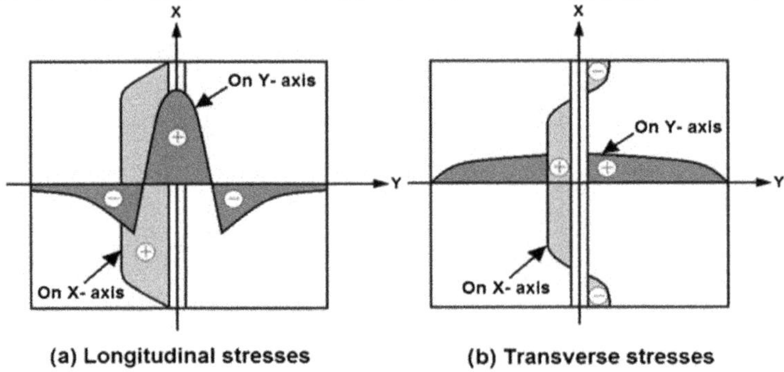

Fig. 17.7 Typical residual stress distribution in a butt weld

ten times larger than the longitudinal plastic strain. As a result, the transverse shrinkage is much larger than the longitudinal shrinkage and therefore the magnitude of transverse residual stress is much smaller than the longitudinal residual stress.

The magnitude of welding residual stress depends on various factors such as rate of heat input, degree of restraint in the structure, coefficient of thermal expansion, melting point of the material, weld joint geometry, preheating, post weld heat treatment, tensile strength of weld metal and parent metal, welding sequence, plate thickness, etc.

17.2.1 Plastic Strain

A simple bar model, as shown in Fig. 17.8 [7], can be considered to explain the mechanical behaviour of a butt-welded joint in the longitudinal direction. The bar is restrained at both the ends. The analogy of this model with a welding situation is as follows: The high temperature welded zone and metal adjacent to it corresponds to the restrained bar. This restraint replicates the restraining effect of the surrounding cold metal which is at near room temperature condition. Through this bar model, the effect of thermal expansion and contraction of the weld metal is simulated.

The restrained bar is assumed to be subjected to a thermal cycle. The heating cycle leads to a maximum temperature of T_{max} from 0 °C and subsequently cooling down to 0 °C. It is to be noted that the elastic-plastic behaviour of the bar depends on the maximum temperature T_{max}. If T_{max} is more than the yield temperature T_Y, it may cause formation of plastic strains leading to development of residual stress in the restrained bar. T_Y is the temperature at which yielding takes place in the restrained bar. It is given by,

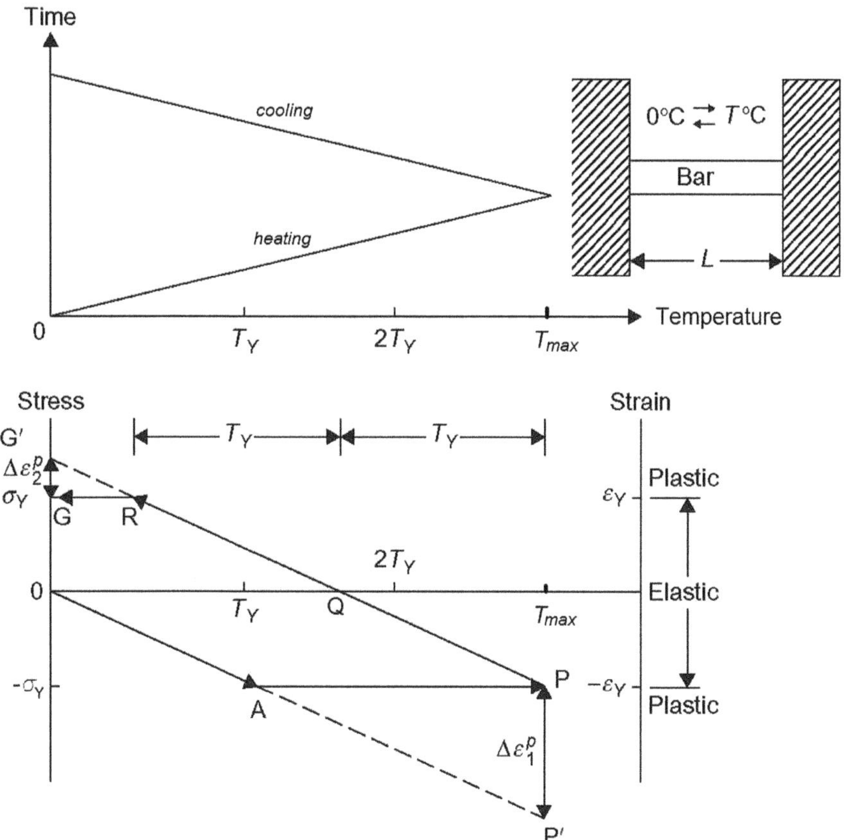

Fig. 17.8 Bar model and stress strain history of the restrained bar [7]

$$T_Y = \frac{\sigma_Y}{\alpha E}$$

Heating Stage (heating from T = 0 °C to T_{max})

In a welding situation, the maximum temperature at the weld zone and its near vicinity is much beyond $2T_Y$. The equivalent bar representing this zone is as shown in Fig. 17.8. It is assumed that this bar has been heated to a temperature T_{max}, which is above $2T_Y$. The corresponding thermal strain is indicated by the point P′ in the figure. As the temperature is much beyond the yield temperature, the bar is therefore in the plastic range. Hence the stress level does not increase and stays at σ_Y and reaches the point P as shown in the figure. The additional thermal strain induced for temperature above T_Y is given by,

$$\Delta\varepsilon^T = \alpha\Delta T = \alpha(T_{max} - T_Y)$$

This strain is suppressed to turn into compressive plastic strain as shown by $P'P$,

$$\Delta\varepsilon^P = -\Delta\varepsilon^T = -\alpha(T_{max} - T_Y) \equiv \Delta\varepsilon_1^P$$

Cooling Stages

Stage 1 (cooling from T_{max} by T_Y, i.e. from P to Q in Fig. 17.8).

At this cooling stage, the shrinkage strain of the bar is $(-\alpha T_Y)$, which reduces the stress by σ_Y and the stress in the bar vanishes. The stress point moves from P to Q.

Stage 2 (cooling further by T_Y to $T = T_{max} - 2T_Y$).

This further cooling results in a shrinkage strain of $(-\alpha T_Y)$. However the bar being not free to shrink because of its restrained ends, tensile stress of yield magnitude is formed. The stress point reaches R as shown in Fig. 17.8.

Stage 3 (cooling to T = 0 °C).

At this stage, the shrinkage strain is $-\alpha(T_{max} - 2T_Y)$, and the strain reaches the point G′ shown in Fig. 17.8. As the stress at this stage is at the yielding point in tension, no further increase of stress occurs. The shrinkage strain thus turns into plastic tensile strain. Stress point R shifts to G, and the plastic strain increment is given by,

$$0 = \Delta\varepsilon = \Delta\varepsilon^P + \Delta\varepsilon^T = \Delta\varepsilon^P - \alpha(T_{max} - 2T_Y)$$

$$\Delta\varepsilon_P = \alpha(T_{max} - 2T_Y) \equiv \Delta\varepsilon_2^P$$

One can observe that, plastic strain develops both during heating and cooling stages, when the peak temperature is well beyond the yield temperature:

Heating stage,

$$\Delta\varepsilon^P = -\alpha(T_{max} - T_Y) \equiv \Delta\varepsilon_1^P$$

Cooling stage 3,

$$\Delta\varepsilon_P = \alpha(T_{max} - 2T_Y) \equiv \Delta\varepsilon_2^P$$

The plastic strain during heating phase is compressive and the one in cooling phase is tensile. The residual plastic strain is therefore the summation of both and is given by,

$$\Delta\varepsilon_P = -\alpha(T_{max} - T_Y) + \alpha(T_{max} - 2T_Y) = -\alpha T_Y$$

This plastic strain causes the residual stress to develop. This plastic strain is often referred to as inherent strain ε^*,

$$\varepsilon^* = \Delta\varepsilon_1^p + \Delta\varepsilon_2^p = -\alpha T_Y$$

The corresponding inherent deformation is given by,

$$\Delta L^* = \varepsilon^* = -\alpha T_Y L$$

Thus one can see that the maximum temperature T_{max} at the heating stage affects the formation of the resulting plastic strain, which is the source of residual stress. Hence the magnitude of residual stress is influenced by the maximum temperature attained. When the base temperature is T °C instead of 0 °C, the yield temperature T_Y should be modified to

$$T_Y = \frac{\sigma_y}{\alpha E} + T$$

17.3 Distortion Mechanism

Fusion welding being a thermal process, it involves application of intense, concentrated heat in localised area. The heat from the welding arc causes an uneven temperature distribution in the welded structure. The weld metal and its adjacent area undergo a thermal cycle, heating followed by natural cooling. It causes restrained expansion during heating cycle, followed by restrained contraction during the cooling cycle resulting in formation of elastic and/or plastic strains. As the structure fully cools down, residual stresses are formed and distortion may occur.

During heating and cooling in the welding process, thermal strains occur in the weld metal and base-metal near the welded zone. The stresses resulting from these strains produce internal shrinkage forces leading to structural deformation, like lateral and longitudinal shrinkage, bending, buckling and rotation.

The mechanism of distortion due to a thermal cycle can be easily visualised through the following example. Consider a small cube of steel that is heated uniformly as shown in Fig. 17.9. It expands in all directions. As the heat source is removed and the cube cools down, it contracts uniformly to its original dimensions. In the next experiment, the cube is placed between the jaws of a vice such that it does not fall off, as shown in Fig. 17.9b. It is again heated as done previously. However, now it is restrained by the jaws of the vice, hence it can expand only in the vertical direction. Thereby compressive reaction forces are generated along the lateral direction as shown in Fig. 17.9c.

Subsequently, on removal of the heating source, as the cube cools down, contraction takes place uniformly, unrestrained from all directions as shown in Fig. 17.9d. The result is, the cube is narrower laterally and thicker in vertical direction. That the cube has become narrower in lateral direction becomes evident

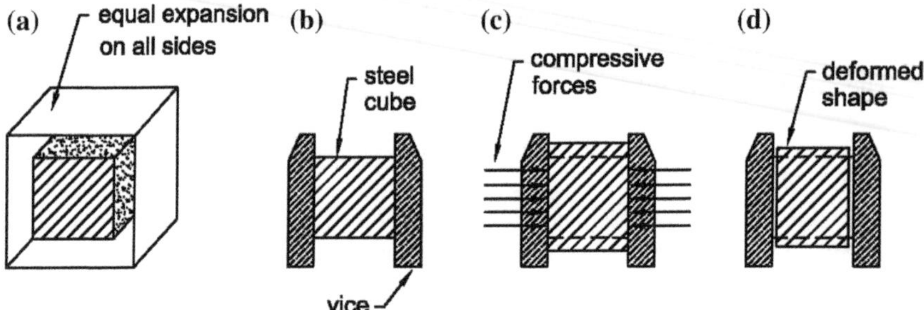

Fig. 17.9 Distortion mechanism

from the fact that the cube slips off from the vice. Hence the residual shape of the cube is permanently deformed or distorted.

The welding of butts and seams in plate panel assemblies may cause an angular deformation along the weld lines. This may also cause shrinkage transverse to the weld seams or butts. Subsequent welding of stiffeners may also add to shrinkage and development of compressive stresses in the plate. If the stress level exceeds the critical buckling stress of the plate panels, buckling may take place. This phenomenon makes fabrication of thin stiffened panels difficult.

The same distortion mechanism also applies to large assembly units and to the ship itself. If welding is carried out on top (above neutral axis) of a long shallow unit having less rigidity, the unit may deflect upwards at the ends. This explains the phenomenon of the bows and sterns of ships to lift off the keel blocks by several millimetres. On the other hand, if the unit is relatively deep and hence rigid, there will be less tendency of such deflection at the ends. This rigidity is achieved by a proper erection sequence like, if an upper unit is installed and tack welded to the lower unit and subsequently the welds are made in the lower unit, the top unit will provide added restraint thus reducing the deflection.

Essentially all the weld induced distortions are caused by the shrinkage force generated due to the thermal loading on the structure. Depending on the pattern of this shrinkage forces (compressive forces) various structural deformation like bending, rotation and buckling take place.

17.3.1 Shrinkage Due to Butt Welds

One of the basic dimensional changes due to welding is in-plane shrinkage of the plates being welded. This shrinkage may take place both in the transverse direction, perpendicular to the weld line and in the longitudinal direction, parallel to the weld line. Hence the resulting shape of the panel after butt welding may become as shown in Fig. 17.10.

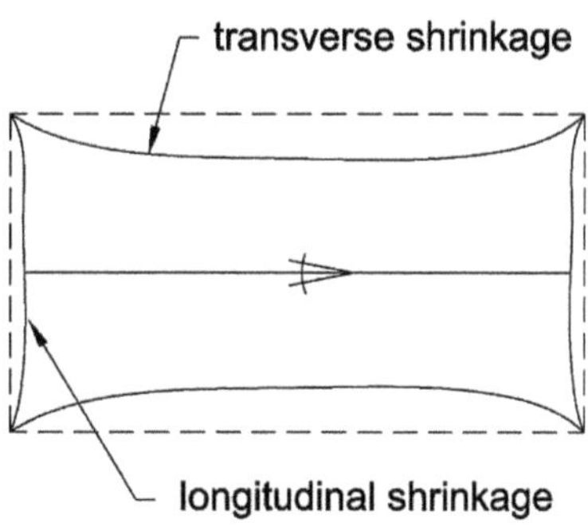

Fig. 17.10 Transverse and longitudinal shrinkage of a butt welded *panel*

Mechanism

Several researchers have studied the mechanism of shrinkage distortion due to welding. The most important finding of these analyses are:

- The major portion of transverse shrinkage of a butt weld is due to contraction of the base plate. Shrinkage of the weld metal itself is only about 10 % of the actual shrinkage.
- Transverse shrinkage is generally of much higher order than longitudinal shrinkage.

During welding as the heat from the molten pool gets conducted, the parent metal expands with simultaneous contraction of the weld deposit as the welding torch moves away. This expansion of the parent metal is resisted by the cold part of the base metal not undergoing sufficient temperature change. Thus the plates undergoing welding can be divided into two zones as shown in Fig. 17.11.

Nearfield: It is directly affected by the thermal cycle and hence experiences substantial expansion and subsequent contraction.

Farfield: Cold part of the plate acting as elastic restraints and not directly affected by the thermal cycle and hence does not undergo a direct expansion or contraction.

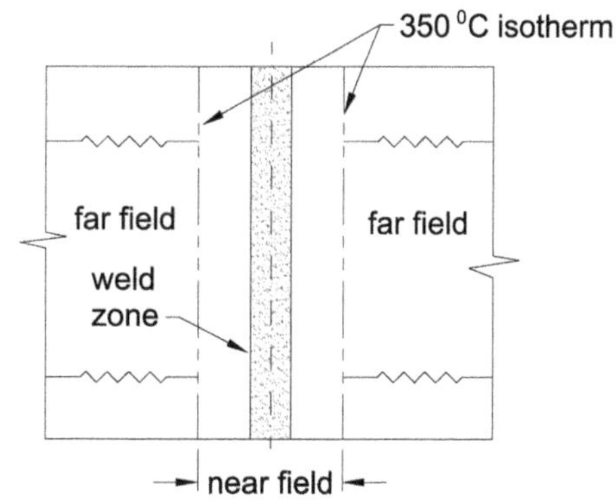

Fig. 17.11 Near field and far field zones of the plates undergoing welding

The boundary of nearfield zone can be the 350 °C isotherm, which means that the thermal expansion in base material below 350 °C is considered to be negligible [8].

The mechanism of transverse shrinkage can be explained as follows:

- The base metal in the nearfield zone expands during the heating cycle.
- The expansion is constrained by the cold part of the base metal, i.e. farfield zone.
- Compressive forces are generated. This can be considered as energy stored in a set of springs along the boundary of the nearfield. This is caused by the elastic compression of the farfield.
- During the subsequent cooling cycle, contraction of the nearfield takes place and simultaneously the energy stored in the said springs gets released thus adding to the contraction forces.
- Under the action of these forces, the base metal along with the weld deposit with low yield point because of elevated temperature yields and plastic flow of material takes place giving rise to transverse shrinkage.

17.3.1.1 Angular Distortion Due to Butt and Fillet Welds

Angular distortion due to butt welding of two plates may take place as shown in Fig. 17.12. This happens due to a bending moment caused by the nonuniform shrinkage forces generated across the plate thickness near the weld zone.

The angular change in a butt weld depends on the joint geometry. A square butt leading to a more uniform heat distribution in thinner plates yields minimum angular deformation. Whereas single V groove welds lead to a higher level of angular distortion. However, thicker plates with double V groove joint geometry, the resulting angular deformation will be much less. This is because of the balancing effect of the top and bottom welding.

When several stiffeners are fillet welded with no external constraints as in the case of a flat plate panel stiffened by longitudinals, the plate bends at each joint and forms a sort of polygon as shown in Fig. 17.13. The angular distortion of a free joint can be characterized by angular change ϕ as shown in Fig. 17.13.

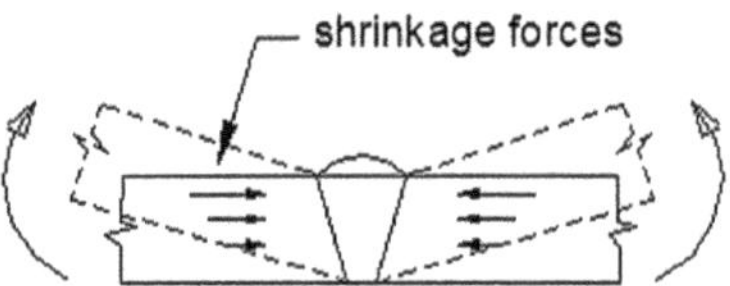

Fig. 17.12 Angular deformation due to butt welding

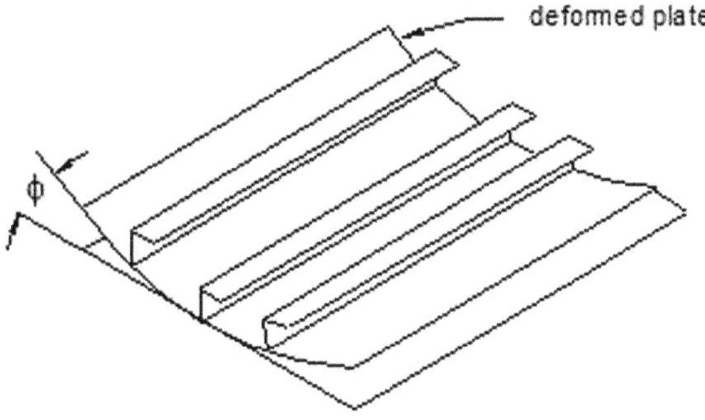

Fig. 17.13 Angular distortion in an unrestrained stiffened *panel*

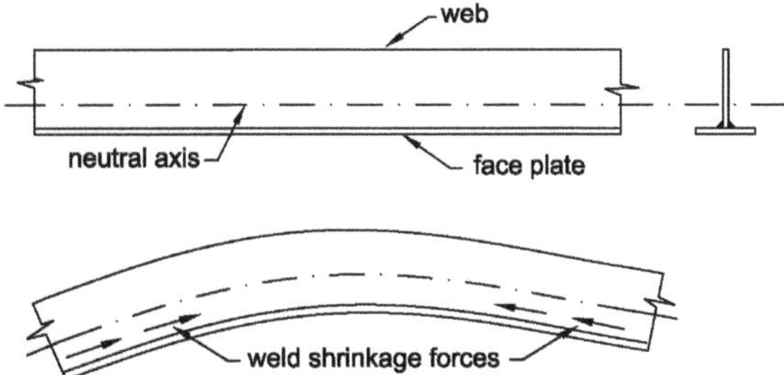

Fig. 17.14 Bowing distortion of built-up girder

17.3.2 Longitudinal Bending (Bowing) in Built-up Girders

In fabrication of built-up girders, it is often observed that bending occurs along the length of the beam. This happens due to the shrinkage of the weld deposit joining the web to the face plate. If the joint is below the neutral axis of the girder, which is normally the situation, the bending takes place as shown in Fig. 17.14. As the fillet deposit shrinks longitudinally, a bending moment is developed about the neutral axis of the tee section causing a "bowing effect".

17.3.3 Buckling Due to Welding

Restrained expansion and contraction of the weld metal and adjacent base metal occurs in a welding process. This causes thermal strains to develop in the weld metal and base metal near the welded zone. The stress resulting from these strains produces internal force, which partially may manifest in structural deformation and rest may remain as residual stress. High tensile stresses are produced in regions near the weld, while compressive stresses are produced in regions away from the weld. This is the usual distribution of residual stresses that remain after welding is completed. If the compressive stress thus developed exceeds the critical buckling stress of the plate panel, buckling of the plate takes place as can be seen in Fig. 17.2.

Ships designed for high speed operation are generally weight sensitive and calls for thin plates and sections. The increased use of thin plates in panel fabrication results in significantly increased distortion. At the same time, the flatness aspect of plates and sections, i.e. the initial deformations, also significantly reduces the critical buckling stress of the stiffened panels. These factors make the stiffened plate members more vulnerable to buckling.

17.3.4 Distortions in Ship Hull Units

In ship construction pre-fabrication is extensively used. First the subassemblies comprising of flat or curved plates and stiffened with longitudinals and transverses are fabricated. These are then put together to get the larger assemblies. Subsequently these assemblies are put together to fabricate the units, followed by blocks. For example, part of the bottom shell and the respective inner bottom panel are fabricated along with the bilge plate. These are then assembled to form the double unit. The side shell panels are prefabricated and erected over this double unit and finally the pre-fabricated deck panel is erected completing the block.

Now this entire erection process is very simple and straight forward, provided, all the pre-fabricated components are correct to their design dimensions. In reality, if adequate preventive measures are not taken, often there will be mismatch between the components, causing extreme difficulty in structural fitup and alignment. For example, weld induced shrinkage may lead to mismatch between the dimensions of the bottom unit comprising of bottom shell and bilge plate and the inner bottom panel. Similarly, once the side shell panels are erected, because of overall shrinkage of the bottom unit, the pre-fabricated deck panel may come out to be dimensionally larger than the distance between the vertical side shell panels. In such a condition either the deck panel has to be trimmed at the edge to match the structure or the side shell panels are to be forcibly stretched out such that the deck can be fitted. In both the cases, the overall dimension of the block will be different from the design dimensions and hence there will be dimensional mismatch with the next block that needs to be connected to this block.

This type of dimensional mismatch takes place primarily because of shrinkage of ship hull units. Hence to mitigate this mismatch, one needs to assess the dimensional shrinkage that may occur in a given hull unit and accommodate that in the initial design as appropriate shrinkage allowance.

17.3.5 Shrinkage of Hull Units

The cumulative effect of distortion resulting from the welding of butts and stiffeners during the fabrication of ship hull units leads to an overall change in dimensions as schematically shown in Fig. 17.15.

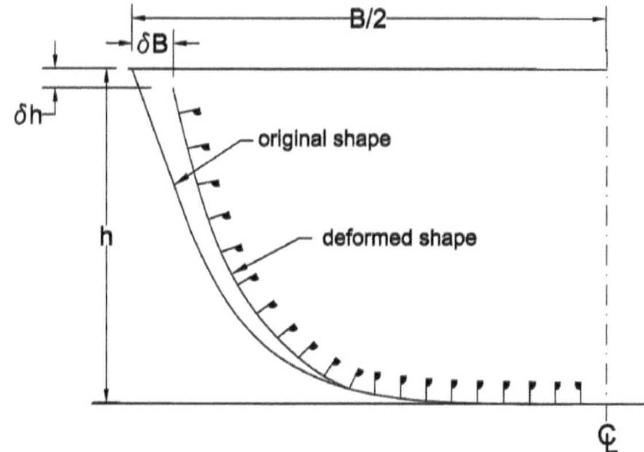

Fig. 17.15 Schematic representation of hull shrinkage

Reduction in breadth due to cumulative effect of shrinkage = δB

Reduction in height caused by cumulative effect of shrinkage = δh.

During the fabrication stage, as a general fabrication practice, the hull plates are laid on the skid and are rigidly connected to it by means of lugs welded to it as can be seen in Fig. 17.16. In this condition all the stiffeners are laid out and welded as shown in Fig. 17.17.

For subsequent erection as this prefabricated unit is released from the skid and shifted to the erection berth for erection of side shell panels and decks, it is often observed that substantial dimensional change has taken place in the bottom hull unit. This dimensional inaccuracy causes further erection difficulties owing to mismatch and misalignment between the side shells and the deck.

This problem can be addressed using the far field–near field concept as described above. Plastic deformation is expected only in the near field, whereas the far field behaves elastically throughout the heating and cooling cycle. The overall shrinkage of a unit can be estimated from the cumulative shrinkage effect of plate butts and

Fig. 17.16 A hull unit connected to the skid using lugs

Fig. 17.17 A hull unit in a skid fitted with longitudinal and transverses

fillet welding of longitudinals. This will help the designer to evaluate the extent of hull shrinkage that may take place while fabricating the hull unit. Suitable shrinkage allowances thus can be incorporated in the structural layout of the unit at the design stage. It would greatly nullify the effect of the weld induced shrinkages and retain the dimensional accuracy of a unit [3].

17.3.5.1 Assessment of Hull Shrinkage

The mathematical model is based on the assumption that the plate is made up of a thermoelastoplastic zone (Near Field) and a fully elastic zone (Far Field). Hence, the plate undergoing welding can be modelled as a combination of a thermoelastoplastic bar and an elastic spring, as shown in Fig. 17.18.

The near field is assumed to extend on both the sides of the weld deposit to a location where the peak temperature attained is about 350 °C [8], and the rest of the plate is far field. Plastic deformation is expected only in the near field, whereas the far field behaves elastically throughout the heating and cooling cycle. The temperature field associated with the moving welding arc is solved assuming a quasi-stationary state using temperature dependent material properties.

The Near Field zone is subdivided into finite strips. The shrinkages of these individual strips are estimated. The algebraic sum of these shrinkages give the total shrinkage caused by the individual butt or fillet welding.

The method followed to calculate the shrinkage can be summarised as follows. The calculations are done using two modules:

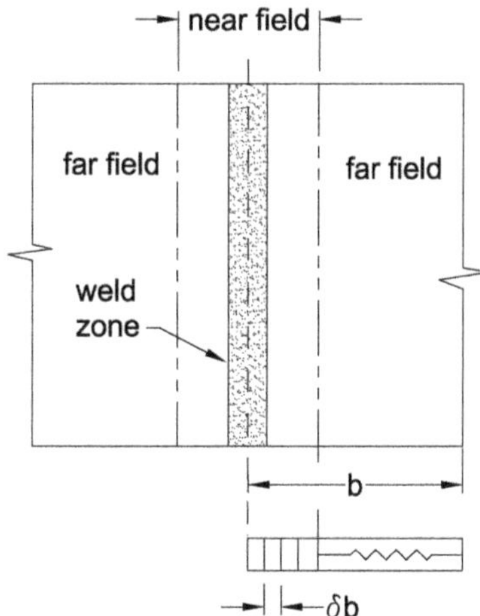

Fig. 17.18 Near field–far field model for estimation of weld induced transverse shrinkage

(i) Module-I calculates the temperature distribution over the plate for given welding parameters and temperature dependent properties of the metal. Based on this temperature distribution, the extent of the near field and far field zones are estimated.
(ii) The near field is further subdivided into a number of finite strips. The temperature across the strips is assumed to be constant.
(iii) For each strip, depending on its peak temperature, the distortion is calculated.
(iv) The algebraic sum of the shrinkages of each strip is taken to obtain the final shrinkage, completing Module-II.

The effect of the transverses on overall deformation of the unit has not been taken into account. Logically these members will have an effect of opposing the transverse deformation of the units. This appears to be one of the main reasons for overestimation of the transverse shrinkage of the units. However, the transverse members can never fully nullify the shrinkage effect caused by the welding of the longitudinal members. The resulting shrinkage is accommodated through elastic deflection of the transverse members.

Calculation of Temperature field

The moving welding arc results in 3-D temperature field. Generally the size of individual stiffened panels are quite large, hence long lengths of welding is involved. Hence solving the heat flow equation a quasi-stationary state is considered. The heat input is given by,

$$Q = \eta VI \quad \text{J/s}$$

where,

V arc voltage (V)
I welding current (A)
η arc heat transfer efficiency.

In case of butt welding, $0.5Q$ is taken for individual plates, whereas in case of fillet welds in longitudinals and transverses, $(1/3)Q$ is taken for individual sides of the base plate and stiffener web [3].

The modified Rosenthal formula, given in Eq. (17.1), incorporating temperature dependent material properties, can be used to calculate the temperature field 'T' for a given welding heat input.

$$T - T_0 = \frac{Q}{2\pi\lambda} e^{\left(-\frac{v}{2\alpha}\right)w} \left\{ \frac{e^{\left(-\frac{v}{2\alpha}\right)R}}{R} + \sum \left[\frac{e^{\left(-\frac{v}{2\alpha}\right)R_n}}{R_n} + \frac{e^{\left(-\frac{v}{2\alpha}\right)R'_n}}{R'_n} \right] \right\} \tag{17.1}$$

where,

$$w = x - vt$$

$$R = \sqrt{(w^2 + y^2 + z^2)}$$

$$R_n = \sqrt{\left[w^2 + y^2 + (2nt_h - z)^2\right]}$$

$$R'_n = \sqrt{\left[w^2 + y^2 + (2nt_h + z)^2\right]}$$

$$\alpha = \lambda/(\rho C_p)$$

Here 'w' is the moving coordinate, x, y, and z are the Cartesian coordinates, 't' the time, n is an integer, taken to be 50 in this calculation, 't_h' is plate thickness, and ρ, λ and C_p are the density, thermal conductivity and specific heat of steel respectively and 'v' welding speed. T_0 is the ambient room temperature. The flow chart for the calculation of the temperature field is given in Fig. 17.19.

Calculation of strip width in Near Field

The near field is further subdivided into finite strips. The temperature across the strip width is considered to be constant. The correlation between welding heat input, welding speed, plate thickness and width of the strips is given by Eq. (17.2), [8],

$$\delta b = -0.9 + (7.402 \times 10^{-4}\acute{Q}) - (10^{-8}\acute{Q}^2) \tag{17.2}$$

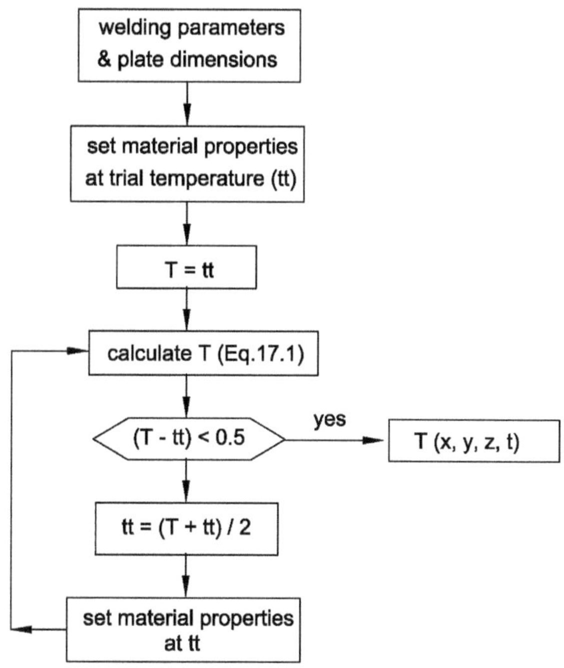

Fig. 17.19 Flow chart for temperature field calculation

where

δb strip width (mm)
$\acute{Q}$ $\eta VI/vt_h$
η arc efficiency
V arc voltage (V)
I welding current (A)
v welding speed (cm/s)
t_h plate thickness (cm).

For calculating shrinkage, a suitable strip width (δb) is chosen on the basis of the Eq. (17.2).

Calculation of shrinkage
For each strip, depending on its peak temperature, the shrinkage (δp mm) is calculated based on the following equation Eq. (17.3),

$$\delta p = \left[\frac{\alpha T_m T_p}{2T_m - T_p} - \frac{\sigma_y}{E}\right]\delta b \tag{17.3}$$

where,

α coefficient of thermal expansion
σ_y Yield stress at room temperature (MPa)
E Modulus of elasticity (MPa)
T_m Melting temperature (°C)
T_p Peak temperature attained in each strip (°C).

The yield stress and modulus of elasticity are assumed to vary linearly with temperature as given below [3]:

$$\sigma_y = \sigma_{y0}[1 - (T/T_m)]$$

$$E = E_0[1 - T/T_m]$$

σ_{y0} and E_0 are the yield stress and modulus of elasticity at ambient room temperature respectively.

The algebraic sum of the shrinkage of each strip (δp) within the near field zone is taken to obtain the actual shrinkage of the plate.

17.3.6 Buckling of Shell and Deck Panels

It has been already shown that residual tensile stress develops in and around the weld zone. These stresses being of self-equilibrating nature, the tensile stress, thus generated, is balanced by compressive stresses in areas away from the weld zone. If the compressive stress exceeds the critical buckling stress of the plate panel in between the stiffeners, buckling of the panel takes place.

Compressive Stress Estimation
The stress field occurs because of the welding heat. At the design stage, the designer may not have information concerning the net weld heat input for any given size of weld. The welding heat is directly related to the leg length in case of fillet welding and to the plate thickness in case of butt weld. Therefore the following formulae are suggested for some common welding processes [9],

For fillet welds
For fillet welding leading to a weld deposit of leg length '*l*' mm, the rate of heat input is given by,

$$\eta Q/S_w = l(l+1)/C \text{ kJ/mm}$$

The factor C depends on the welding process as shown in Table 17.1.

Table 17.1 Values of factor C for various welding processes

S. No.	Welding process	C
1	GMAW—CO_2	33
2	MMAW—Rutile (horizontal)	17
3	MMAW—Rutile (horizontal with iron powder)	20
4	MMAW—Rutile (vertical)	14
5	MMAW—Rutile (downhand with iron powder)	30
6	SAW	33

For butt welds

The rate of heat input for butt welding of plate of thickness 't_h' mm is given by,

$$\eta Q/S_w = t_h(t_h + 3)/C \text{ kJ/mm}$$

The heat input causing the expansion followed by contraction of the heated metal, will cause compressive shrinkage forces to develop in the plate as given by Eq. (17.4).

$$F = f\,\eta(Q/S_w)\,\text{kN} \tag{17.4}$$

where, $f = 200$ for a wide range of welding process variables for steel [9], S_w is welding speed in (mm/s) and Q is welding heat input in (kJ/s). Therefore the compressive stress (σ_r) in the panels of thickness 't_h' mm bounded by longitudinals at a spacing of 's' mm will be as given by Eq. (17.5),

$$\sigma_r = \frac{200\,\eta(Q/S_w)}{s\,t_h}\,\text{kN/mm}^2 = \frac{2\times 10^5\eta(Q/S_w)}{s\,t_h}\text{N/mm}^2(\text{MPa}) \tag{17.5}$$

Critical buckling stress estimation

The critical buckling stress for the stiffened panel with frame spacing 's' under inplane compression along the length is given by Eq. (17.6),

$$\sigma_{cr} = \frac{k\pi^2 E}{12(1-\gamma^2)(s/t_h)^2} \tag{17.6}$$

where,

k buckling coefficient
E modulus of elasticity
t_h plate thickness
γ Poisson's ratio

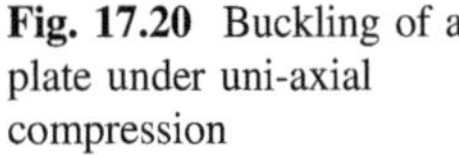

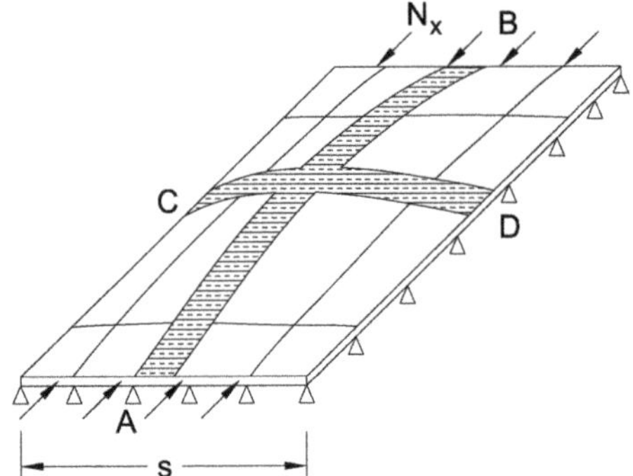

Fig. 17.20 Buckling of a plate under uni-axial compression

The expression for the critical buckling stress is a function of the width to thickness ratio s/t_h. Now, why should the critical buckling stress for compression along x-direction be a function of the width 's' in the y-direction?

As the compressive load Nx on the plate is increased and reaches the critical buckling load N_{cr}, the central part of the plate, such as the strip AB tends to buckle as shown in Fig. 17.20. Now, if a plate strip CD in the transverse direction is considered, one can observe that this strip resists the tendency of the strip AB to deflect out of the plane of the plate, i.e. in the z-direction. The shorter the width 's' more will be the resistance offered by CD to AB. Hence the strip AB, till it buckles, behaves like a column on elastic foundation, whose stiffness depends on 's'. That is why the plate width 's' figures in the expression for critical buckling stress.

The transverse stiffeners are spaced 3 times the spacing of longitudinal as per general shipbuilding practice. The plate panel in between the stiffeners tend to buckle under the compressive stresses developed due to the welding of the longitudinals. The longitudinals provide for sufficient restraint to the panel edges and thereby restrain them from rotation. The edges more or less behave like fixed edges. Therefore for fixed-fixed condition of the panel edges the buckling coefficient k is 6.97 [10, 11]. Whereas for free-free condition it is equal to 4. However, depending on the scantling of the longitudinals, there might be some deflection of these longitudinals. Keeping this in view, the buckling coefficient k was taken equal to 5.5 [12]. With increasing plate thickness and decreasing frame spacing the critical buckling stress will increase. However it will also lead to increase in weight of the stiffened panel. For a given plate panel as σ_r exceeds σ_{cr}, buckling is expected.

References

1. Huang, T. D., Dong, P., Decan, L. A., & Harwig, D. E. (2003). *Residual stresses and distortions in lightweight ship panel structures* (pp. 1–26). Spring/Summer: Northrop Grumman Technical Review Journal.

2. Huang, T. D., Dong, P., Decan, L., Harwig, D., & Kumar, R. (2004). Fabrication and engineering technology for lightweight ship structures, part 1: distortions and residual stresses in panel fabrication. *Journal of Ship Production, 20*(1), 43–59.
3. Mandal, N. R. (1999). Prediction of dimensional changes during fabrication of thin shell ship plate hull bottom units. *Science and Technology of Welding and Joining, 4*(5), 290–294.
4. Mcpherson, N. A., & Crow, A. (2006). Plate requirements for current naval vessel builds. In *Proceedings, Achieving Profile and Flatness in Flat Products Conference*, January, Birmingham, UK.
5. Spicknall, M. H., Kumar, R., & Huang, T. D. (2005). Dimensional management in shipbuilding: a case study from the Northrup Grumman ship systems lightweight structures project. *Journal of Ship Production, 21*(4), 209–218.
6. McPherson, N. A. (2007). Thin plate distortion—The ongoing problem in shipbuilding. *Journal of Ship Production, 23*(2), 94–117.
7. Ueda, Y., Murakawa, H., & Ma, N. (2012). Welding deformation and residual stress prevention. ISBN 978-0-12-394804-5, Elsevier Inc.
8. Mandal, N. R., & Sundar, C. V. N. (1997). Analysis of welding shrinkage. *Welding Journal, 76*(6), 233s–238s.
9. Ractliffe, M. A. (1983). The basis and essentials of thermal residual distortion in steel structures. *The Royal Institution of Naval Architects.*
10. Bulson, P. S. (1970). *Theory of flat plates*. London: Chatto and Windus.
11. Timoshenko, S. P., & Gere, J. M. (1961). *Theory of elastic stability.* New York: McGraw Hill Book Company.
12. Mandal, N. R., Prabu, Sree Krishna, & Kumar, Sharat. (2014). Buckling of stiffened panels and its mitigation. *Journal of Ship Production and Design, 30*(4), 201–206.

Chapter 18
Distortion Control and Mitigation

Abstract Out-of-plane distortion results from buckling and angular deformations of panel edges as well as in the panel in between the stiffeners. The most effective way of reducing distortion is to control the formation of plastic stains produced in regions near the weld. If the appropriate distortion control is applied, the final distortion will be reduced. On the other hand improper control may lead to higher distortion. Control of distortion needs to start from implementation of appropriate design and fabrication techniques. Reducing welding heat input reduces all types of weld-induced distortions. Distortion mitigation through heat sinking involves removal of heat from the welded region such that it does not get spread in the plates away from the weld zone thus reducing residual stresses and distortion. Mitigation of buckling distortion of stiffened panels can be done by applying an active distortion mitigation technique known as Thermo-Mechanical Tensioning (TMT). In this method through the use of restraining and tensioning lugs, the tensile residual stress field generated by welding of stiffeners is reduced. This leads to reduction of the balancing compressive stress field.

Weld-induced distortion is a problem well known in shipbuilding industry. Over the years, shipbuilders have come to expect some degree of distortion and have learned to live with its consequences. This can be observed from the prevailing practice of flame straightening to remove distortion, leaving stiffener ends unwelded to facilitate fitting and over sizing of panels for later onsite adjustment and trimming.

In recent years, greater use of thin (10 mm and below) plates in panel fabrication, has resulted in significantly increased distortion. The net result has been increased man-hours for fitting, fairing by line heating and rework.

In general two forms of weld-induced distortions are of major concern to shipbuilders:

- In-plane distortion.
- Out-of-plane distortion.

In-plane distortion is a result of shrinkage in the overall panel dimensions, resulting in dimensional inaccuracy. To overcome this difficulty, green material is kept at the

N.R. Mandal, *Ship Construction and Welding*, Springer Series
on Naval Architecture, Marine Engineering, Shipbuilding and Shipping 2
DOI 10.1007/978-981-10-2955-4_18

panel boundaries. Thus the implementation of "shipbuilding without tolerances" technology becomes difficult. Out-of-plane distortion results from buckling and angular deformations of panel edges as well as in the panel in between the stiffeners. Many a time out-of-plane distortion also results in shortening of panel dimensions simply because a deformed plate is shorter than a flat plate. The most effective way of reducing distortion is to control the formation of plastic stains produced in regions near the weld [1]. The necessary control must be made in real time, during welding. If the appropriate distortion control is applied, the final distortion will be reduced. On the other hand improper control may lead to higher distortion.

The basic structural components of ships are orthogonally stiffened plate panels. They are fabricated by welding longitudinal and transverse stiffeners to plate panels. This welding causes a stress field to develop in the stiffened panel. When the magnitude of the compressive component of the residual stress exceeds the critical buckling strength of the panels, it results in buckling of the same. With increasing use of higher tensile steel, it has led to reduction in plate thickness, making them more vulnerable to buckling. At the same time to reduce overall weight and to lower the centre of gravity, thinner plates are being widely used for superstructures.

Buckling of shell and superstructure panels apart from possible loss of strength, it may cause mismatch during block assembly, difficulty in erection of seatings for equipment and aesthetic loss of structures. On a rough estimate, about 3–5 % of labour cost is incurred in the correction of this buckling distortion. In the following sub-section various techniques and methodologies will be discussed to control and mitigate these distortions.

18.1 Distortion Control Through Design

At the design stage, it would be appropriate to study the effect of frame spacing and plate thickness on the buckling strength of the stiffened panels. By increasing plate thickness and/or reducing stiffener spacing, the critical buckling stress of a panel can be increased. Thus out-of-plane distortion due to buckling caused by weld induced stresses can be avoided to a great extent by suitably choosing plate thickness and/or stiffener spacing.

However this may lead to increase in the weight of the structure. Hence these parameters are to be so chosen that the increase in weight is minimised. The critical buckling stress of a plate panel, bounded by longitudinal and transverse stiffeners, depends on stiffener spacing, plate thickness and the boundary restraints caused by the stiffeners. Therefore the buckling strength of the stiffened panels can be changed by changing either of these parameters.

Now if the stiffener spacing and plate thickness are so chosen that the critical buckling stress of the panel becomes higher than the compressive stress developed due to welding of the stiffeners, then buckling of the panels is not expected. However for ship structures, it should comply with classification requirement as well as should not lead to unwanted increase in weight of the stiffened structural panel.

Example Twenty one test cases were considered with 3 different plate thicknesses, 3 different frame spacing and 4 different fillet leg lengths [2]. The variation in fillet leg length were taken keeping in view the cases of possible over welding that may take place during panel fabrication. For different fillet leg length 'l' mm, plate thickness 't_h' mm, frame spacing 's' mm, the compressive stresses (σ_r) developed due to welding heat input of 'Q' kJ/s at a welding speed of 'S_w' mm/s and the corresponding critical buckling stresses (σ_{cr}) of the stiffened panels were calculated based on equations as given in (18.1), (18.2) and (18.3).

$$\sigma_r = \frac{200\,\eta(Q/S_w)}{s\,t_h}\,\mathrm{kN/mm^2} = \frac{2\times 10^5\,\eta(Q/S_w)}{s\,t_h}\,\mathrm{N/mm^2(Mpa)} \tag{18.1}$$

For fillet welds

For fillet welding leading to a weld deposit of leg length 'l' mm, the rate of heat input is given by,

$$\eta Q/S_w = l(l+1)/C\ \mathrm{kJ/mm} \tag{18.2}$$

The factor C depends on the welding process as shown in Table 18.1.

$$\sigma_{cr} = \frac{k\,\pi^2 E}{12(1-\gamma^2)\,(s/th)^2} \tag{18.3}$$

where,

k buckling coefficient = 5.5
E modulus of elasticity (MPa)
t_h plate thickness (mm)
s frame spacing (mm)
γ Poisson's ratio.

Table 18.1 Values of factor C for various welding processes

S. No.	Welding process	C
1	GMAW—CO_2	33
2	MMAW—Rutile (horizontal)	17
3	MMAW—Rutile (horizontal with iron powder)	20
4	MMAW—Rutile (vertical)	14
5	MMAW—Rutile (downhand with iron powder)	30
6	SAW	33

Table 18.2 Weld induced compressive stress and critical buckling stress for varying plate thickness, stiffener spacing and fillet leg length

S. No.	Plate thickness (mm)	Frame spacing (mm)	Leg length (mm)	Net heat input (J/mm)	σ_r (MPa)	σ_{cr} (MPa)	Remarks
1	4	400	3	705.9	88.2	104.3	No buckling
2	4	500	3	705.9	70.6	66.7	Buckling
3	4	600	3	705.9	58.8	46.3	Buckling
4	4	400	4	1176.5	147.1	104.3	Buckling
5	4	500	4	1176.5	117.6	66.7	Buckling
6	4	600	4	1176.5	117.6	46.3	Buckling
7	5	400	5	1176.5	117.6	162.9	No buckling
8	5	500	5	1176.5	94.1	104.3	No buckling
9	5	600	5	1176.5	78.4	72.4	May buckle
10	5	400	5	1764.7	176.5	162.9	Buckling
11	5	500	5	1764.7	141.2	104.3	Buckling
12	5	600	5	1764.7	117.6	72.4	Buckling
13	6	400	4	1176.5	98.0	234.6	No buckling
14	6	500	4	1176.5	78.4	150.2	No buckling
15	6	600	4	1176.5	65.4	104.3	No buckling
16	6	400	5	1764.7	147.1	234.6	No buckling
17	6	500	5	1764.7	117.6	150.2	No buckling
18	6	600	5	1764.7	98.0	104.3	No buckling
19	6	400	6	2470.6	205.9	234.6	No buckling
20	6	500	6	2470.6	164.7	150.2	Buckling
21	6	600	6	2470.6	137.3	104.3	Buckling

The calculated values are given in Table 18.2. The cases in Sl No. 4–6, 10–12 and 19–21 are situations of definite over welding. In such a situation when the plate thickness is 4 or 5 mm, in all cases buckling is expected. However if the thickness is increased to 6 mm having frame spacing of 400 mm, even with excess welding, buckling will not take place as can be seen from the results in Sl No. 19. Also one can observe that with marginal over welding, which is very likely in shop floor condition, with plate thickness of 6 mm and frame spacing as high as 600 mm, buckling will not take place as shown in Sl Nos 16–18 in Table 18.2.

For 4 mm thick panels, frame spacing of 400 mm with fillet leg length of 3 mm is recommended for no-buckling condition. Whereas if fillet deposition cannot be controlled to 3 mm then to have a no-buckling condition, the plate thickness needs to be increased to 5 mm and frame spacing can be equal to or less than 500 mm provided the fillet leg length remains 4 mm. If weld metal deposition cannot be controlled to 4 mm but will be within 5 mm, then for no buckling condition, the plate thickness has to be minimum 6 mm even with frame spacing of 600 mm.

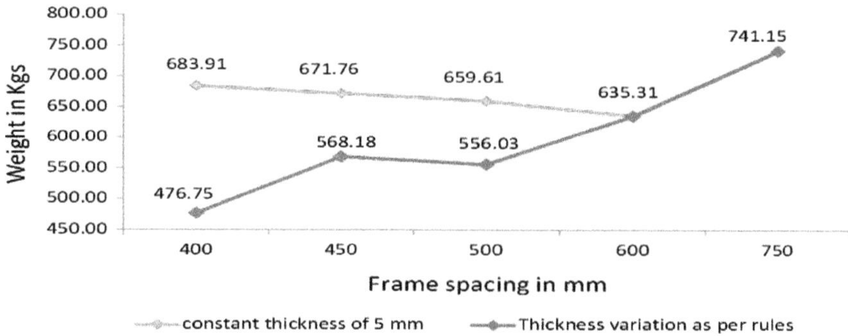

Fig. 18.1 Variation of weight of stiffened *panel* with stiffener spacing

Thus it can be seen that the buckling of such stiffened panels could be effectively avoided by suitably choosing the panel thickness and frame spacing keeping in mind the possibility of over welding.

As far as the effect of changing frame spacing and plate thickness on weight of stiffened panel is concerned, it was observed that, for a constant 5 mm plate thickness, reducing the frame spacing from 600 to 450 mm will cause an increase in weight of the panel by about 5.7 % as shown in Fig. 18.1 [3]. This increase in weight will incur more material cost but at the same time rework will reduce which has direct impact on productivity and quality.

Increasing plate thickness along with reducing scantlings of stiffening members, desired panel strength can be maintained. However it may cause an increase in weight of the panel, e.g. increasing plate thickness from 3 to 4 mm may cause an increase in weight of the stiffened panel by about 30 %. The weight penalty may be significant, however the tradeoff could be the necessary advantages of buckle free panel. One can see from Table 18.3 that buckling is expected in the 3 mm deck panel on welding of stiffeners, whereas buckling could be avoided by increasing plate thickness to 4 mm.

Table 18.3 Example of increase in buckling strength with increase in plate thickness

S. No.	Plate thickness (mm)	Frame spacing (mm)	Leg length (mm)	Net heat input (J/mm)	σ_r (MPa)	σ_{cr} (MPa)	Remarks
1	3	390	3	364	62.2	61.7	Buckling
2	3	475	3	364	51.0	41.6	Buckling
3	4	390	4	606	77.7	109.7	No buckling
4	4	475	4	606	63.8	74.0	No buckling

18.2 Distortion Control Through Fabrication Technique

Appropriate fabrication techniques and procedures are to be followed for controlling weld induced distortion. Joint mismatch, excessive restraint in unit fabrication, high heat input, etc. are to be avoided to minimise structural distortions.

Joint Mismatch
Joint mismatch due to inherent distortions in plates and sections will give rise to increased overall distortion of a structure after fabrication. Improperly fitted joints will require increased quantity of deposited weld metal which in turn tends to increase distortion. Hence the plates and sections used for fabrication should be first relieved of inherent distortion and residual stresses, if any. This is carried out by rolling the plates and sections in suitable straightening machines. The thinner the plate higher the number of rollers required. For plates up to 6 mm in thickness a 21 roller straightening machine should be used. For block alignment, lugs with nut and bolt arrangement may be used. The lugs are to be welded on the stiffeners as shown in Fig. 18.2 and not on the plates.

Restraint
Restraining structures against possible distortions leads to accumulation of stress which results in more distortion when the restraining fixtures are released. It is a known phenomenon that on releasing a unit from skid after its completion, the overall breadth of the unit may become shorter due to shrinkage action. Therefore better alternative is to provide required amount of shrinkage allowances and to allow the structure to shrink freely without any restraint. This dimensional inaccuracy leads to further erection difficulties because of mismatch and misalignment. This phenomenon takes place as a cumulative effect of weld induced shrinkage and angular deformation of plate panel. The welding of each longitudinal member progressively adds to the shrinkage and angular deformation leading to an overall shrinkage of the assembled unit as shown in Fig. 18.3.

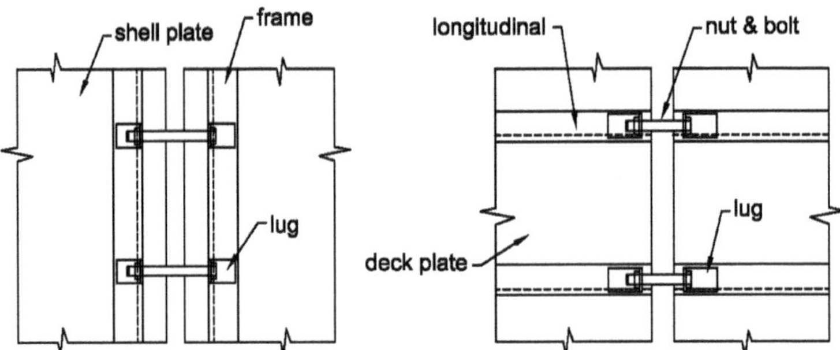

Fig. 18.2 Fixtures for block alignment

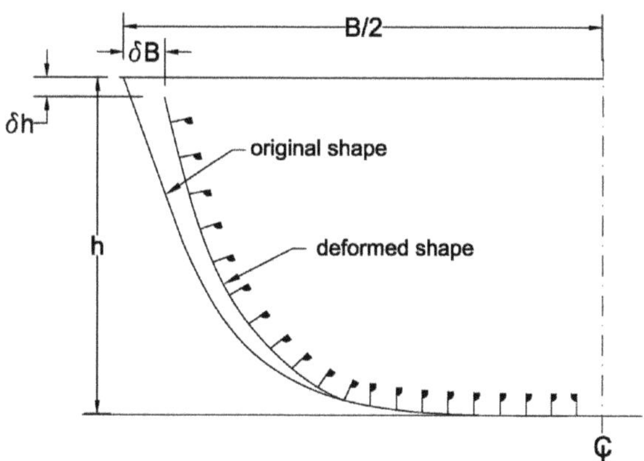

Fig. 18.3 Dimensional change of a unit on releasing from the skid

To overcome this phenomenon suitable shrinkage allowances have to be incorporated in the design [4]. However the required shrinkage allowances have to be estimated for every individual unit being assembled over skid as has been explained in Chap. 17.

18.3 Heat Input

One of the most influential parameter in weld-induced distortion is the heat input. Reducing welding heat input reduces all types of weld-induced distortions. Unfortunately heat input cannot be reduced enough to completely eliminate distortion, however its magnitude can be reduced by controlling heat input. The heat input can be controlled through the implementation of the following:

Reduce fillet weld size
This implies over welding has to be avoided. With decreasing plate thickness, the fillet size also reduces. Hence preferably pulsed GMAW with flux cored wire should be used for depositing fillets of appropriate sizes.

High speed low heat input welding
Increasing welding speed reduces heat flow in the welded components leading to reduced deformations. SMAW, which is inherently a slow speed welding process should be replaced by higher speed automatic SAW or semiautomatic GMA welding.

Intermittent welds
By welding intermittently, the overall weld metal deposition is greatly reduced thereby reducing the total heat input. Hence for controlling weld-induced distortions wherever possible, staggered welds should be used. Weld segments should be 40 mm minimum and the maximum unsupported length should not exceed 12 times the plate thickness as shown in Fig. 18.4.

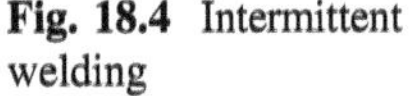
Fig. 18.4 Intermittent welding

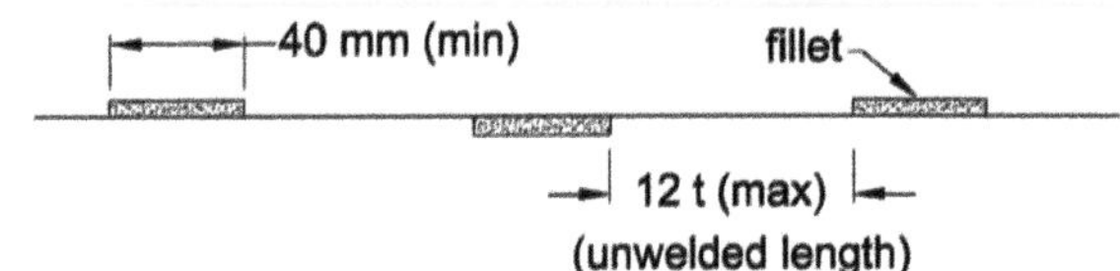

18.4 Heat Sinking

Distortion mitigation through heat sinking involves removal of heat from the welded region such that it does not get spread in the plates away from the weld zone. Therefore by keeping the base metal cool, the modulus of elasticity and yield strength of the base metal is not lowered and also the residual stresses and distortion are reduced. A schematic representation of heat sinking through water cooling below the plate surface undergoing welding is shown in Fig. 18.5. Theoretical analysis and experimental investigation of deformation with and without water cooling as the medium of heat sinking as schematically shown in Fig. 18.5 are presented in Figs. 18.6 and 18.7 for 6 and 8 mm thick plates respectively [5].

One can observe that with the proposed heat sinking, the overall deflection at the plate edges reduced by more than 50 %. This indicates a promising possibility of devising a suitable heat sinking arrangement which will be able to reduce the weld

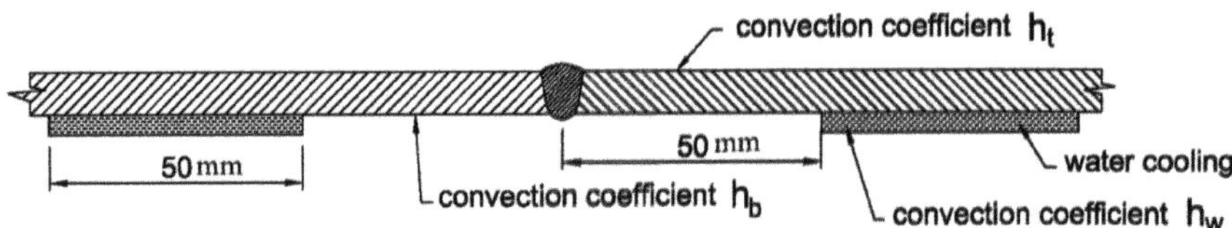

Fig. 18.5 Schematic representation of heat sinking

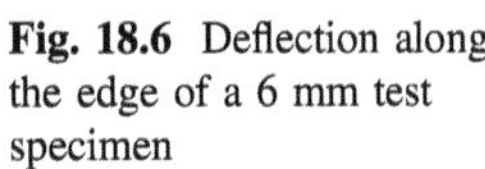
Fig. 18.6 Deflection along the edge of a 6 mm test specimen

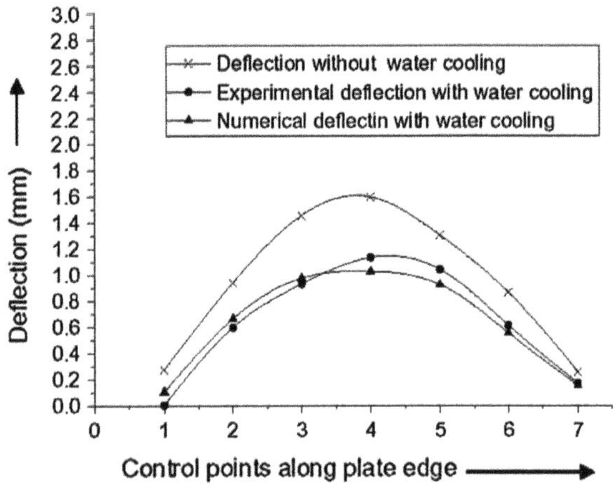

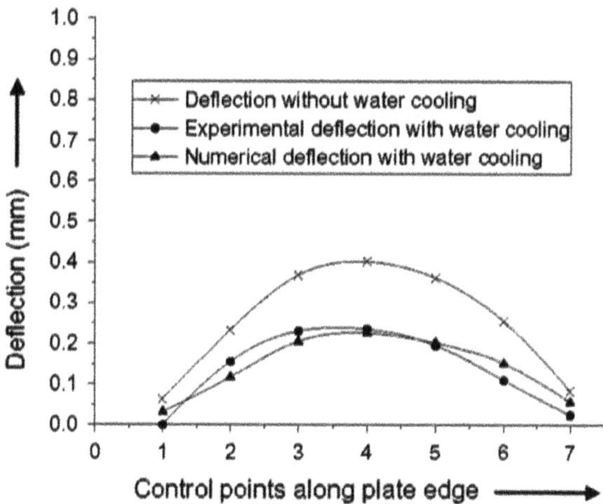

Fig. 18.7 Deflection along the edge of a 8 mm test specimen

induced deformation quite significantly. However, the metallurgical consequences that are associated with changes in the temperature gradients and cooling rates must be accounted for in practical situations.

18.5 Thermo-Mechanical Tensioning

The basic structural components of ships are orthogonally stiffened plate panels. The increased use of thin plates in panel fabrication results in significantly increased buckling distortion. Mitigation of such buckling distortion of stiffened panels can be done by applying an active distortion mitigation technique known as Thermo-Mechanical Tensioning (TMT).

This is based on the mechanism of counteracting the compressive forces developed due to welding of stiffeners. Tensile residual stress is generated in and around the weld zone, which is balanced by a compressive stress field away from this tensile zone. If the compressive stress thus developed exceeds the critical buckling stress of the plate panel in between the stiffeners, buckling of the plate takes place.

Hence, buckling can be avoided if this compressive stress could be reduced below the critical buckling stress of the given panel. This can be achieved by reducing the tensile stress level which will automatically lead to reduction of the compressive stress field. Therefore more effective means of mitigating the effect of residual compressive stress would be by reducing the level of residual tensile stress caused by welding of the stiffeners.

As tensile residual stress develops along the welding zone, therefore one of the ways to reduce it would be by applying compressive forces along that zone. In the proposed Thermo Mechanical Tensioning (TMT) method, tensioning and

Fig. 18.8 Tensioning lug being heated prior to welding

restraining lugs are placed along the stiffener location both for longitudinal and transverse stiffeners. The lugs (restraining) at one of the plate edges, are welded to the plate as well as to the base plate thereby completely restraining the movement of the plate at that end. The lugs (tensioning) welded to the plate at the other end are then heated up by a gas flame and as it cools down to about 500 °C, in that condition, they are welded to the base plate as shown in Fig. 18.8. Subsequently as these lugs cools down, they shrink and exert tensile force to the plate panel along the stiffener locations.

Thus the plate panel remains under tension both in longitudinal and transverse direction by the tensioning lugs welded all around as shown in Fig. 18.9. It has additional advantage of making the plate panel absolutely flat nullifying initial deformation, if there was any. Thus subsequent positioning of stiffeners and welding them becomes very easy. Thus the fitting up time of stiffeners reduces to a great extent.

With these TMT lugs in position, the longitudinal and transverse stiffeners are welded to the plate panel. The lugs are so designed that they generate sufficient tensile force to have elastic elongation of the panel. That means the panel will be under pre-tension. Once stiffener welding is completed, these TMT lugs are then released. The plate will then shrink back because of release of the elastic tensioning force, causing compressive stress to generate. This compressive stress will partly/fully nullify the tensile stress that developed due to welding of the stiffeners. Thus the residual tensile stress will get significantly lowered resulting in lowering of compressive stress field. As this compressive stress field goes below the critical buckling stress of the panel, the buckling of the same can be avoided.

Fig. 18.9 The plate *panel* with lugs fitted all around

18.5.1 TMT Model

A three bar model is adopted to explain the concept of TMT [6]. The model has been schematically shown in Fig. 18.10. The plate is put under TMT and then the stiffener is welded. The tensile pul shown in Fig. 18.10 is generated by the TMT lugs after they have cooled down to room temperature. An infinitesimal thin strip of width 'dx', perpendicular to the welding direction is considered for three bar model. If 'dx' is small the temperature in the welding direction is assumed to be constant. All thermal as well as all mechanical transverse interactions between the bars are neglected. The longitudinal stresses and strains of the three bars are coupled by two rigid bars at the top and bottom. The model is assumed to be fixed at one end and connected to a movable rigid bar at the other end. For simplicity the material is assumed to be elastic perfectly plastic. The longitudinal stresses will be developed with corresponding heating and cooling cycles and they should be in equilibrium continuously.

In Fig. 18.11 it can be seen that, initially tensile stresses were provided in the 'Bar 1' through TMT. This results in formation of elastic tensile stress at 'Bar 1' and a comparatively lower magnitude of elastic tensile stress at 'Bar 2'. At time 't_1'

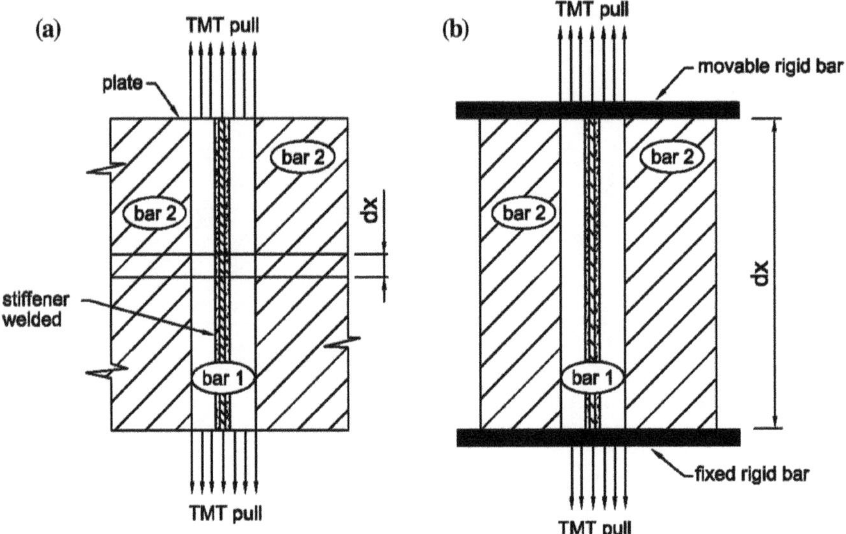

Fig. 18.10 **a** Stiffener welding is done under TMT pull. **b** Three bar model under TMT condition

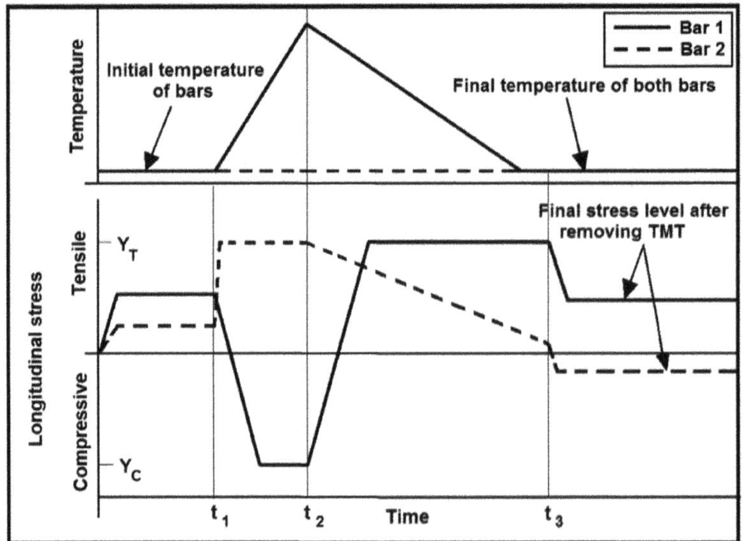

Fig. 18.11 Temperature plot and resulting longitudinal stress for three bar model under TMT condition

welding of stiffener is commenced with TMT applied condition. From time 't_1' the temperature rises to the melting temperature at time 't_2' and thereafter gradually cools down to the room temperature. As the temperature gradually rises beyond time 't_1', the 'Bar 1' tries to expand. But this expansion is constrained by 'Bar 2' which is at a much lower temperature and can be considered as the cold part in the 3 bar model.

The stress field that develops due to the thermal cycle of the bar 1 is shown in Fig. 18.11. As the stresses in both the bars reach their yield limits, the stress remains constant as a elastic and purely plastic material model has been considered.

'Y_T'and 'Y_C' represent tensile and compressive yield stresses respectively ass shown in Fig. 18.11. As for TMT applied condition, 'Bar 2' will also be in elastic tension. During welding this cold part will attain tensile yield stress from its pre tensile state more quickly than 'Bar 1' attains its compressive yield magnitude. This is depicted in Fig. 18.11 as the gradient of 'Bar 2' is steeper than 'Bar 1'. At time 't_2' the temperatures in both the bars reach their maximum limit and from this point onwards the bars start to cool down. During cooling, the constrained shrinkage in 'Bar 1' leads to the formation of tensile stresses and a balancing compressive stress occurs in 'Bar 2'. But as both the bars are in pre tension, the stress in 'Bar 2' will not reach its compressive value. Rather its value will gradually reduce from time 't_2' until the room temperature is attained at around time 't_3'.

At this stage a lower tensile stress (than its initial pretention) may remain in 'Bar 2' until the TMT is cut. At time 't_3' when both the bars reach room temperature, the lugs are cut and thus the pretension in 'Bar 1' and 'Bar 2' gets released. This will lead to reduction in the tensile stress level in Bar 1, causing reduction in the balancing compressive stress level in Bar 2, thereby reducing the chances of buckling.

18.6 Weld Sequencing

The extent of distortion of large orthogonally stiffened panels normally used in ships and offshore structures depend very much on the welding sequence of the stiffening members. The extent of distortion depends on various aspects, such as welding parameters, plate thickness, plate material properties, restraints applied to the job while welding as well as sequence of welding of the stiffeners. Keeping all other parameters constant a poorly designed welding sequence may lead to excessive shrinkage and buckling distortion. In fabrication of stiffened panels normally longitudinal stiffeners are fillet welded to the plate followed by the transverse members. Such a sequence of erection is convenient with the available automated welding systems. However, this method of fabrication results in a higher level of buckling distortion of the panels.

Therefore the effect of welding sequence of the stiffening members on the distortion pattern and its magnitude in fabrication of orthogonally stiffened plate panels needs to be analysed. A computational strategy, based on finite element analysis and quasi-stationary nature of welding, was developed to study the effect of welding sequence on distortion of stiffened panel [7].

Table 18.4 Plate thickness and welding parameters used in the study

Plate No.	Thickness (mm)	Current (A)	Arc voltage (V)	Welding speed (mm/s)
1	6	370	24.5	10

Effect of welding sequence in stiffened panel fabrication

It is observed from the distortion pattern of the welded structure that the sequence of welding of stiffeners has a significant effect on structural distortion. The welding parameters used in the present example are shown in Table 18.4.

Three different sequences of welding of stiffeners as shown in Figs. 18.12, 18.14 and 18.15 are examined. The panels comprised of 6 mm thick steel stiffened by 3 longitudinal members and 2 transverse members. The spacing between the transverse members is 4 times that of longitudinal members. The depth of web of the transverse members is twice that of the longitudinal members. In the FE model, the intersections of the webs of longitudinals and transverses are considered to be rigidly connected. This implies that the vertical intersections of the stiffening members are welded and then the full grillage of longitudinal and transverses is welded following different welding sequences as depicted by numbers 1, 2, 3… in Figs. 18.12, 18.14 and 18.15.

The arrows in the following figures indicate the sequence followed for welding of the stiffeners. In these examples, the stiffeners were assumed to be welded by single run as indicated by the arrows. Three different cases of welding sequence were studied.

In case 1, the sequence followed is that of welding of longitudinals starting from one end of the panel moving towards the other end. Once the longitudinals are completely welded, the transverses are welded. The welding direction and the sequence of welding are indicated by the arrows and the numbers respectively.

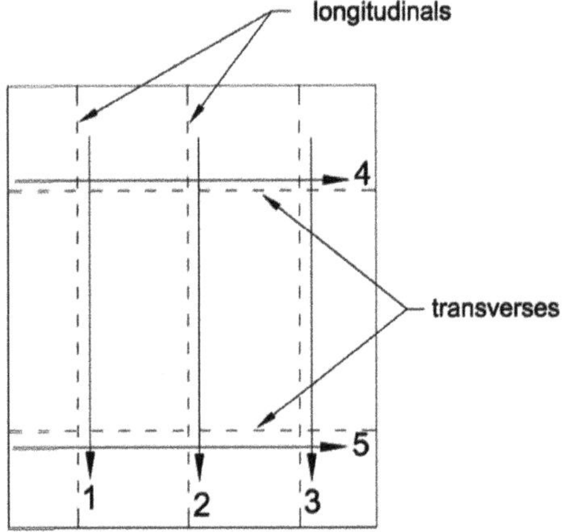

Fig. 18.12 Welding sequences for case 1

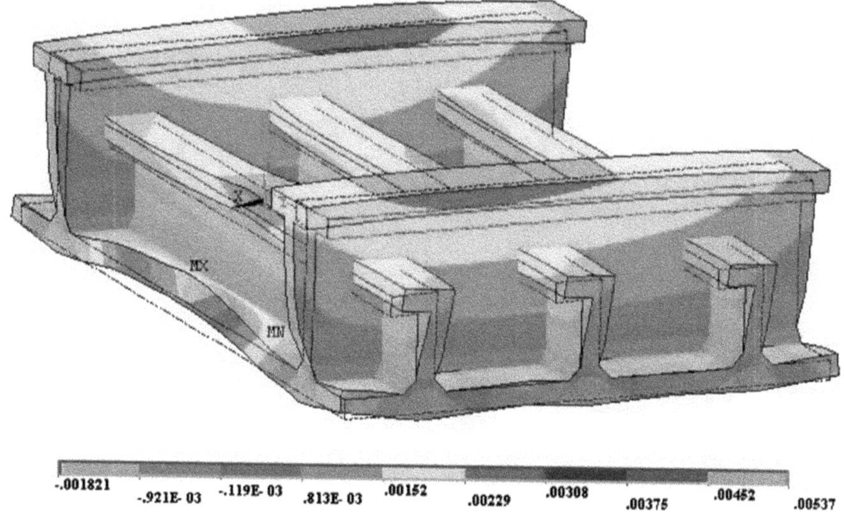

Fig. 18.13 Typical distorted shape for welding sequence—case 1

The typical distorted shape of the stiffened panel following the sequence as shown in case 1 was numerically calculated and is shown graphically in Fig. 18.13.

In case 2, the welding was carried out starting from the central longitudinal followed by the side longitudinals. However the welding of the longitudinals was done starting from one transverse-longitudinal intersection to the next intersection as shown in Fig. 18.14. Subsequently the portions of transverses in between the

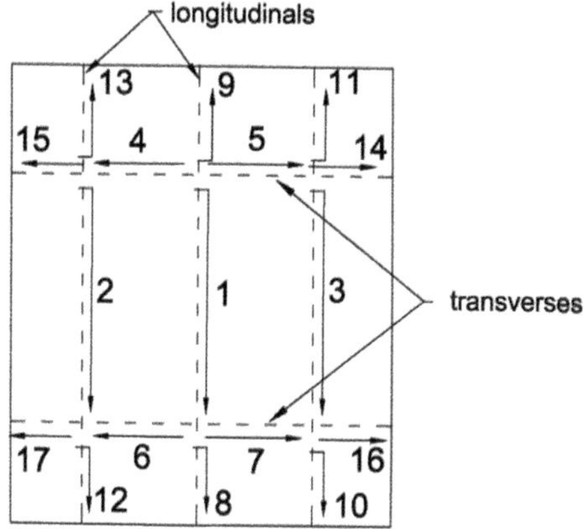

Fig. 18.14 Welding sequences for case 2

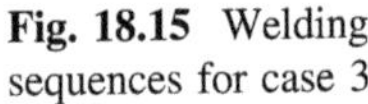

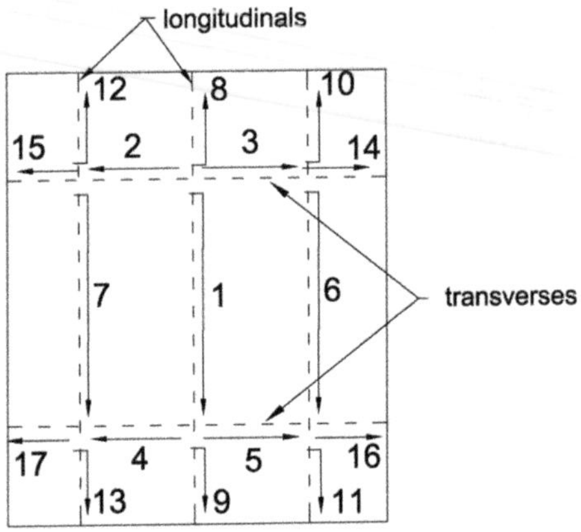

Fig. 18.15 Welding sequences for case 3

intersections were welded as depicted by the arrows in the Fig. 18.14. The welding of the transverses was carried out again from the central position moving outwards. In this sequence of welding, it resulted in a more even distribution of stiffening of the plate and balancing of stresses due to welding as compared to that of Case 1.

In case 3, the central longitudinal was welded first followed by the top and bottom transverses starting from the centre moving outwards as shown in Fig. 18.15. This sequence of welding resulted in progressive stiffening of the plate starting from centre moving outwards. At the same time it further enhanced balancing of stresses caused by welding as compared to that of Case 2.

As a result one finds that the resulting distortions are maximum in Case 1, where the stresses developed due to welding was left fully unbalanced. As expected a significant drop in distortion was observed in case 2. Whereas minimum distortion was achieved in Case 3 is evident from the Figs. 18.17 and 18.19.

A comparison of the deformation data as obtained for the control points shown in Figs. 18.16 and 18.18 are plotted in Figs. 18.17 and 18.19 respectively. From this comparison it can be observed that the extent of deformation tends to reduce with a specific pattern of welding sequence.

From these results one can observe that the sequence of stiffener welding as depicted in Fig. 18.15 led to minimum residual deformation of the entire stiffened panel. Hence this sequence of welding of stiffeners, if followed in real life structures, will result in stiffened panels having minimum deformation.

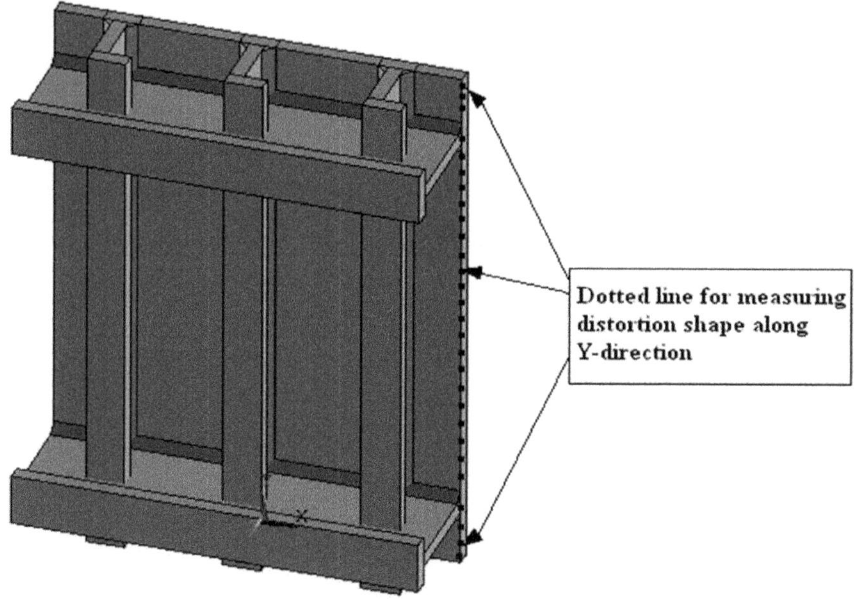

Fig. 18.16 Control points at the *panel* edge for distortion measurement

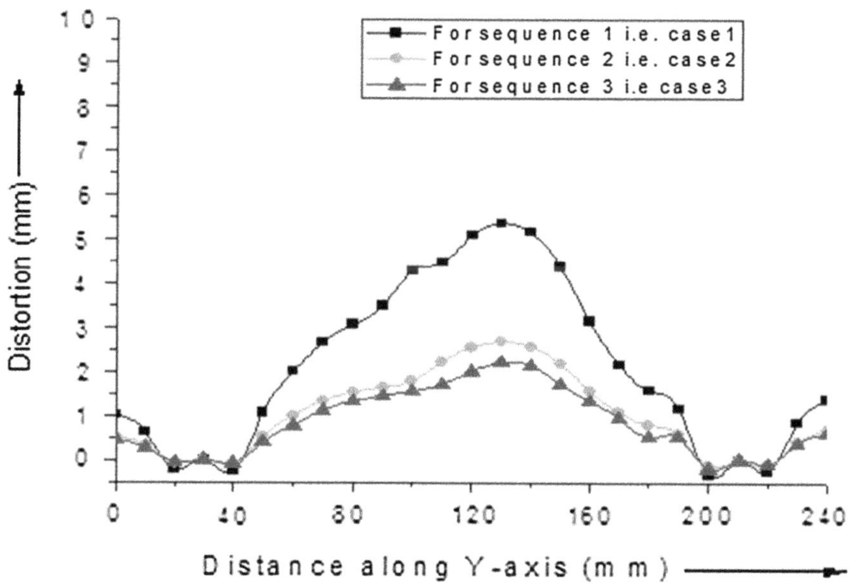

Fig. 18.17 Comparison of distortion at the *panel* edge fabricated with 3 different sequence of welding

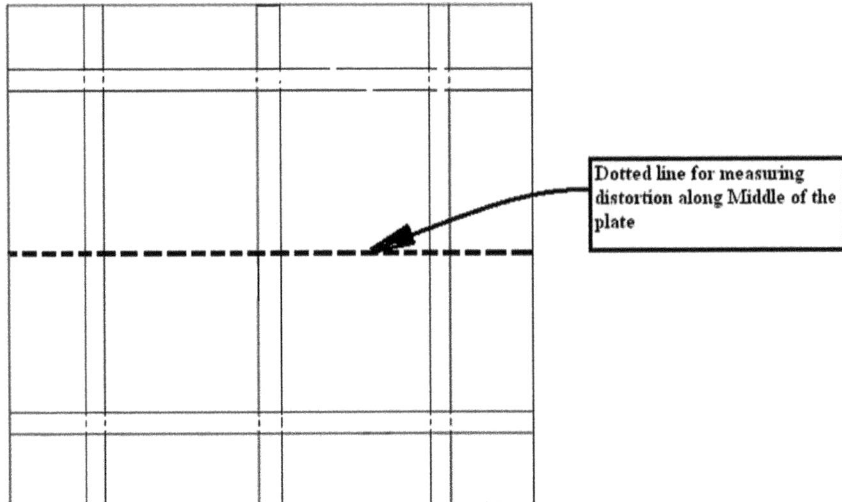

Fig. 18.18 Control points at the middle of the stiffened *panel* for distortion measurement

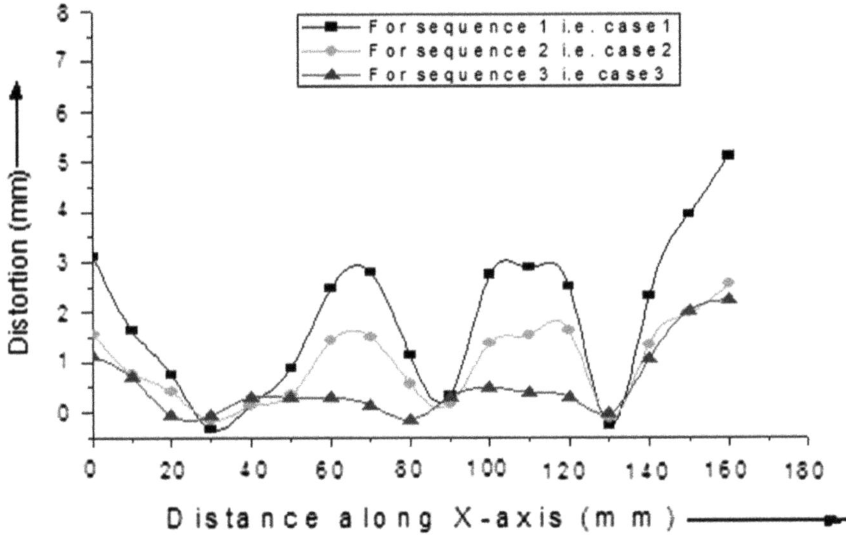

Fig. 18.19 Comparison of distortion at the middle of the *panel* fabricated with 3 different sequence of welding

References

1. Masubuchi, K. (1991). Research activities examine residual stress and distortion in welded structures. *Welding Journal* 41–47.
2. Mandal, N. R., Krishna Prabu, Sree, & Kumar, Sharat. (2014). Buckling of stiffened panels and its mitigation. *Journal of Ship Production and Design, 30*(4), 201–206.
3. Joshi, A. D. (2014). Experimental study on buckling and its mitigation in thin stiffened panels used for ship construction. Thesis, Master of Technology, Department of Ocean Engineering and Naval Architecture, IIT Kharagpur, May 2014.
4. Mandal, N. R. (1999). Prediction of dimensional changes during fabrication of thin shell ship hull bottom units. *Journal of Science and Technology of Welding and Joining, 4*(5), 290–294.
5. Adak, M., & Mandal, N. R. (2010). Numerical and experimental study of mitigation of welding distortion. *Applied Mathematical Modelling, 34*, 146–158.
6. Podder, D. (2016). Numerical and experimental study of fusion welding on residual stresses. Distortions and its Mitigation, Ph.D. thesis.
7. Pankaj, B., & Mandal N. R. (2007). Finite element analysis to study the effect of welding sequence in fabrication of orthogonally stiffened plate panels. In National Conference on "*Welding – Productivity & Quality*", Feb.20–21, NMRL, Ambernath, Mumbai.

Chapter 19
Welding Defects

Abstract In a welding process, there can be flaws in the welded joint. The flaws depending on their, size, location and type may be considered as defects. Once these defects are detected, remedial measures are to be implemented to remove them, as because a structure cannot be put to service with defects. Flaws are nothing but imperfections in a welded joint. Welding procedure, joint features, access and welding technique will have direct effect on fabrication imperfections. Incorrect procedure or poor technique may produce imperfections leading to premature failure in service. The majority of the defects encountered in welded structures are primarily due to improper welding procedure. Once the causes are established, the operator can easily correct the problem. Most encountered welding imperfection/defects are lack of penetration (lack of root fusion), lack of fusion, slag inclusion, undercutting, porosity, and weld cracks.

Welding, as a very convenient joining method, is extensively used in the fabrication of marine structures. This extensive use of welding also has led to many welding related catastrophic failures, causing economic and human loss. Offshore platform 'Alexander Kielland' capsized in the North Sea on March 27, 1980. The rig capsized in heavy seas killing 123 of the 212 on board. The cause of the disaster was attributed to a 6 mm fillet weld which joined a non-load bearing flange to one of the main braces. Thus the importance of ensuring defect free welded structures becomes very evident.

In a welding process, there can be flaws in the welded joint. These flaws are not necessarily qualified to be defects. The flaws depending on their, size, location and type may be considered as defects. Once these flaws are termed as defects, they need to be rectified, or in other words remedial measures are to be implemented to remove them, as because a structure cannot be put to service with defects.

Flaws are nothing but imperfections in a welded joint. However they may not render the structure unsuitable for its application to service. As mentioned the size, type and location of these imperfections may render the structure defective. For example, stray gas entrapped pores of diameter within permissible limits in a weld deposit remain as a flaw or imperfection but not a defect. Whereas an imperfection

N.R. Mandal, *Ship Construction and Welding*, Springer Series on Naval Architecture, Marine Engineering, Shipbuilding and Shipping 2,
DOI 10.1007/978-981-10-2955-4_19

of the type crack of certain minimum length is not acceptable and has to be rectified, i.e. this imperfection renders the structure to be defective.

Welding procedure, joint features, access and welding technique will have direct effect on fabrication imperfections. Incorrect procedure or poor technique may produce imperfections leading to premature failure in service. The majority of the defects encountered in welded structures are primarily due to improper welding procedure. Once the causes are established, the operator can easily correct the problem. Most encountered welding imperfection/defects are lack of penetration (lack of root fusion), lack of fusion, slag inclusion, undercutting, porosity, and weld cracks.

19.1 Lack of Penetration

This defect can be found in either of the following ways as shown in Fig. 19.1a–c:

- The weld bead does not penetrate the entire thickness of the plate being welded.
- Two opposing weld beads do not interpenetrate.
- The weld bead does not penetrate the toe of a fillet weld but only bridges across it.

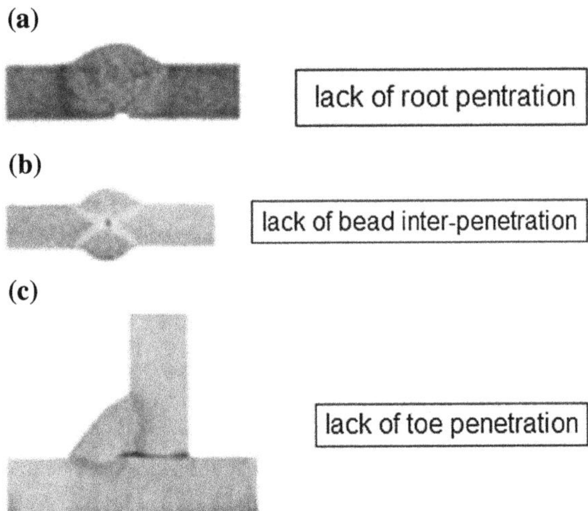

Fig. 19.1 **a**, **b**, **c** Examples of lack of penetration

This defect in case of lack of root fusion, is an external imperfection, however in other case as shown above it is an internal defect. Welding current has the greatest effect on penetration. Too low a welding current usually leads to Lack of penetration. It can be eliminated by simply increasing the welding current. Other causes can be, welding speed being too low or an incorrect torch angle. Both will cause the molten weld metal to flow in front of the arc, acting as a cushion preventing penetration. That is why the welding arc should be kept on the leading edge of the weld pool.

19.2 Lack of Fusion

It is primarily an internal defect. Lack of fusion occurs when there is no fusion between the weld metal and the surfaces of the plate being welded as is clearly visible in Fig. 19.2. The most common cause of lack of fusion is a poor welding technique. Either the weld pool is too large because of too slow welding speed or because of a very wide weld joint.

In a wide weld joint if the arc is directed to the center of the joint, the molten weld metal will only flow and cast against the side walls of the parent plate without melting them. The heat of the arc should melt the parent plate. This is accomplished by making the joint narrower or by directing the arc towards the side wall of the parent plate. For welding thick plates with wider V-groove, deposits in the form of large weld beads bridging the entire gap should be avoided. Instead after the root run, a split bead technique should be used.

However, it is also very often caused by too low a welding voltage. As a result, the wetting of the bead will be poor causing lack of fusion in the parent metal. In aluminium welding, lack of fusion may occur due to the presence of aluminum oxide. This oxide has a very high melting point of approximately 2500 °C. If this oxide is present on the surfaces to be welded, lack of fusion of the parent metal with the weld metal may occur. The oxide layer should always be removed from the plate surface in the weld vicinity just prior to welding.

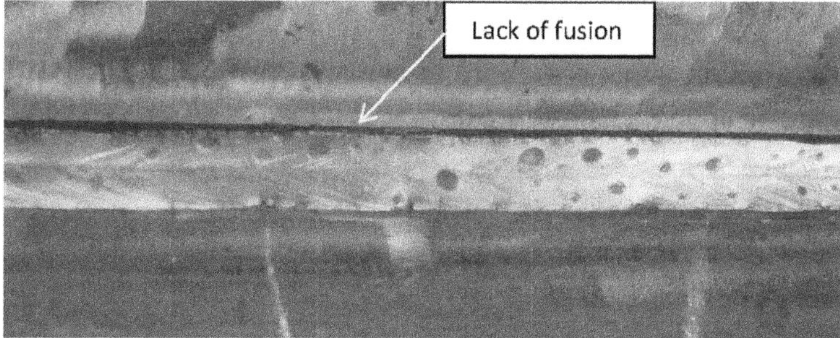

Fig. 19.2 A typical example of lack of fusion

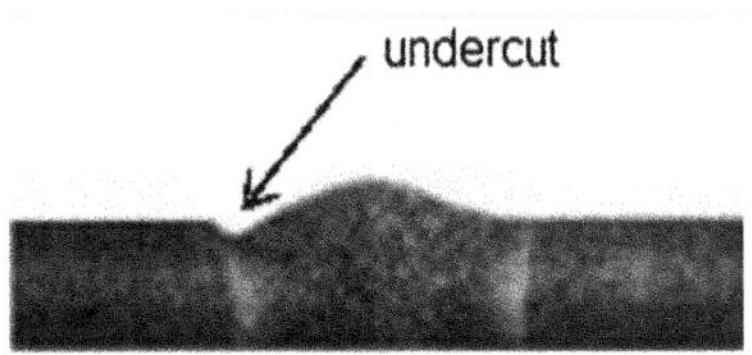

Fig. 19.3 Example of undercut

19.3 Undercutting

This defect appears as a groove in the parent metal running along the edges of the weld bead as shown in Fig. 19.3. Therefore it is an external surface imperfection. It is mainly encountered in butt welds. It is caused by improper welding parameters, particularly the travel speed and arc voltage.

This type of defect occurs because of molten metal from the weld edges is drawn towards the weld centre by surface tension forces and rapid solidification takes place because of high welding speed. Here the molten metal is drawn towards the weld centre and cannot flow back for adequate wetting of the edges because of fast solidification of the molten metal.

The remedy is to decrease the welding speed, it will gradually reduce the size of the undercut and eventually eliminate it. When only small or intermittent undercuts occur, raising the arc voltage or using a leading torch angle solves the problem of undercut. In both cases, the weld bead will become flatter and wetting will improve.

However, with excessive increase in arc voltage, undercutting may again appear. This happens particularly in arc welding having spray transfer of metal. With increase in arc voltage, it increases the arc length and makes the arc very wide. However, the heat transfer in a long arc is rather poor. Therefore the areas along the weld bead edge cool of quickly and adequate wetting does not take place causing undercutting. Even excessive welding current can also lead to undercutting. It is therefore always advisable to set welding current within the specified range for a given wire diameter.

19.4 Porosity

Porosity generally occurs as a cluster of gas pores in the solidified weld bead as shown in Fig. 19.4. Generally this forms just under the weld bead or at the surface. Hence this can be either internal or external imperfection.

Porosity is generally caused from atmospheric contamination. The gases in the air that are primarily responsible for porosity in steel are nitrogen and excessive oxygen. However, in absence of nitrogen, only oxygen does not lead to porosity. However in aluminium welding, oxygen can cause severe problems because of its rapid oxide formation.

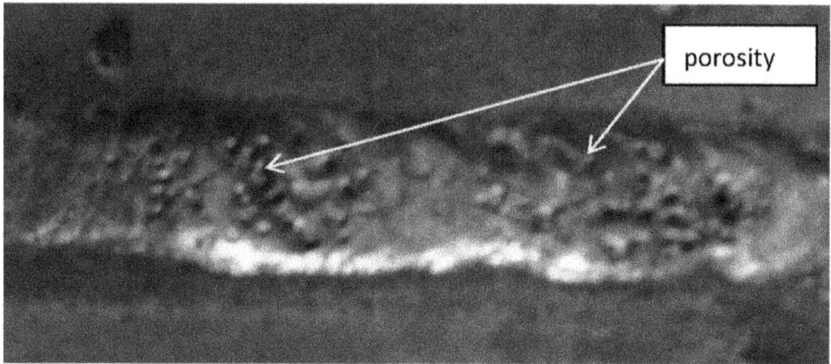

Fig. 19.4 Typical example of porosity in weld bead

In gas metal arc welding, the shielding gas supply lines need to be checked at regular intervals to insure that leakage has not taken place. In addition to nitrogen contamination, excessive moisture in the atmosphere can cause porosity in steel and particularly in aluminum. Hence in humid climates appropriate measures are to be taken to avoid water condensation in water cooled welding torches that may cause contamination of the shielding gas.

The following can be the possible sources of atmospheric contamination of the shielding gas causing porosity in the weld deposit:

- Inadequate shielding gas flow.
- Excessive shielding gas flow. This can cause aspiration of air into the gas stream.
- Severely clogged gas nozzle, leaking hoses, fittings, etc.
- Strong winds in the welding area in outdoor welding situation. This can blow away the gas shield.

Welding of mill scale coated or heavily rusted steel plates may result in porosity in the welded joint because of the obvious source of oxygen as well as entrapped moisture in the mill scale and the rusted layer on the steel plate. In aluminium it is particularly true because of the hydrated oxide layer that naturally forms on the surface.

Foreign matter in the welding consumable can be a source of porosity, e.g. excessive lubricant on the welding wire. These lubricants being hydrocarbons are sources of hydrogen which is particularly harmful for aluminum.

Other causes of porosity:

- Extremely fast weld solidification rates causing entrapment of gas that otherwise may have normally escaped. To reduce the rate of solidification, extremely high welding speeds and low welding current levels should be avoided.
- Erratic arc characteristics. This is caused by poor welding conditions, e.g. too low or too high arc voltage and fluctuating wire feed speed. These cause severe weld pool turbulence. This turbulence will tend to break up the shielding gas envelope exposing the molten weld metal to atmosphere.

19.5 Weld Cracks

Cracks due to welding may occur in various forms and in various ways and at different time instant with respect to the temperature of the weldment. The cracks that develop during welding are mostly due to the shrinkage strains that occur with the cooling of the weld metal. The hot metal tries to shrink as it cools, however this shrinkage is opposed by the surrounding cold metal causing significant tensile forces to develop in the weld metal. In case of highly restrained structures these stresses become even more severe. These shrinkage forces and stresses increase with increase in the volume of the shrinking metal. That means in case of large weld size and in deep penetration welding, the shrinkage stresses will be higher. It will also be higher where higher tensile strength filler and parent metal are used. Therefore welding of highly restrained structures and where higher tensile strength materials are used, extra care is to be taken in terms of welding sequence, pre-heating and interpass temperatures to avoid cracking tendencies.

The various types of cracks due to welding and their characteristic features [1] are listed in the Table 19.1.

Table 19.1 Types of weld induced cracks and their characteristic features

Sl. No.	Description of cracks	Features
1.	Micro crack	Visible under magnification by almost 6 times
2.	Macro crack	Normally visible or with magnification less than 6 times
3.	Intergranular crack	Crack propagates through grain boundaries
4.	Transgranular crack	Crack propagates through the grains
5.	Hot crack	Cracks forms through a low melting phase
6.	Solidification crack	Crack occurs during solidification of weld metal
7.	Liquation crack	Crack occurs when only the low melting phase is in molten state as at a grain boundary
8.	Cold crack/Delayed crack	It is hydrogen induced crack develops through an increase of residual stress; hydrogen precipitates as it cannot effuse out of the material due to microstructure changes
9.	Shrinkage crack	Crack forms due to restrained shrinking
10.	Toe crack	Develops due to high tension concentration at the weld toe of fillet joint
11.	Precipitation induced crack	It occurs due to precipitation of brittle phase during welding
12.	Lamellar tearing	It develops due to the welding shrinkage forces acting in the through thickness direction. The crack lies within the parent metal often below the HAZ and is generally parallel to the fusion boundary

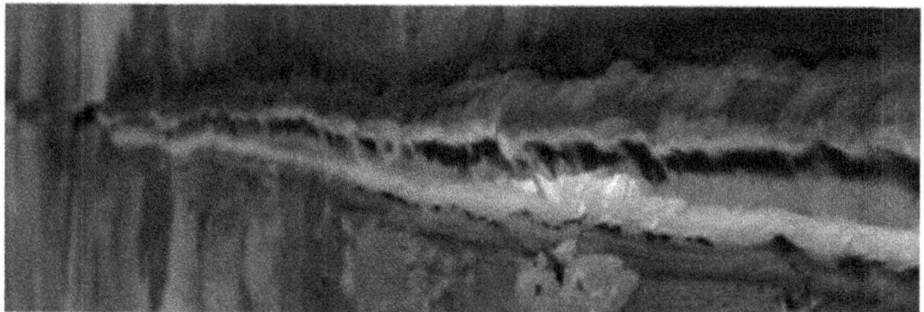

Fig. 19.5 A severe case of longitudinal crack

Hot and cold cracks

The various cracks that are caused due to welding and not by the effect of service loads, occur as a result of weld metal solidification, its cooling and the stresses that develop due to shrinkage of the hot metal. All these cracking take place within the period of fabrication. The cracking that takes place while metal is still at elevated temperature is referred to as hot cracks. They occur while the weld bead is in a mushy condition between the liquidus (melting) and solidus (solidifying) temperatures. These are essentially solidification related. This type of defect usually results from the use of an incorrect alloy composition of filler electrode particularly in welding of aluminum and stainless steel alloys. Hot cracking generally is promoted by presence of sulphur and phosphorous. These elements form low melting alloys with iron and they tend to segregate very much.

Whereas cracks may occur after the metal has fully cooled down to room temperature. These are known as cold cracks, also often referred as delayed cracking. These cracks are mainly caused by a phenomenon known as hydrogen embrittlement.

Longitudinal or centerline cracking (macro crack) of the weld bead can be either hot crack or cold crack. A severe case of longitudinal crack is shown in Fig. 19.5. Usually such cracks once formed, it continues with the weld all along the length. It is primarily driven by the transverse shrinkage forces causing tensile stress to develop in the weld material, finally rupture takes place either due to hot cracking or cold cracking phenomena.

Hydrogen embrittlement

Hydrogen embrittlement often also referred to as hydrogen assisted cracking as it is associated with hydrogen. Atomic hydrogen may form during welding. The sources of this could be from the moisture present in the welding consumables, organic compounds in the form of grease or dirt in the weld vicinity and alike. The dissociation of these produces atomic hydrogen which dissolves in the liquid metal of the weld. The hydrogen then migrates to the HAZ. This takes time and therefore, weld inspection is done after a delay of 48 h after the completion of welding, because delayed cracking may take place. With time hydrogen diffuses out from weld deposits. However its concentration is maximum during the welding period and hydrogen induced crack, if it forms, will occur within few days of fabrication.

The cooling rate of the weldment is important as it determines the time for the hydrogen to diffuse out of the weld metal. As the weld metal cools below about 100 °C, the hydrogen effusion stops. Hence slower cooling rate helps in avoiding cold cracking by hydrogen embrittlement. The cooling rate is measured by the parameter $t_{8/1}$, i.e. the time taken to cool from 800 to 100 °C. The longer the time taken for the weld metal to cool, the more time is available for the absorbed hydrogen to diffuse out of the weld metal.

This hydrogen concentration in the HAZ causes an increase of microscopic stresses accompanying martensitic transformation of the HAZ. It often leads to formation of time delay cracks in the heat affected zone adjacent to the weld.

In order to counter this cold cracking susceptibility, the source of hydrogen should be eliminated or reduced. Low hydrogen processes are to be used, like using low hydrogen electrodes, dehumidifying shielding gas, baking of electrodes prior to welding, cleaning of the weld surface from oil, grease and dirt, avoiding outdoor welding during rains, etc. At the same time care should be taken to reduce the structural restraint to reduce the stress levels as well as suitable measures like pre-heating and interpass temperature are to be maintained to avoid/reduce martensitic transformation.

Cold crack development can be monitored through sound emission measurement. A suitable microphone can be attached to the structural component to detect and measure the sound pulses generated by crack development. The pulse intensity gives a qualitative assessment of the crack size.

Lamellar Tearing

Lamellar crack or tearing as it is often referred to, occurs due to development of stress across the plate thickness, i.e. in a plane perpendicular to the plate rolling plane. In welded structures where the shrinkage of the weld metal is greatly obstructed, there stresses near equal to yield stress may get generated as in the case shown in Fig. 19.6 causing typical cracks known as lamellar tearing. This happens across the plate thickness as the strength of the rolled plates at certain areas can be low along the thickness direction because of local weaker microstructure.

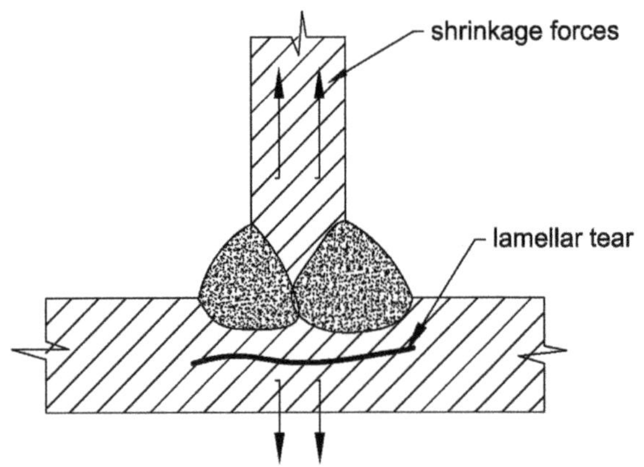

Fig. 19.6 A schematic of lamellar tearing

Table 19.2 Locations of different form of cracking in welding zone

Cracking phenomenon	Position in weldment
Hydrogen attack	Weld deposit
Porosity	Weld deposit
Solidification cracking	Weld deposit
Liquation cracking	Fusion zone
Lamellar tearing	HAZ
Cold cracking or hydrogen embrittlement	HAZ

Some of the metallurgical effects associated with welding leading to cracking are listed in the Table 19.2.

Crater Crack

This type of welding imperfection is generally observed at the end of a weld run. It occurs if the welding is improperly stopped at the end of a welding pass without giving a proper finishing pass. It forms a situation of lack of deposition in the form of a depression at the weld end. Small cracks may form at this location. Even if they are small as they form, however they may propagate into the weld bead causing serious damage.

To avoid such crater cracks, the weld finishing has to be done properly. The crater that forms needs to be filled by backfill technique, i.e. the welder needs to weld backwards slightly and fill in the area at the end of the weld.

19.6 Slag Inclusion

This type of weld imperfection is naturally associated with welding processes using flux for shielding of welding arc and molten metal pool, e.g. SMAW, SAW and FCAW. It is an internal imperfection. As such depending on the size of a slag particle, some slag inclusions are tolerated, i.e. no rectifying measures are taken. However in case of larger slag particles or slag entrapment constitute a weld defect and it has to be removed either by arc gouging or by deep grinding and rewilding.

As the name suggests, slag gets entrapped in between weld deposits. The entrapped slag remains as it is. It does not melt under the arc heat in the subsequent weld run. In multipass welding, slag may also get trapped in cavities or in excessive undercut in the weld toe or in the uneven surface profile of the preceding weld run. To avoid slag inclusion, slag removal has to be done meticulously prior to each overlapping welding run in case of multipass welding.

The composition of flux, joint geometry, welding position and access restrictions all have an influence on the risk of slag inclusion. Slag needs to be removed manually, hence adherence of solidified slag to the weld bead plays an important role. This is highly influenced by flux composition as well as joint geometry and quality of weld deposition. For example, flux containing higher percentage of silicate produces a glass-like slag, which is more or less self-detaching. Whereas fluxes containing lime produce an adherent slag that is difficult to remove.

Effect of different types of fluxes on slag formation and its removal [2]:

- Rutile or acidic flux
  It contains large amount of titanium oxide (rutile) along with some silicates. This provides for higher oxygen level of the weld pool to produce flat or slightly convex weld bead. The fluidity of the slag is determined by the calcium fluoride content in the flux. For positional welding (i.e. other than down hand welding) higher slag fluidity is required, however the more the slag is fluid more it becomes difficult to remove. Whereas fluoride-free flux used for down hand welding produces smooth bead profiles and the slag is also easily removed.
- Basic flux
  It comprises of high proportion of calcium carbonate (limestone) and calcium fluoride (fluospar). This reduces the oxygen content of the weld pool and therefore its surface tension. The slag is more fluid than that produced with the rutile coating. The slag thus produced is of fast freezing type. It helps in vertical and overhead welding. However with these advantages its drawback is, the slag removal is difficult.

Consequently, slag removal is comparatively more difficult with basic flux particularly in multi-pass welds.

To achieve a weld deposit without slag inclusion, slag should be removed thoroughly before depositing the next run. In between the welding runs the slag can be removed by grinding, light chipping or wire brushing.

Best practice to avoid slag inclusion

To prevent slag inclusion, the following may be implemented:

- Suitable welding techniques and welding parameters are to be chosen to produce smooth weld profile with adequate fusion in between overlapping runs to avoid pocket formation trapping slag.
- Undercutting of side wall should be avoided by appropriately choosing welding current and welding speed. Slag removal from such undercutting is difficult.
- Slag should be meticulously removed in between consecutive overlapping runs, particularly any slag trapped in crevices, if any.

References

1. Welding Defects. (2005). *Classification of cracks to DIN 8524 Part 3*. ISF: Aachen.
2. Lucas, B., Mathers, G., & Eileens, C. *Defects/imperfections in welds—slag inclusions, Job Knowledge 43*, The Welding Institute.

Chapter 20
Nondestructive Testing

Abstract Welding being a process of joining, the end user would like to assure himself of the quality of the joint that has been achieved though welding. There should be some mechanism which will give assurance to this aspect of no-defect in the welded joint. Hence the necessity of non-destructive testing (NDT). Through this the welded joints are inspected to determine the presence of any flaw, its type, extent and its location, such that, if it qualifies to be an unacceptable flaw, i.e. a defect, then necessary remedial measures could be undertaken to eliminate the defect. Thus it will ensure that the structural component is free from any welding defect. Various standards have been worked out by various regulatory authorities. The non-destructive testing (NDT) is carried out to verify compliance to the standards by suitable methods of examination of the surface and subsurface of the welded joint. The NDT methods that are commonly used to examine finished welds in shipbuilding are: visual, dye penetrant, magnetic particle, radiographic and ultra-sonic testing.

Welding being a process of joining, the end user would like to assure himself of the quality of the joint that has been achieved though welding. A good quality welding should have acceptable external appearance and the structural component should be able to perform its intended functions, ideally for an infinite period. However the life of the structural component will depend not only on the quality of welding but also on many other factors, like, loading pattern (stress levels), work environment (corrosive), material property (fatigue life), etc.

However if the weld quality is not up to the mark, i.e. there are some inherent defects in the welded joint itself, then automatically, the component is likely to fail in service much earlier that it was intended for. Hence there should be some mechanism which will give assurance to this aspect of no-defect in the welded joint. There could be two ways to establish this aspect, one through destructive testing of samples taken from the structural component and the other is through non-destructive testing. Now once the structure has been fabricated, it cannot be cut into pieces to extract testing samples. Hence the necessity of non-destructive testing. Through this the welded joints are inspected to determine the presence of any

N.R. Mandal, *Ship Construction and Welding*, Springer Series
on Naval Architecture, Marine Engineering, Shipbuilding and Shipping 2,
DOI 10.1007/978-981-10-2955-4_20

flaw, its type, extent and its location, such that, if it qualifies to be an unacceptable flaw, i.e. a defect, then necessary remedial measures could be undertaken to eliminate the defect. Thus it will ensure that the structural component is free from any welding defect.

Various standards have been worked out by various regulatory authorities. The non-destructive testing (NDT) is carried out to verify compliance to the standards by suitable methods of examination of the surface and subsurface of the welded joint. There are various NDT methods available. The NDT methods that are commonly used to examine finished welds in shipbuilding are: visual, dye penetrant, magnetic particle, radiographic and ultra-sonic testing.

20.1 Visual Inspection

One of the basic methods of NDT is Visual inspection. It is a simple human skill based method. It does not require any special equipment or instrument. Visual inspection is often aided with magnifying glasses as well as computer imaging. This is an effective method to identify almost all kinds of surface defects. It is a recommended practice to have visual inspection of all welded joints in a ship structure. The first inspection ideally is done by the welder himself. He can take rectification measures, if any flaw/defect is detected. The drawback of this method from documentation point of view, it does not automatically create any record of testing.

20.2 Dye Penetrant Testing

Dye penetrant testing is another simple method of NDT, that is widely used to primarily detect surface cracks. The surface is first cleaned and the dye is spread over the weld zone. It penetrates the cracks due to surface tension effects. The dye is subsequently wiped out from the weld surface. In case there are cracks the dye bleeds out thus indicates presence of cracks. The basic steps [1] in the process are:

Surface Preparation
In dye penetrant testing, surface preparation is an important part of the process. The testing surface should be thoroughly cleaned of any kind of contamination like oil, grease, water, dirt and dust that would prevent dye penetrating the surface cracks.

Dye Application
After cleaning and drying the surface thoroughly, the dye is sprayed over the surface being tested for surface cracks.

Dye Dwell Time
The dwell time is generally recommended by the dye manufacturer. It is the time required for the dye to penetrate into the cracks, typically it ranges from 5 to

60 min. Increased dwell time does not cause any harm to the process however the dye should not dry up.

Removal of Excess Dye
Excess dye is to be removed carefully such that it does not get removed from the cracks where it has seeped in.

Developer Application
In some of the dye penetrant systems, another liquid is prayed over the dye on the test piece. These developers are available in liquid or powder form. It is applied either by dusting or by spraying depending on whether it is in powder form or liquid.

Indication Development
Depending on the type of cracks, the bleeding time of the dye may vary significantly. Therefore sufficient time should be given to the developer to extract the dye from the cracks. This time can be as high as 10 min.

Inspection
Visual inspection is carried out often using florescent light to detect bleeding of dye indicating a crack.

Clean Surface
Finally the test surface is thoroughly cleaned to remove the traces of developer as well as dye.

20.3 Magnetic Particle Testing

Magnetic particle testing (MPT) is applicable to only ferromagnetic materials. It is used to detect surface and near-surface flaws that may have formed due to welding. In this method a magnetic field is applied around the welded joint using a permanent horse shoe magnet or any other equivalent hand-held device. In case there is no crack or any material discontinuity is present, the magnetic flux remains mostly concentrated below the surface of the material. However, in case of a discontinuity in the form of crack or any other foreign inclusion, like slag inclusion, the magnetic flux as if 'leaks' at the location of the discontinuity. Fine magnetic particles used in MPT get attracted to the area where this flux leakage takes place indicating the presence of a discontinuity or a flaw.

The Principal

When a magnet is cut in two pieces, each will behave like an individual magnet having its own north and south poles. Similarly if a magnet is cracked, then north and south poles will form at the cracked edges. As the magnetic flux encounters the cracked edges, flux leakage takes place. This happens as the fine gap in the cracked region fails to support the magnetic field per unit volume same as that of the material without any discontinuity [1]. It is schematically shown in Fig. 20.1.

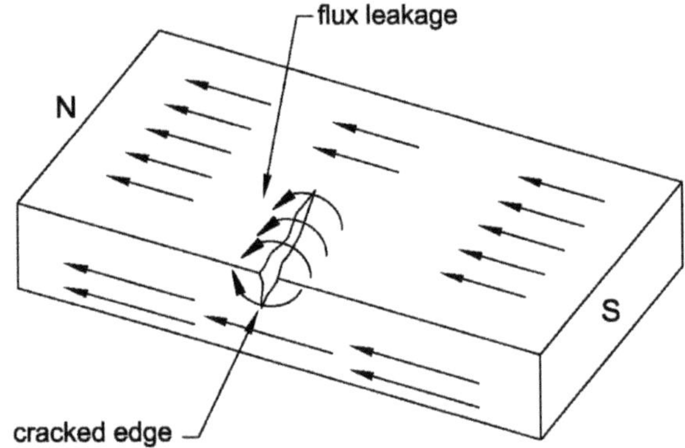

Fig. 20.1 Schematic representation of magnetic flux leakage in a cracked region

If magnet is brought near to a zone sprinkled with magnetic (iron filings) particles, the particles will get clustered at the magnetic poles. Thus when they are spread on a surface having a crack and a the zone is magnetised, the magnetic particles will get clustered at the cracked region. Even if it is a very fine crack, which may not be visible in naked eyes, this cluster of particles will be easily visible as shown in Fig. 20.2, indicating the presence of a discontinuity. It can be a surface or a sub-surface crack or subsurface blow hole or slag inclusion. The limitation of the method is it can only detect surface discontinuity and sub-surface up to a limited depth of about 6 mm depending on the magnet strength.

The most commonly used magnetic particles are black iron particles and red or yellow iron oxides. Often these particles are sprayed with a fluorescent material, thus improving the visibility under a UV illumination in dark areas.

20.4 Radiographic Testing

In radiographic testing, gamma or X-radiation is used to examine the welded joints of any form of discontinuity external or internal. Gamma or X-radiation is obtained from a X-ray machine or radioactive isotope. The test surface is exposed to this

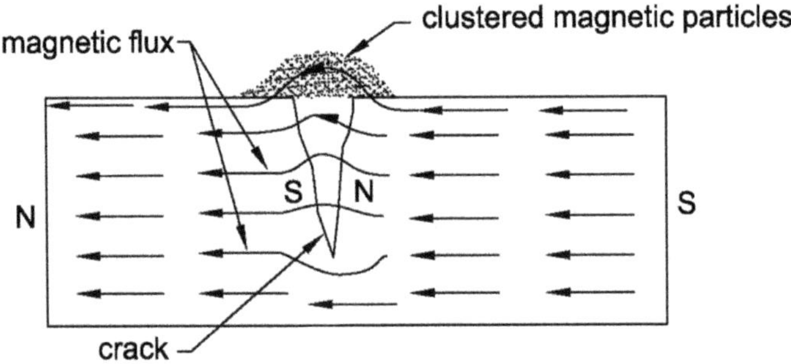

Fig. 20.2 Schematic view of clustered magnetic particles near the crack zone

radiation. The radiation penetrates the material and exposes the photographic film kept below the test surface. Any form of external or internal discontinuity or flaw present in the welded zone becomes visible in the resulting shadowgraph (photograph). These are indicated as density changes in the film as one is more familiar with X-ray of broken bones.

The objective of radiographic testing is to produce images of welded region exhibiting maximum possible of detail of surface as well as the internal region of the fusion zone. This is very much dependent on the image quality. This quality is measured by the parameter *Radiographic Sensitivity*. Two sets of independent variables, (i) variables affecting *contrast*, and (ii) variables affecting the *definition* of the image define the radiographic sensitivity.

The degree of difference in density between two areas of a radiographic image gives the radiographic contrast. Radiographic definition gives the sharpness of an abrupt change from one area to another in an image. Images with high definition and high contrast make it easier to identify minute details in the radiographs.

Radiograph Interpretation of Welds
It is not enough just to produce high quality radiographs, what is extremely important is to have a radiographer who is skilled to interpret these radiographic images. Interpretation of radiographs is a very important part of radiographic testing. It is a skill that is acquired over time. The skill of interpreting these radiographs to identify the presence of any flaw or defect improves with time and experience.

There are three basic steps in interpretation of radiographs: (i) detection, (ii) identification, and (iii) assessment. In all these the visual skill of the radiographer to resolve a spatial pattern in an image is most important. It naturally depends on the level of experience for identifying various features in an image.

This method of NDT has the advantage of generating test documents which can preserved for future reference if required.

20.5 Ultrasonic Testing

Beams of high-frequency sound waves are used in ultrasonic testing. Subsurface (internal) flaws in the material can be detected by this method of NDT. The sound waves being focused on the test surface, penetrates the same and travel through it. However if it encounters any material interface in the form of a discontinuity because of any crack or any foreign inclusion, the sound wave reflects back. Any discontinuity in the medium acts as an interface because of sharp change of material density. For example, as the wave meets a crack in its path, it will actually encounter a medium having much less density (air in the cracked zone) compared to the metal. This reflected sound wave is then captured and appropriately interpreted to detect the source of reflection, i.e. to detect if there is any flaw or defect in the tested region. Pulse echo method of ultrasonic testing technique is widely used. In this ultra sound wave is made incident on the test object and the reflections (echoes), if any, are captured in a receiver.

Ultrasonic testing has several advantages over other NDT methods:

- It can detect both surface and subsurface discontinuities.
- It has a higher penetrating power for flaw detection compared to other NDT methods.
- It has high accuracy level in detecting and assessing the location, size and shape of the flaw.
- Preparation time for carrying out testing is minimum.
- No need for post processing, instantaneous results are obtained.
- Computer aided systems can produce detail images of the flaws.
- It can be used for other purpose also e.g. thickness gauging.
- To carry out ultrasonic testing only one side access is enough.
- No health hazard.
- Equipment is portable and can be easily handled.
- The ultrasonic probe can be very easily taken to structurally congested areas.
- Test reports can be well documented.

The limitations of this technique include:

- Test surface should have required accessibility to reach out with the ultrasonic probe.
- The training and skill required is much more exhaustive than with other methods.
- It is difficult to conduct ultrasonic testing on materials that are rough and irregular in shape.
- One of the major drawbacks of this method is, in case of weld defects those may lie parallel to the sound beam may not get properly detected as to its complete shape and size and its extent of discontinuity.
- To effectively use this method, suitable equipment calibration is required along with appropriate characterization of flaws.

20.6 Acoustic Emission Testing

Acoustic emission is an effective, versatile and at the same time a non-invasive method of monitoring a material or a structure. Any abrupt redistribution of stress within a material generates transient elastic waves. In the event of any cracking phenomenon, like crack initiation and/or crack growth, opening and/or closure of a crack, any slip and/or dislocation movement, etc. triggers sudden release of energy causing generation of the elastic waves. These waves get transmitted to the surface and are captured by acoustic emission sensors. This is referred to as acoustic emission. Right kind of equipment and setup can capture very sensitive and precise motions in the order of picometers (10^{-12} m). It can also monitor fiber breakage and fiber-matrix debonding in composites [1].

Acoustic emission testing fundamentally differs from other NDT techniques in two ways; (i) It captures energy from the structure whereas in other methods, energy is supplied to the structure, (ii) it deals with events/processes or changes taking place in the material in real time.

The limitations of AE system is it can only qualitatively gauge the damage in a structure. This method indicates about the occurrence of a damage (e.g. crack), however for obtaining quantitative information about the damage other NDT methods are to be applied. AE system generally operates in a high ambient noise level, hence to acquire useful information suitable noise reduction is very crucial.

Often the main focus of an AE system is to monitor and locate the source of any significant acoustic emission. Using multiple sensors it is possible to identify the location of the source of any acoustic emission, thereby the location of a possible damage is obtained.

Reference

1. NDT Resource Centre. https://www.nde-ed.org/index_flash.htm.

Chapter 21
Accuracy Control

Abstract Accuracy Control is based on the basic fact that there is no such thing as absolute accuracy. Whatever be the method of production, variations from design dimensions are unavoidable. However these variations are measurable and at the same time they are anticipated. Hence it is necessary to know the ranges of variations so that one can quantitatively target the end product accuracy. A system is needed to monitor and control the accuracy of interim products. Otherwise, work in succeeding stages of production will be adversely affected by inaccurate interim products. The reasons for dimensional variations in any work process can be attributed to either some C*ommon Cause* or some *Special Causes*. Variations caused by the so-called Common Causes reflect the status of accuracy level of the existing manufacturing process, which includes production process, machineries used etc. Whereas the so-called Special Cause variation points directly to some fault in the production line. For implementation of accuracy control system, it is necessary to have an accuracy control data base. This gives the statistical history of the accuracy level of the work processes employed in the shipyard.

Accuracy Control can be defined as "The use of statistical techniques to monitor, control and continuously improve shipbuilding design details, planning, and work methods so as to maximize productivity" [1]. Accuracy Control essentially is based on the basic fact that there is no such thing as absolute accuracy. Whatever be the method of production, variations from design dimensions are unavoidable. The reasons of which can be multifarious. However these variations are measurable and at the same time they are anticipated. Hence it is necessary to know the ranges of variations so that one can quantitatively target the end product accuracy.

Shipbuilding process based on group technology, using concurrent hull construction, requires that accuracy be controlled. Otherwise, work in succeeding stages of production is adversely affected by inaccurate interim products. Consequently, a system is needed to monitor and control the accuracy of interim products. The justification of such a system may initially be the need to monitor the construction of interim products to minimize delays and rework during erection. It involves the regulation of accuracy as a management technique for improving the

N.R. Mandal, *Ship Construction and Welding*, Springer Series on Naval Architecture, Marine Engineering, Shipbuilding and Shipping 2,
DOI 10.1007/978-981-10-2955-4_21

productivity of the entire shipbuilding system by focusing attention on individual areas where improvements offer significant benefits.

The reasons for dimensional variations in any work process can be attributed to either some ***Common Cause*** or some ***Special Causes***. It is the result of the production system, including raw materials, incoming parts or interim products, worker training, work environment, tools and machinery etc. Variations caused by the so-called Common Causes reflect the status of accuracy level of the existing manufacturing process, which includes production process, machineries used etc. It can only be changed when the production process or the machineries or both are changed.

Whereas the so-called Special Cause variation points directly to some fault in the production line. These dimensional variations are due to some fault that may have taken place in the machine or may be due to ignorance of the worker or even can be due to some kind of accident or any such reasons. Hence in such a situation these causes ought to be identified and removed as a regular part of monitoring a work process.

In shipbuilding, for example, spacing between longitudinals or the dimensions of plates cut in a NC machine or any other such work of repetitive nature will have dimensions varying from designed dimensions. For implementation of accuracy control system, it is necessary to have an accuracy control data base. This gives the statistical history of the accuracy level of the work processes employed in the shipyard. It is the quantitative measure of normal performance at every work stations in a shipyard. Each piece of data in accuracy control data base is associated with a particular work process.

Statistical Principles

The principle is based on the observation that there is no such thing as absolute accuracy. Irrespective of technique or control measures, there will always be measurable and anticipated variation from specified dimensions. Working with ranges of variations is crucial for production control and for achieving specified end product accuracy.

Any repeatable work process i.e. a work process which is consistent in facilities and workers skills, produces products that have variations in characteristics. For example, spacing between longitudinals will vary from the designed frame spacing. These variations, when plotted by the number of times they occur, approximate a normal distribution. Two parameters describe the relative shape of a normal distribution:

- Mean, $\overline{X}$, the arithmetical average of variations in a sample. It describes the central tendency of distribution.
- Standard deviation, σ, which classifies the sizes of variations from the mean value by their frequencies of occurrence. This is a measure of the relative scatter of points around the mean value.

For a normal distribution, 67 % of the values fall within one standard deviation of the mean, 95 % fall within two standard deviations, 99.7 % fall within three standard deviations, and so on.

21.1 Accuracy Control System

An accuracy control system can be considered to have two primary goals, one short term and one long term. The short term goal is to monitor the construction of interim products to minimize delays and rework during erection. The more important long term goal is the establishment of a management system that permits the development of quantitative information that can be used to continuously improve productivity.

Accuracy control system includes three major components [2]:

- Planning
- Execution
- Evaluation

Accuracy control planning prepares for accuracy control work to be performed on a specific shipbuilding project.

Accuracy control execution is the actual work to be done, including development of specific check sheets, methods of measuring and recording data.

Accuracy control evaluation is the analysis phase. It provides data for use in planning, execution and evaluation of next shipbuilding project. The results of evaluation are used to plan for future work. Thus the results of accuracy control evaluation continuously help to improve shipyard productivity.

To implement such accuracy control system, the basic requirement is to have well-defined work processes, procedures and coding system so that required data can be acquired through proper measurements. These data are then interpreted using statistical theory. The Group Technology approach to shipbuilding provides for standardization of various work processes within a yard, and through this repeatability of processes is achieved. This makes application of accuracy control techniques possible.

The second requirement to full scale application of accuracy control is to have an accuracy control data base. This gives the statistical history of the accuracy of the work processes employed in the shipyard. It is the quantitative measure of normal performance at every work stations in a shipyard. Its preparation takes time, and it may yield some short term benefits, but its primary aim is to build the base for implementation of an Accuracy Control System.

A large data sample of each work process over a set period of time is required to develop the accuracy control data base. The actual data that are to be recorded for a particular process are the differences between design dimensions and actual dimensions that are produced by the machine or the work process. This data sample is then used to determine the distribution of variations from design or target dimensions (mean and standard deviation of an assumed normal distribution). Using this data, the regular performance for the work process can be monitored.

Once the data base is ready, it serves two purposes:

- Provides basis for making the control charts for monitoring of performance of an individual machine or a process.
- Provides the necessary information to do process analysis.

The objective of process analysis is productivity improvement through cutting costs, improving quality, and shortening lead times. In fact the impact of alteration of any work process on the overall production process can be predicted and analyzed using the accuracy control data base.

21.2 Control Charts

The most common tool used in statistical quality control is the Shewhart control chart [1]. Control charts apply the concept of expected and measurable variation in work processes. These are used to distinguish between common cause variations and special cause errors. When only common cause variations are detected, the process can be taken as running normally and no adjustment is necessary. It implies that the machine or the process is operating within the known or expected accuracy level. The dimensional deviations are within the expected range of the machine or the process.

On the other hand if special cause variations are detected, logically the production in that machine should be immediately stopped. It implies that the dimensional variations of the interim product are beyond the normal range of that particular machine. The functioning of the machine in question should be checked to establish the cause of such deviation such that it can be rectified and the machine returns to normal performance.

The control charts are plots of mean ($\overline{X}$) and range (R) of random samples of measurements from a specific work process over time. R is the difference between the largest and the smallest value in the sample, and although it is a less rigorous measure of variability than the standard deviation, its simplicity has led to its widespread use in control charts [1]. The $\overline{X}$ chart is used to control the central tendency and R chart is used to control the dispersion.

Control chart theory is based on the statistical central limit theorem. The central limit theorem states that the distribution of the means of random samples taken from a normal distribution is another normal distribution itself. It has the same mean of the original distribution and a standard deviation equal to the standard deviation of the original distribution divided by the square root of the sample size.

The technique involves initially determining regular performance of a work process i.e. its normal distribution, using a large data sample. This normal performance is then used to establish the expected range of variation for the given process. Subsequently random product samples from the work process are monitored to check, whether there is any deviation in the performance of the process.

Control charts give us the limits on the variation of the mean and the range of these random samples. The limits are commonly set at three standard deviations below and above the process mean and the average range. Three standard deviations are taken because they provide 99.7 % assurance that exceeding these limits is a result of a change in the normal performance of the process indicating a special cause i.e. an error has occurred in the process.

Thus control chart is a tool to monitor a production process to evaluate whether the process is under control or some error/fault has occurred. When a process is under control and no special causes are present, variations on the $\overline{X}$ and R charts are due to common causes and they lie within the limits. When the points fall outside the control limits, they indicate a special cause of variation. The control limits indicate the level of accuracy and the variation that is expected for the given process.

In developing the $\overline{X}$ and R control charts, six values are required, three for each chart. These values are the centreline, the upper control limit (UCL), and the lower control limit (LCL). The control chart values are determined using the accuracy control database for specific operations.

The values of mean process variation $(\overline{X})$ are taken from the accuracy control data sheets for each subgroup in a work process. A sample data sheet [4] showing measurements of one subgroup is shown in Fig. 21.1. The mean of process variation $(\overline{X})$ and the range of process variation (R) are used to prepare the $\overline{X} - \text{R}$ control charts. For a sub group size of 'k', the central line is calculated as follows:

$$CL = \overline{\overline{X}} = \frac{\sum_{j=1}^{k} \overline{X}_j}{k}$$

After setting the central line, the upper control limit (UCL) and lower control limit (LCL) are to be calculated. By definition these are calculated from the following equations:

$$UCL = \overline{\overline{X}} + 3\sigma_{\bar{x}}$$
$$LCL = \overline{\overline{X}} - 3\sigma_{\bar{x}}$$

For ease of calculation and at the same time keeping the level of accuracy of calculation, the following equations are used in practice [5]:

$$UCL = \overline{\overline{X}} + A_2\overline{R}$$
$$LCL = \overline{\overline{X}} - A_2\overline{R}$$

DATA SHEET									
Machine No : Plasma II									
Process : Plasma cutting									
Instrument used : Steel Tape (837)									
Sl No	Ship No.	Block	Nesting No	Plate thk	Part No.	Type of measurement	Dimensions		Variation
							Required, R (mm)	Actual, A (mm)	X = A-R (mm)
Date : 10/07/2013									
1	93	5	5N8	11	05-TT1-W5-S	Length	7780	7780	0
2	93	5	5N8	11	05-TT1-W5-S	Width	2460	2453	-7
3	93	5	5N2 (Normal)	8	05-BS-W5-P	Length	3590	3588	-2
4	93	5	5N2 (Normal)	8	05-BS-W5-P	Width	1395	1395	0
5	93	5	5N2 (Mirror)	8	05-BS-W5-S	Length	3590	3588	-2
6	93	5	5N2 (Mirror)	8	05-BS-W5-S	Width	1395	1395	0
7	93	5	5N7	11	05-TT1-W6-P	Length	6998	6998	0
8	93	5	5N7	11	05-TT1-W6-P	Width	2460	2458	-2
9	93	5	5N18 (Normal)	15	05-KNY1730-W1-P	Length	2440	2440	0
10	93	5	5N18 (Normal)	15	05-KNY1730-W1-P	Width	1041	1043	2
11	93	5	5N18 (Normal)	15	05-KNY3082-W1-P	Length	1665	1667	2
12	93	5	5N18 (Normal)	15	05-KNY3082-W1-P	Width	1041	1042	1
13	93	5	5N18(Mirror)	15	05-KNY1730-W1-S	Length	2440	2440	0
14	93	5	5N18(Mirror)	15	05-KNY1730-W1-S	Width	1040	1044	4
15	93	5	5N18(Mirror)	15	05-KNY3082-W1-S	Length	1665	1667	2
16	93	5	5N18(Mirror)	15	05-KNY3082-W1-S	Width	1040	1043	3
17	93	5	5N04	10.5	05-Y1000-W2-P	Length	2072	2073	1
18	93	5	5N04	10.5	05-Y1000-W2-P	Width	900	900	0
19	93	5	5N04	10.5	05-991-W2-P	Length	4501	4493	-8
20	93	5	5N04	10.5	05-991-W2-P	Width	873	868	-5
Mean of process variation = - 0.55 mm								ΣX =	-11
Range = 12 mm								X_{max} =	4
								X_{min} =	-8

Fig. 21.1 A sample accuracy control data sheet

These are calculated using statistical constant A_2 as a function of sub group size 'k' as given in Table 21.1, [3].

The range of variation in measured data is given by R = $X_{max} - X_{min}$. The R chart is made using $\overline{R}$ as the centre line (CL) and the UCL and LCL are calculated using the following equations:

Table 21.1 Statistical constants

k	A_2	D_3	D_4
2	1.880	0	3.267
3	1.023	0	2.574
4	0.729	0	2.282
5	0.577	0	2.114
6	0.483	0	2.004
7	0.419	0.076	1.924
8	0.373	0.136	1.864
9	0.337	0.184	1.816
10	0.308	0.223	1.777

$$CL = \overline{R} = \frac{\sum_{j=1}^{k} R_j}{k}$$

By the definition of control limits, UCL and LCL are calculated using the following equations:

$$UCL = \overline{R} + 3\sigma_R$$
$$LCL = \overline{R} - 3\sigma_R$$

However in order to simplify the calculations, the following derived equations are used:

$$UCL = D_4\overline{R}$$
$$LCL = D_3\overline{R}$$

D_3 and D_4 are statistical constants depending on subgroup size 'k' as given in Table 21.1.

Sub group - 1

Avarage variation =	0.11111
Range =	11
Sample size =	18

Sub group - 2

Av variation =	-0.32143
Range =	8
Sample size =	28

Sub group - 3

Av variation =	-0.875
Range =	8
Sample size =	16

Sub group - 4

Av variation =	-0.33333
Range =	6
Sample size =	12

Sub group	**Av variation** (mm)	**Range** (mm)
1	0.1111	11
2	-0.3214	8
3	-0.875	8
4	-0.3333	6

ΣX = -1.4186   Av ΣX = -0.35 mm

ΣR = 33   Av ΣR = 8.25 mm

For sample size of 4

A2 = 0.729

D3 = 0

D4 = 2.282

For X Chart

UCL =	5.66	mm
LCL =	-6.37	mm

For R Chart

UCL =	18.83	mm
LCL =	0	mm

Fig. 21.2 A typical data set of 4 subgroups analysed from accuracy control data sheets

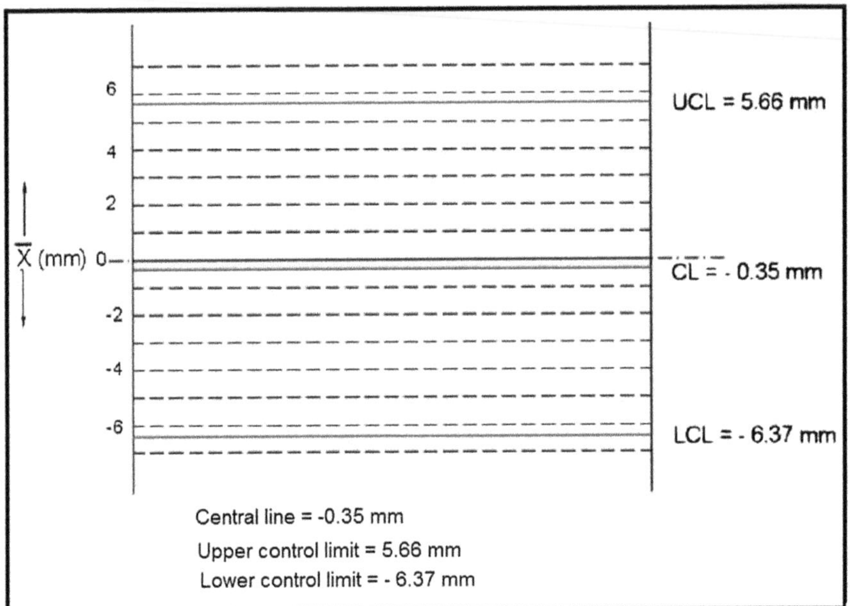

Fig. 21.3 A typical $\overline{X}$ control chart

A typical set of analysed data is shown in Fig. 21.2 having 4 subgroups. The corresponding $\overline{X}$ and R control charts are shown Figs. 21.3 and 21.4 respectively.

Such control charts can be made for each work station and they can be used by the shop supervisors to routinely monitor performance at each work station. The charts serve as a visual signal to the workers and supervisors to see whether their workstation is functioning within the expected accuracy level or is not "in control". The randomly measured data from a given workstation are plotted in the control charts and if they fall within the control limits, it can be concluded that the work is progressing in normal fashion. However if some of the data points fall outside the prescribed limits, it points to some error in the functioning of the machine or the work process. The production at the concerned workstation should be stopped. The cause must be determined and necessary corrective measure is to be implemented to eliminate the cause of variation.

The control charts are quantitative management tools. With help of these any work process that has gone beyond the normal accuracy level can be easily identified and appropriate corrective action can be initiated immediately. Thereby a continuous check can be maintained on all production processes.

One of the advantages of the use of control charts is that production workers get directly and actively involved in managing their own work. This can be a source of pride and motivation for workers. It also actively involves them in problem solving, and may stimulate them to suggest creative and workable process improvements.

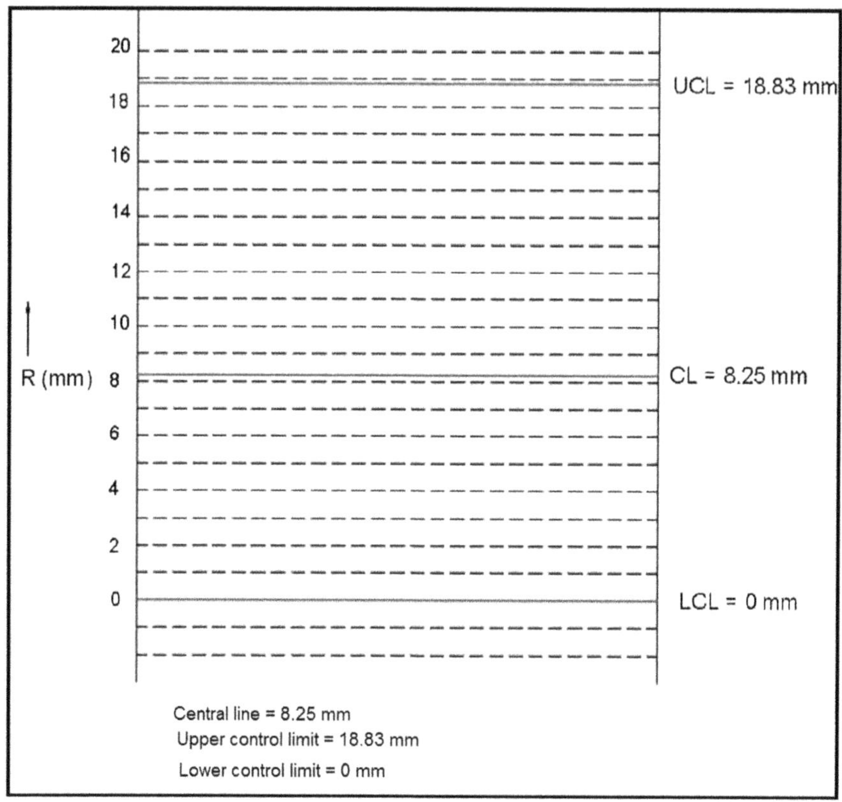

Fig. 21.4 A typical R control chart

Such an expanded role for production workers can promote greater job satisfaction, and produce tangible rewards for the organization.

21.3 Accuracy Control Planning

Accuracy control planning is essential for proper functioning of the system. Since deviation from design dimensions is expected to occur at each stage of production, it needs to be decided at what stage of construction, necessary corrective action must be taken to minimize rework at erection. Therefore the role of accuracy control planning is to:

- Pinpoint the stage of construction and dimensions that are critical to the dimensional and geometrical accuracy of the units and blocks.
- Appropriate checkpoints and reference lines are to be earmarked in subassemblies and assemblies from which the units and blocks are assembled.

- Estimation of shrinkage allowances and its location of application.
- Extent of green material, if any and deciding the stage at which the green material needs to be removed.
- Identification of work processes during which accuracy control measurements will be made.
- Determine the numbers of interim products that should be measured based upon random sampling.
- The working drawings to show the shrinkage allowances, tolerance limits and margins.

21.4 Accuracy Control Standardization

Standards related to accuracy control can be classified into two groups:

- Work standards.
- Accuracy standards.

Work standards relate to things like:

- Green material.
- Shrinkage allowances.
- Checking procedure.
- Part fabrication and assembly schemes.

Deciding on green material the following questions need to be answered:

- Why is green material required?
- Where such green material is required?
- How much green material is necessary?
- During what work process will the green material be removed?

Green material is kept as buffer to compensate for accuracy variations in all hull construction processes. This leads to a difficulty in detecting the true causes of accuracy variations. Thus improvement in fabrication process becomes difficult. Where no systematic statistical approach is taken, large amount of green material is provided based on feedback from production. This method hides the actual causes and reduces opportunities for improvements.

With proper implementation of accuracy control system, the extent of green material reduces to just the excess allowances needed to compensate for variations. This is characterized by finish cutting, based on a high probability that no rework will be necessary.

To further improve situation, the following questions are to be looked into:

- Why is excess needed?
- Where is excess needed?

- How much excess is necessary?
- If needed, during what stage should rework take place?

This type of questioning becomes the motivation for continuous improvements in work standards.

An appropriate manual is to be made specifying the standardized procedures for recording data for analysis. Standards for parts fabrication and assembly schemes are to be developed to achieve specified accuracy during each work process. Specific allowances are to be included in working drawings along with the usual information regarding structural details and edge preparation, etc. Working drawings should contain total instructions and information for how to construct a ship's hull.

Accuracy Standards

Accuracy standards are needed to control accumulation of dimensional deviations at the final stage from preceding work processes or operations. Data from past experience within a given shipyard are used to set accuracy standards for a new construction. However as the new construction progresses new data are acquired and the old set of data are reviewed in light of these new data. Standard range of variation with tolerance limit should be established for every work process.

21.5 Setting Accuracy Standards

Through quantitative analysis of data, accuracy standards can be set. For example, when plate butts are aligned, the achieved distribution of root gap variations at its extremities may indicate requirements of rework, which are:

- Cutting is required, where the root gap is small or negative.
- Building up on the edge when root gap is too much.

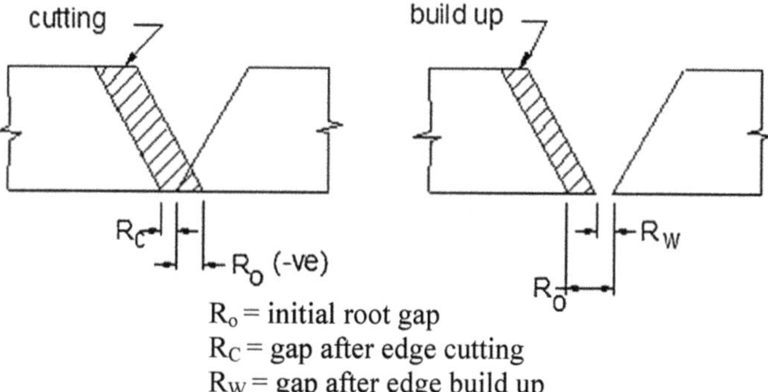

Fig. 21.5 Rework requirements in root gap alignment

As shown in the Fig. 21.5, when R_0, the initial root gap is negative, i.e. plate overlap is there, edge cutting will be required to achieve the desired root gap R_C. Whereas for bigger than the required root gap, edge build up will be required to obtain the desired gap R_W. Now retaining the original material as much as possible works out to be cheaper, hence the tendency is to cut as less as possible in case of R_0 being negative, whereas, building up of edge being a costlier process, the tendency will be to build up as less as possible. Therefore it leads to R_C always less than R_W.

Hence the condition for avoiding rework works out to be:

$$R_C \leq R_0 \leq R_W$$

Therefore by definition the lower tolerance limit is R_C and the upper tolerance limit is R_W.

21.6 Factors Leading to Dimensional Error

To build precise, dimensionally correct assemblies, units and blocks should be the main focus of a good shipbuilding practice. As one can observe in functioning of various shipyards, ***rework*** is one of the major contributors to the total production time. At various stages of assembly, factors like, welding distortion, improper erection schedule, lack of proper information feedback, etc. result in dimensionally inaccurate structures. This leads to the requirement of substantial rework elongating the production cycle.

The various factors which contribute to possible dimensional errors are as follows:

(a) Material

- Accuracy of plate flatness and its thickness.
- Straightness of rolled steel sections.

(b) Cutting and Marking

- Cutting accuracy of plates.
- Marking accuracy for fitting.

(c) Sub-assembly

- Accuracy of web plate connection.
- Fitting error of face plate.
- Fitting error of tripping brackets and flat bars on floor or web plates.
- Welding deformation.
- Residual deformation after fairing.

(d) Flat Panel assembly

- Marking and cutting errors of plate.

- Fitting errors of stiffeners.
- Deformation due to welding of stiffeners.
- Deformation due to welding of stiffened panels.

(e) Curved Panel assembly

- Dimensional inaccuracy of skid.
- Marking and cutting errors of plate.
- Accuracy of plate bending.
- Accuracy of stiffener bending.
- Fitting errors of stiffeners.
- Deformation due to welding of stiffeners.

Therefore, there are five distinct factors affecting the accuracy of an assembly:

- Accuracy of materials.
- Accuracy of marking and cutting.
- Accuracy of fitting.
- Welding shrinkage.
- Residual deformation after fairing.

21.7 Self Check

A self-check system that can be operated by the concerned worker and/or supervisor is very much essential to make an accuracy control system effective. Until and unless the jobs are checked to assure compliance with the accuracy standards by the workers or the work groups, it is to be considered that the said job is not yet completed. Self-checks thus are to be made part of the production work.

Start and finish schedules are usually posted at each work station for parts fabrication, subassembly, and block assembly. Self-checks and other subsequent checks, and accuracy control data recordings are to be regarded as part of the work processes that should adhere to these schedules. The status of block assembly is to be recorded on a day to day basis.

Accuracy checks are to be performed daily. For each work stage, items are to be checked for conformance with accuracy standards. Supervisors from each work group should record necessary data on day to day basis then only the accuracy control system becomes effective.

Typically following items are to be checked, measured and recorded:

- **Tame plate preparation**: overall dimensions, including shrinkage allowances and marking required for part fabrication, assembly and checking work.
- **Part fabrication**: Overall dimensions of cut plates, edge preparations, deformation, and the curvature for bent parts.

- **Subassembly, unit and block assembly**: Positioning of parts or subassemblies, their fitup, gaps for welding, etc.
- **Erection**: Fit-up, gaps for welding, and component alignment.

Measuring each and every part is impractical, therefore random sampling is employed to monitor dimensional accuracy levels. Maintenance checks on cutting machines should be done on a regular basis. The accuracy of bent parts is particularly important. Inaccurately bent parts are generally forced to fit and thus results in stress build up which adds to deformation during welding.

By introducing Accuracy Control, a shipyard aims to totally eliminate the requirement of adjustment and rework, which accounts to about half of the total fitting man-hours. Downstream production is adversely affected by inaccurate interim products. So it is necessary to monitor the construction of interim products to minimize delays and rework during erection.

References

1. Storch, R. L. (1995). *Ship production* (2nd edn.) SNAME.
2. Storch, R. L. (1985). Facilitating accuracy control in shipbuilding. In *Computer Applications in the Automation of Shipyard Operations and Ship Design V*. North-Holland: Elsevier Science publishers B.V.
3. Trietsch, Dan. (1999). *Statistical quality control—A loss minimization approach*. Singapore: World Scientific Publishing.
4. Mandal, N. R., Harikrishnan, S., Mathew, T., Jimmy, S., George, F. (2013). *Introduction of accuracy control in steel fabrication*. Cochin Shipyard Ltd.
5. Storch, R. L. (2002). *Accuracy control implementation manual*. Shipyard Production Process Technologies Panel, NSRP.

www.ingramcontent.com/pod-product-compliance
Ingram Content Group UK Ltd.
Pitfield, Milton Keynes, MK11 3LW, UK
UKHW021448280726
14060UKWH00001BA/297